THE END *of* ERROR

Unum Computing

John L. Gustafson

CRC Press
Taylor & Francis Group
Boca Raton   London   New York

CRC Press is an imprint of the
Taylor & Francis Group, an **informa** business

A CHAPMAN & HALL BOOK

Chapman & Hall/CRC
Computational Science Series

SERIES EDITOR

Horst Simon
Deputy Director
Lawrence Berkeley National Laboratory
Berkeley, California, U.S.A.

PUBLISHED TITLES

COMBINATORIAL SCIENTIFIC COMPUTING
Edited by Uwe Naumann and Olaf Schenk

CONTEMPORARY HIGH PERFORMANCE COMPUTING: FROM PETASCALE
TOWARD EXASCALE
Edited by Jeffrey S. Vetter

DATA-INTENSIVE SCIENCE
Edited by Terence Critchlow and Kerstin Kleese van Dam

THE END OF ERROR: UNUM COMPUTING
John L. Gustafson

FUNDAMENTALS OF MULTICORE SOFTWARE DEVELOPMENT
Edited by Victor Pankratius, Ali-Reza Adl-Tabatabai, and Walter Tichy

THE GREEN COMPUTING BOOK: TACKLING ENERGY EFFICIENCY AT LARGE SCALE
Edited by Wu-chun Feng

GRID COMPUTING: TECHNIQUES AND APPLICATIONS
Barry Wilkinson

HIGH PERFORMANCE COMPUTING: PROGRAMMING AND APPLICATIONS
John Levesque with Gene Wagenbreth

HIGH PERFORMANCE PARALLEL I/O
Prabhat and Quincey Koziol

HIGH PERFORMANCE VISUALIZATION:
ENABLING EXTREME-SCALE SCIENTIFIC INSIGHT
Edited by E. Wes Bethel, Hank Childs, and Charles Hansen

INTRODUCTION TO COMPUTATIONAL MODELING USING C AND
OPEN-SOURCE TOOLS
José M Garrido

INTRODUCTION TO CONCURRENCY IN PROGRAMMING LANGUAGES
Matthew J. Sottile, Timothy G. Mattson, and Craig E Rasmussen

INTRODUCTION TO ELEMENTARY COMPUTATIONAL MODELING: ESSENTIAL
CONCEPTS, PRINCIPLES, AND PROBLEM SOLVING
José M. Garrido

PUBLISHED TITLES CONTINUED

INTRODUCTION TO HIGH PERFORMANCE COMPUTING FOR SCIENTISTS
AND ENGINEERS
Georg Hager and Gerhard Wellein

INTRODUCTION TO REVERSIBLE COMPUTING
Kalyan S. Perumalla

INTRODUCTION TO SCHEDULING
Yves Robert and Frédéric Vivien

INTRODUCTION TO THE SIMULATION OF DYNAMICS USING SIMULINK®
Michael A. Gray

PEER-TO-PEER COMPUTING: APPLICATIONS, ARCHITECTURE, PROTOCOLS,
AND CHALLENGES
Yu-Kwong Ricky Kwok

PERFORMANCE TUNING OF SCIENTIFIC APPLICATIONS
Edited by David Bailey, Robert Lucas, and Samuel Williams

PETASCALE COMPUTING: ALGORITHMS AND APPLICATIONS
Edited by David A. Bader

PROCESS ALGEBRA FOR PARALLEL AND DISTRIBUTED PROCESSING
Edited by Michael Alexander and William Gardner

SCIENTIFIC DATA MANAGEMENT: CHALLENGES, TECHNOLOGY, AND DEPLOYMENT
Edited by Arie Shoshani and Doron Rotem

CRC Press
Taylor & Francis Group
6000 Broken Sound Parkway NW, Suite 300
Boca Raton, FL 33487-2742

© 2015 by Taylor & Francis Group, LLC
CRC Press is an imprint of Taylor & Francis Group, an Informa business

Printed on acid-free paper
Version Date: 20141112

International Standard Book Number-13: 978-1-4822-3986-7 (Paperback)

Visit the Taylor & Francis Web site at
http://www.taylorandfrancis.com

and the CRC Press Web site at
http://www.crcpress.com

Contents

Part 1
A New Number Format: The Unum

Chapter 4 The complete unum format................................35

Chapter 5 Hidden scratchpads and the three layers................55

Chapter 6 Information per bit................................85

Chapter 7 Fixed-size unum storage.....................93

Chapter 8 Comparison operations.....................103

Chapter 9 Add/subtract, and the unbiased rounding myth..111

Chapter 10 Multiplication and division..................127

Part 2
A New Way to Solve: The Ubox

Preface

"The purpose of mathematics is to eliminate thought." —*Philip G. Saffman (1931 – 2008)*

The class laughed hard when Professor Saffman said this in a classroom lecture in 1978. The class, a third-year undergraduate course in applied math that was required for many majors at Caltech, had a well-earned reputation as being a difficult one. It certainly seemed to us at the time that mathematics was failing miserably at "eliminating thought."

But Saffman's observation was wise and profound.The elaborate abstract structures mathematicians build make it unnecessary to "think" about the problems they address. Once the math is done, problems can be solved just by following the rules. With the advent of automatic computing, we do not even have to follow the rules; we can let computers do that. This assumes that computer arithmetic follows mathe-matical rules, so that it is safe to build a vast structure on top of the strong foundation.

However, the rules for performing simple arithmetic on computers have *not* been mathematical, so as a result, we *still* have to think. Pick up any pocket calculator and divide 1 by 3, and you will see something like this:

That is not the same as 1/3, of course. We recognize the repeating decimal, and perhaps hope that the calculator knows the right answer but simply cannot display a string of 3s that goes on forever. Multiply its result by 3, however, and it often becomes clear that the calculator returned an *incorrect* answer:

Some calculators return what looks like 1.0000000, but if you then subtract 1 from that result, up pops something like -1×10^{-10} or -1×10^{-23}. Oops.

People sometimes say, "Computers don't lie." Actually, they lie all the time, and at incredibly high speeds. At the time of this writing, the world's fastest computer system proudly produces over 30 quintillion lies per second like the one above: Computers have no way to store the correct answer, so they substitute a different one that—you hope—is close enough. *Is* it close enough? Because you have to ponder that, existing computer math fails to "eliminate thought."

Why have we learned to tolerate such unrigorous behavior in devices as fundamentally logical as computers? The rationalization goes something like this: "Computers have finite storage, but the number of real numbers is infinite, so you *have* to give up at some point. If we use high precision, it will probably be good enough. We hope."

The phrase "correctly rounded" is an oxymoron, since a rounded number is by definition the substitution of an incorrect number for the correct one. Symbolic mathematics programs like Macsyma, *Mathematica*®, and Maple™ offer a great solution but at a very high computational cost. If you've got the time, God's got the integers. *Interval arithmetic* is a seductive answer, promising to compute with rigorous bounds on the answer instead of rounded numbers. But when one tries to use it to solve basic problems in physics, instead of making rounding errors it often produces bounds on the answer that are much looser than they need to be, and too loose to be useful. Why spend twice the storage and more than twice the arithmetic work for a technique that takes even more expertise to wrangle than rounding errors?

When the United States agreed to end nuclear testing in the 1990s, the national laboratories that develop and maintain nuclear weapons thought they could replace experiment with an ambitious federal program relying on computer simulation: The Accelerated Strategic Computing Initiative (ASCI). However, simulations with rounded arithmetic are simply guesses and guidance, not *proof* of anything. The late Nobel laureate Kenneth Wilson was the first to declare computational methods a third branch of science, complementing experimental and theoretical methods.

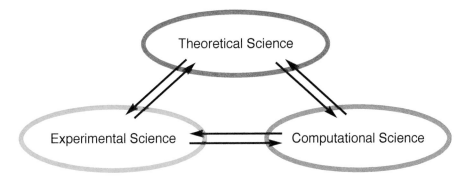

The ASCI program tried to achieve firm scientific conclusions in the absence of experiments, but because floating point computation lacks mathematical rigor, the results have still been guesses. About nuclear weapons.

While a *small* number of people care about provably valid answers, a *large* number of people care about how much electricity computers use, and how much memory it takes to solve a problem. Executives at computer companies have said, "No one is complaining about rounding errors," yet everyone is now acutely aware of the need for better performance-per-watt, or ultimately computations-per-dollar. What if there were a rigorous form of computer arithmetic that could guarantee answers, freeing the programmer to use *fewer* bits to store numbers? By the early 2000s, computers were increasingly limited in performance by memory speed and energy consumption. The users and programmers of technical applications insisted on 64-bit precision as insurance against rounding errors, but perhaps there is a way to get *more accurate*, provably bounded answers with *fewer* bits. If it can be done with *less* numerical expertise, *not more*, then programmer productivity can be improved at the same time.

Formulas, numbers, and computer code all make a book harder to read. The densest code is relegated to the appendices to relieve this, and in general the reader does not need to read the code sections to understand what follows. This book was written in *Mathematica*, a good way to make sure the math is sound and the graphs are correct. In one example, a four-bit "unum" value gets the right answer whereas a quadruple precision (128-bit) float fails; the main reason the code is shown is because some skeptics would otherwise not believe such calculations are possible, and would suspect smoke-and-mirrors. Everything needed to define a complete, working, rigorous computing environment is here.

None of the new ideas presented here are patented or pending patents, and that is quite intentional.

The purpose of the unum approach is to eliminate thought.

Acknowledgments

One of the best editors I have ever worked with, and one of the most supportive during the creation of this document, has been my son David. He has done so much to raise its quality that I wondered if I should list him as a coauthor.

Robert Barton, the Burroughs architect who invented the stack architecture in the 1960s, was a mentor to me in the 1980s when I worked at Floating Point Systems. He pointed me to interval arithmetic via the book by Miranker and Kulisch, and gave me an education in computer architecture I would never have otherwise received. He also taught me that the crux of good computer design was performance per watt, decades before everyone else figured that out.

What kind of nut cares this much about computer arithmetic? I have had the honor and pleasure of working with and meeting several members of the equally obsessed. Bill Walster has been an enormous help on this journey, with enthusiasm just short of that of a tent revivalist preacher. At Intel, a virtual team that included Tim Mattson, Helia Naeimi, Marius Cornea, and Peter Tang helped make major strides in studying flexible precision and new ways to represent numbers. Roger Golliver was also on that team, and he has been a great help in polishing this book. Thomas Lynch, who developed the high-radix online arithmetic concept that may be an excellent way to perform scratchpad calculations for unum math, added many crucial insights to the text, including noticing the tie-in between "information" as defined by Claude Shannon and "information" as defined here. Others I am indebted to for their very careful reading and constructive suggestions include David Bailey, Phyllis Crandall, Martin Deneroff, Vassil Dimitrov, Jürgen Wolff von Gudenberg, John A. Gunnels, Ken Jacobsen, Mary Okocha, and Andrew Shewmaker.

I save my deepest appreciation for last. In 2012, I visited a legendary Berkeley professor named William Kahan just after discovering the ubox method. We spent six hours talking about everything from the IEEE Standard to n-body dynamics to car repair. The cool thing about the visit is that I got to thank him for introducing me to numerical analysis. You see, Professor Kahan was the man who created the software for the HP calculators that introduced me to the hazards of rounding error. He has therefore been my tutor in numerical analysis, one way or another, for almost forty years.

Thank you, Professor Kahan.

How to read this book

Except in a few sections, which are flagged, high school math is all that is needed to understand everything described here. Calculus in particular has been avoided. In fact, not even high school math is needed to understand over 95% of this material. Some of the symbols that are introduced, like $\otimes$ and $\hat{\omega}$, may *look* cryptic and bizarre, but they turn out not to be frightening or hard to grasp at all when you read what they mean. This book is intended not for mathematicians, but for *everyone* who uses computers for technical calculations of any kind.

Even the use of Greek letters has been resisted, which is a very hard thing for a mathematician to do. An exception is made for π, which plays a starring role in many of the examples.

When jargon is introduced (some of it industry-wide and some of it coined in this work), the first use of the jargon is explained in a $\boxed{\textbf{Definition}}$ box, like this:

> **Definition**: The *h-layer* is where numbers (and exception quantities) are represented in a form understandable to humans.

All the jargon terms in Definition boxes are also collected into a Glossary at the end, so if you find yourself forgetting what some word means and where in the text it was introduced, just look it up alphabetically in the Glossary.

Background color is often used to signal the reader. A boxed paragraph with light red background is usually a $\boxed{\text{formula}}$ or something mathematical, like

$$x = b_{n-1}\, 2^{n-1} + \ldots + b_2\, 2^2 + b_1\, 2^1 + b_0\, 2^0 - \left(2^{n-1} - 1\right).$$

Text that has a light green background is computer code, a line command, or a description of a program:

```
Module[{x = 65 536.},
    While[x ≠ 1, {Print[x], x = √x }]]
```

The code is **boldface** if it actually executes and generates what follows, and plain if it is describing code written in C or similar language.

Formulas and code make for slow reading, and the reader can generally skim those without missing the thread of the text. Really large blocks of computer code are relegated to the appendices so they do not get in the way.

Except for a few examples written in C or an abstract language, any communication to and from the computer is in *Mathematica,* which makes the code look about as close to conventional math as any programming language in existence. It can still be cryptic, so please do feel free to ignore it unless you are interested in creating your own version of a computing environment with the new capabilities presented here.

On the other hand, boxed paragraphs with light blue backgrounds, like

> **If you think rounding error, overflow, and underflow are only obscure academic worries, think again. The errors caused by floats can kill people, and they have.**

are important, not to be skipped . Boxed paragraphs with light yellow backgrounds are also for emphasis, but as a warning that the paragraph expresses a bad or fallacious idea , even if the idea seems reasonable enough at first glance:

> If a result is too large to represent, the Standard says to substitute *infinity* for the correct answer.

Different fonts are also used, for clarity. Since monospaced fonts like `Courier` always remind us of old-fashioned computer output, `Courier` is used to display binary numbers like `01101101`. Decimal numbers are left in the same font as the main text. `Courier` is also used for computer function and variable names, like `setenv[{3,5}]` and `posbig`. Raw computed output is also shown in `Courier`, but not bold. (Some computed output is not raw, but instead formats the text as part of the computation.) Scattered throughout are **Exercises for the reader** , highlighted in light gray:

> **Exercise for the reader**: Find the closure plot for six-bit rational arithmetic, with various allocations of bits to the numerator and the denominator. Does rational arithmetic yield more results that are representable within the format? Does it help to make the first bit a sign bit?

The publisher has graciously agreed to full-color printing, so every effort has been made to use color throughout to increase the clarity of what may seem like a very dry and hard-to-read topic.

Part 1

A New Number Format: The Unum

1 Overview

"You can't boil the ocean." —Former Intel executive, reacting to the idea of unums.
Incidentally, the pronunciation is "you-num," and it is spelled with lower-case letters.

The universal number, or *unum*, encompasses all standard floating point formats, as well as fixed point and exact integer arithmetic. Unums get more accurate answers than floating point arithmetic, *yet use fewer bits* in many cases, which saves memory, bandwidth, energy, and power. Unlike floating point numbers ("floats"), unums make *no rounding errors*, and *cannot overflow or underflow*. That may seem impossible for a finite number of bits, but that is what unums do. Unums are to floats what floats are to integers: A superset.

Unum arithmetic has more rigor than *interval arithmetic*, but uses far fewer bits of storage. A unum computing environment dynamically and automatically adjusts precision and dynamic range so programmers need not choose which precision to use; it is therefore easier and safer to use than floating point arithmetic. Unlike floats, unum arithmetic guarantees bitwise identical answers across different computers. Its main drawback is that it requires more gates to implement in a chip, but gates are something we have in abundance. *Memory bandwidth*, *power efficiency*, and *programmer productivity* are precious and limiting. Unums help with all three issues.

It is time to supplement the century-old idea of floating point arithmetic with something much better: Unum arithmetic. It is time to boil the ocean.

1.1 Fewer bits. Better answers.

Happy 100th birthday, floating point. We are still using a number format that was designed for the computing technology of the World War I era.

The idea of automatic computing with numbers that look like *scientific notation* dates back to 1914 (Leonardo Torres y Quevado). Scientific notation expresses a number as the product of two parts, like "6.022×10^{23}", where the 6.022 is the fractional part and the 23 in "10^{23}" is the exponent. If you had to use an *integer* to express that value, it would take up far more space, both on paper and in a computer system:

$$602\,200\,000\,000\,000\,000\,000\,000$$

The exponent can be negative or positive, so the location of the decimal point can "float" right or left to represent very large or very small numbers. Torres y Quevado's 1914 proposal for such numbers in automatic computers started showing up in working machines around 1941 (Konrad Zuse's Z3), using base two (binary) instead of base ten (decimal). Zuse's computer could store numbers with a 16-bit fraction and 6-bit exponent, like

$$1.0001110111010001 \times 2^{110100}$$

and the "binary point" instead of the "decimal point" is what floats left and right.

However, every early computer design was different in how it represented floating point values and how arithmetic behaved. After decades of incompatibility, an Institute of Electrical and Electronics Engineers (IEEE, usually pronounced "I triple E") committee standardized the format and some of the behavior in the mid-1980s. That standard is referred to as IEEE 754.

The chemists reading this introduction will recognize 6.022×10^{23} as the physical constant known as "Avogadro's number," a standard number of atoms or molecules of an element or compound. Chemists almost always insist on using double precision for their computer simulations. In IEEE double precision, Avogadro's four-decimal-digit approximate number looks like *this* sesquipedalian pile of bits:

$$(-1)^0 \times 2^{10\,001\,001\,101-1023} \times \texttt{1.1111111000010101010011110100010101111110101000010011}$$

> Astonishingly, the IEEE Standard does *not* require that different computers using the same format produce the same results! The use of hidden "guard digits" was encouraged to improve the accuracy of rounded answers, even at the price of causing baffling *inconsistency between results from different computer systems.* The developers of the Java programming language were surprised to find out, the hard way, that specifying floats as IEEE Standard provides *no* assurance that results will be identical from any computer. The Java workaround has been to disable all hidden methods that improve accuracy.

We now have several different IEEE Standard binary floating point precisions: 16-bit (half), 32-bit (single), 64-bit (double), and 128-bit (quad). Currently, programmers must choose the precision they think is right for their application. If they choose too little, the resulting accuracy might be too low or the dynamic range might not be large enough to represent the largest and smallest values (overflow and underflow). If they choose too much precision, their application will consume unnecessary storage space, bandwidth, energy, and power to run, and of course will also run slower.

Since energy consumption or power consumption limits many computers, from mobile devices (where we always want more battery life) to giant data centers (which run out of available megawatts before they run out of space or money for equipment), maybe it is time to re-examine this century-old idea of floating point format to see if there is something better. "Better" means:

- easier to use
- more accurate
- less demanding on memory and bandwidth
- less energy- and power-hungry
- guaranteed to give bitwise identical results on different computers
- in most situations, faster.

Suppose we want to build on the IEEE principles, but be able to vary the precision and dynamic range to the optimum number of bits, and also **record whether the number is exact or lies within a range, instead of rounding the number**. We can do that by attaching additional bit fields that make the number *self-descriptive*. Call this a "universal number," or *unum.*

Definition: A *unum* is a bit string of variable length that has six sub-fields: Sign bit, exponent, fraction, uncertainty bit, exponent size, and fraction size.

As a quick preview, here is what the unum format looks like:

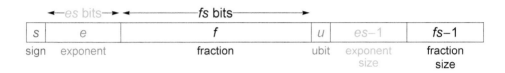

The left three fields are like IEEE floating point, but with better rules about Not-a-Number (NaN) and infinity, and a much better way to handle overflow, underflow, and rounding. It does so with the uncertainty bit, or ubit for short (rhymes with "cubit").

Definition: The *ubit* is a bit in the unum format that is **0** if a unum is exact and **1** if the unum represents the open interval between two exact unums.

The ubit is exactly like the "···" in an expression like "$2 \div 3 = 0.666 \cdots$". It means there are more bits after the last bit of the fraction, not all 0 and not all 1, but unspecified. Instead of rounding, the ubit stores the fact that a result lies between representable exact values.

The ubit also marks the case of a value being between the largest representable real number and infinity, or between the smallest magnitude representable number and zero. The ubit allows unums to handle cases that floats would overflow to infinity or underflow to zero, instead treating them as precisely defined ranges that are *never* confused with infinity or zero. In short, a unum is totally honest about what it does and what it does not know about a value. Unums manage uncertainty by making it explicit, and storing it in the number self-description.

The last two fields indicate the number of bits in the exponent and the fraction, allowing both to change with every calculation, much the way the exponent in a float changes the binary point location with every calculation. (They are offset by one, so if the bits in the exponent size are **010** = 2, for example, the exponent field has a width of 3 bits, and similarly for the fraction size.) The number of bits for the fields that describe the exponent size and fraction size are set in the *environment.*

Those fields can be as small as *zero* bits wide. In that extreme case, all the unums have only four bits: A bit for the sign, a one-bit exponent, a one-bit fraction, and the ubit indicating whether the fraction is an exact value or a range of values. Ultra-short unum bit strings turn out to be surprisingly powerful tools.

That was the fire-hose version of the explanation of unums. The chapters that follow work up to the unum format systematically. The unum approach will prove to have some startling advantages over the way computers usually store a real number, just as floating point format has advantages over fixed point. For one thing, unums are terse compared to floats, which saves storage and bandwidth as well as reduces power and energy requirements of the hardware. More importantly, **a unum never substitutes a nearby number for the mathematically correct one the way floating point does**. They are mathematically honest, and give bitwise identical answers on different computers.

Unums require more logic than floats for a hardware implementation, but transistors have become so inexpensive that we welcome anything that gives them useful new things to do. Wouldn't it be better if the computer could do most of the decision-making about precision and dynamic range, so the programmer does not have to?

1.2 Why better arithmetic can save energy and power

Heat-removal structures for processor chips are now hundreds of times larger than the chips themselves. A 140-watt chip has about the same power density as the coils on an electric stove top.

Many people are surprised when told that the *arithmetic* on a modern computer uses very little energy, and that far more energy goes into *moving* data. That is because transistors have gotten faster, cheaper, and more energy-efficient, but the wires that connect them remain relatively slow, expensive, and energy-inefficient.

Here are some measurements for 2010-era technology, where "pJ" is picojoules:

Operation	Approximate energy consumed	Where
Perform a 64–bit floating point multiply–add	64 pJ	on–chip
Load or store 64 bits of register data	6 pJ	on–chip
Read 64 bits from DRAM	4200 pJ	off–chip

The message is clear: The big energy drain is for *off-chip* data motion, not the math inside the chip.

If a processor has enough cores and multiplier-adder units to perform 32 float operations per clock, at 3 GHz, that still only adds up to about 6 watts. Moving data around within a chip is fairly low-power because the wires are microscopic, but moving data to and from external DRAM means driving wires that are big enough to see, gigantic by chip standards. That is why it can take over 60 times as much energy to move a double-precision float into a processor as it takes to perform a multiply and an add. That ratio is *increasing* as transistors improve.

An analogy: Suppose you purchase car insurance, but you have no idea how much your car is worth. So you make a conservative guess and declare that the car is worth *ten times* its actual value, just to make sure you are covered in case of an accident. Later, you get the bill and are outraged at the shockingly high cost of car insurance these days.

The fifteen-decimal numbers computer users ask computers to shovel back and forth are the same kind of over-insurance against disaster. Users have *no idea* what the right number of decimals to demand is, so they ask for far too many. Later, they are outraged that their application is running too slowly and consuming too much electrical power.

The intent here is not to scold computer users into doing a better job of assessing their precision needs; they should *not* have to do that. They should instead specify the accuracy of the input values and the accuracy they wish to have in the output values. Demanding double precision, or any precision for that matter, is a confusion of means with goals. The computer should manage its own precision needs to achieve requested accuracy goals, which it can do better than any programmer. But first the computer needs a number format *that allows it to do just that.*

2 Building up to the unum format

Positional notation and the concept of zero are at least as ancient as the abacus. The oldest known forms of the abacus were round stones in troughs, not beads on rods. The Latin word for stone is "calculus," which makes it clear where the word "calculate" comes from. Europeans learned enough about the positional system from Arab traders that by 1202, Fibonacci published a description of it in *Liber Abaci*. The woodcut above is from *Margarita Philosophica* by Gregor Reisch (1503), and depicts a competition between an "algorist" using Arabic numerals and an "abacist" using a four-decimal abacus. Apparently this rivalry was the 16th-century equivalent of a Mac versus Windows argument.

2.1 A graphical view of bit strings: Value and closure plots

In the sections that follow, we start with the simplest way to represent counting numbers, and gradually increase the complexity to make them more and more expressive, ending in the definition of the universal number (unum) format. Even if you already know all about integer and floating point storage formats, the visualizations are novel and you may find them of interest.

In practice, integers and floats have 16, 32, or 64 bits on most computers, but long bit strings are hard to read and we do not need them to explain each part of the representation. Also, using fewer bits highlights the problem of finite precision that we need to overcome, so the following sections will use *very* short bit strings.

From the earliest days, there has been little debate about how to represent *counting numbers* in binary. We use the same positional notation as with decimal numbers, but with each location representing a power of two instead of a power of ten, and with only digits **0** and **1** allowed. (We use **this typeface** to help distinguish **binary** digits from decimal digits.) The word "bit" is a shortening of "binary digit," just as "unum" is a shortening of "universal number."

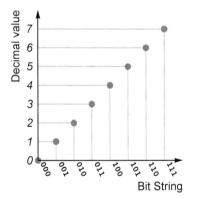

We can show the value represented by bit strings with a *value plot*. In a value plot, the horizontal axis shows the bit strings, sorted in numerical order as binary integers. The vertical axis shows the value *expressed* by the string, shown in decimal form, for a particular format. Since the bit strings represent specific numbers, the plot is a series of discrete points, not a continuous line or curve.

Some people prefer to understand concepts via pictures, others prefer symbolic formulas, and still others like to see actual numbers. We will use all three in the sections that follow, and let the reader choose. As a formula, suppose the bit string is n bits long, $b_{n-1} \ldots b_2\, b_1\, b_0$ where each b_i is a **0** or **1**. The bit string $b_{n-1} \ldots b_2\, b_1\, b_0$ represents the numerical value

$$b_{n-1}\, 2^{n-1} + \ldots + b_2\, 2^2 + b_1\, 2^1 + b_0\, 2^0.$$

If we use n bits, the integers represented range from 0 (all bits set to **0**) to $2^n - 1$ (all bits set to **1**). Interesting question: Why do the powers of the base go *down* to zero from left to right, instead of starting at zero and counting up? Answer: These are Arabic numerals, and Arabic writing goes from right to left! So it *does* actually start with 2^0 (one) and increases in the power of the number base, if you read things as the Arabs did when they invented the notation, about 1500 years ago.

When children in countries that read left-to-right are first introduced to multi-digit numbers, they often ask the teacher why the digits do not stand for the ones, then tens, then hundreds, etc. from left to right instead of having to first count the total number of digits to figure out what the first digit means. (For the fussy historians: Yes, the Arabs got the idea of digit representation from India, which had a well-developed form of positional mathematics centuries earlier. That is why it is properly called the Hindu-Arabic numeral system.)

Binary numbers are more "dilute" than decimals, in that you need more symbols to represent a number. In general it takes about 3.3 times as many digits to write a counting number in binary as it does in decimal. For example, a decimal value like $999\,999$ turns into **11110100001000111111**. To be literate (or perhaps the right word is "numerate") in binary, it is helpful to be able to recognize at a glance the three-bit binary number values, the way you would recognize common three-letter words:

Decimal	0	1	2	3	4	5	6	7
Binary	000	001	010	011	100	101	110	111

If you think about how to use a fixed-size representation for basic arithmetic, a shortcoming appears: The result of an operation is often not expressible in the format! To visualize this, *closure plots* show what happens for addition, subtraction, multiplication, and division on five-bit strings.

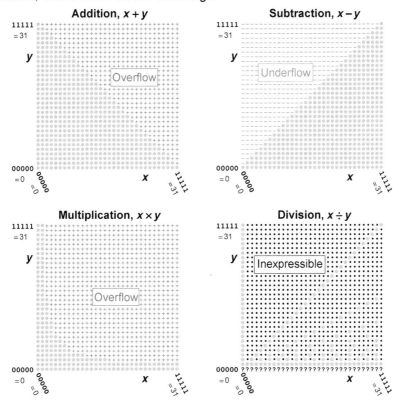

The larger "•" dots show cases where the answer can be represented; everywhere else, something goes wrong. Ideally, *every* dot would be "•", indicating the format is *closed* under arithmetic operations. With addition, the result can *overflow* (marked with "+" signs) for about half the cases, meaning the result is larger than the largest representable number. With subtraction, the result can *underflow* (marked with "-" signs) about half the time, which means the result is less than the smallest representable number, zero. Multiplication overflows for almost all input values. And with division, either the numbers do not divide evenly (marked with small "·" dots), or worse, we divide by zero (marked with "?"); either way, there is no way to express the exact result except for a few special cases.

2.2 Negative numbers

The operator's console and frame of the IBM 701, introduced in 1952, the first commercial computer to have sign-magnitude binary representation. It is also the source of the popular misquotation that Thomas J. Watson said "I think there is a world market for maybe five computers." He actually said at a stockholder meeting that after a sales tour to twenty customers he thought there would be five orders for the IBM 701, but the company instead received eighteen.

There are several ways to represent *negative* numbers, and we will need a couple of them. For example, the *offset-m* or *excess-m* method says to subtract a fixed number *m* from the usual meaning of the bits so that the first *m* numbers represent negative values.

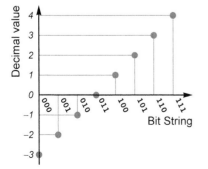

Sometimes this is called *biased* representation, since the represented value has a bias of negative *m*. If we are using *n* bits, and want about half the values to be negative and half to be positive, the best offset *m* might be an integer close to half the maximum unbiased value. That nearly balances the number of positive and negative values.

Say, $m = 2^{n-1} - 1$. So for three-bit numbers, we subtract $2^{3-1} - 1 = 3$ from the conventional meaning of the bit string, as shown in the value plot above.

The bit string $b_{n-1} \ldots b_2\, b_1\, b_0$ therefore represents the following value in this format:

$$x = b_{n-1}\, 2^{n-1} + \ldots + b_2\, 2^2 + b_1\, 2^1 + b_0\, 2^0 - \left(2^{n-1} - 1\right).$$

Does this help make more of the results of arithmetic representable? Here again are the closure plots for five bits with a bias of $2^{5-1} - 1 = 15$, using the same symbol meanings as in the previous plots for results of operations that are representable, overflow, underflow, or unrepresentable.

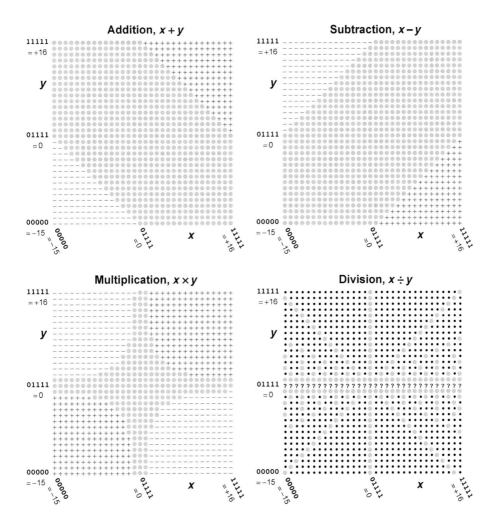

This is an improvement. About 75% of the time, addition and subtraction produce a representable result, compared to the ~50% success rate we had before.

Another way to represent both positive and negative binary numbers is *sign-magnitude,* which is much more like the way we usually write signed decimal numbers: Put a "−" in front of the string if it is negative, or a blank (or a "+") in front of the string if it is positive. We can use a **0** bit for positive and a **1** bit for negative and treat the first bit of the string as the *sign bit* and the other bits as the *magnitude*. This is the beginning of the idea that the bit string can be **partitioned**, with different rules for the meaning of each part.

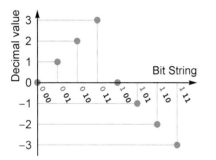

For clarity in the value plot, we color-code the sign bit of each string as red and add a space between the sign bit and the other bits. This notation has the minor drawback that there are two ways to represent the value zero. That redundancy wastes a bit pattern; with a three-bit string, we can represent only seven different numbers instead of eight.

In the early days of computing, IBM favored sign-magnitude format in its first scientific computers, ones that used vacuum tubes as logic elements.

A six-bit sign-magnitude number can be envisioned the following way, where we distinguish between the sign bit s and the b_i magnitude bits that have the same meaning as a simple counting number:

s	b_4	b_3	b_2	b_1	b_0
$\pm$	2^4	2^3	2^2	2^1	2^0

With six bits, we can represent integers from −31 to +31 using sign-magnitude format. In general, the *n*-bit string $s\, b_{n-2}\, b_{n-3} \dots b_0$ represents the value

$$x = (-1)^s \left(b_{n-2}\, 2^{n-2} + \dots + b_2\, 2^2 + b_1\, 2^1 + b_0\, 2^0 \right).$$

An advantage of sign-magnitude format is that it is trivial to negate a number simply by flipping the sign bit, so subtraction is as easy as addition of a negated number. It is also just as easy to take the absolute value of a number, by setting the sign bit to **0**.

Testing whether a number is positive or negative is *almost* as trivial, but you have to test if all bits are **0**, since then the represented value is zero and the sign bit does not matter. Similarly, if you want to know whether two numbers are equal, you cannot simply say "They are equal if all the bits are the same" because there are two ways to represent zero and that exceptional case has to be tested.

The following closure plots show that the ability to represent the results of arithmetic operations with sign-magnitude format is similar to what it is for the biased format, but the quadrants are in different orientations, and now we make a subtle change in definition. "Overflow" and "underflow" now mean that the *magnitude* is too large and too small to express. Division can produce underflow; for example, 1 ÷ 4 is too small to express since the smallest nonzero magnitude is 1.

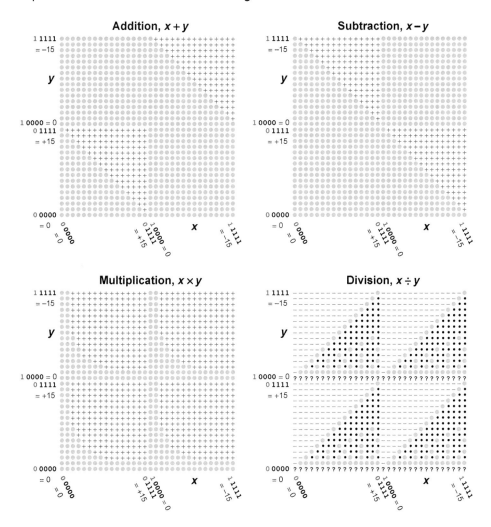

The weakest part of the format now seems to be in the multiplication and division closure results, so we need to introduce something to make those operations more robust. Balancing positive and negative values made it equally likely for addition and subtraction to create larger and smaller numbers. In the plots above, multiplication usually makes numbers that are larger (often overflowing), and division usually creates numbers that are smaller (often underflowing). If we *scale* the value so that about half the time the magnitude is a fraction less than one, then multiplication and division will be similarly balanced, making numbers larger about half the time and smaller half the time.

2.3 Fixed point format

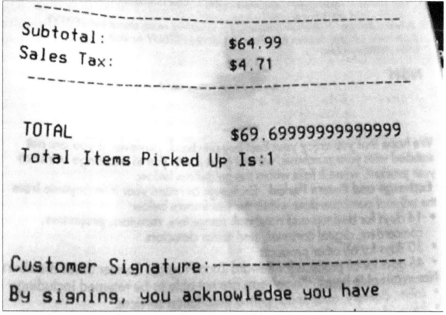

A case where fixed point arithmetic should have been used, but wasn't.

The next step up from signed integers is to represent non-integer *fractional* values. Again there are many ways to do this; for example, we could store the numerator and the denominator separately, like using **011** and **111** to represent $\frac{3}{7}$. There are *rational arithmetic* systems based on this idea, but they are not common, perhaps because the exact addition and subtraction of fractions requires a common denominator, the size of which grows very quickly and overflows.

> **Exercise for the reader**: Find the closure plot for six-bit rational arithmetic, with various allocations of bits to the numerator and the denominator. Does rational arithmetic yield more results that are representable within the format? Does it help to make the first bit a sign bit?

Using the binary analog of a number with a decimal point is simpler. Just as negative numbers can be represented with an offset or bias, we can place the "binary point" somewhere in the bit string. Digits to the right of that point represent negative powers of two, so **0.1**, **0.01**, and **0.001**… in binary means $\frac{1}{2}$, $\frac{1}{4}$, $\frac{1}{8}$,…. just as in decimal they would represent $\frac{1}{10}$, $\frac{1}{100}$, $\frac{1}{1000}$, … . For example, the number $13\frac{1}{8}$ is expressible as $8 + 4 + 1 + \frac{1}{8}$, or **1101.001** in binary. If the format puts the binary point just before the last three digits, that is the same as saying that the integer represented by the magnitude part of the bit string should be divided by $2^3 = 8$.

The value plot shows a two-point shift marked with a "." binary point within each bit string, for clarity:

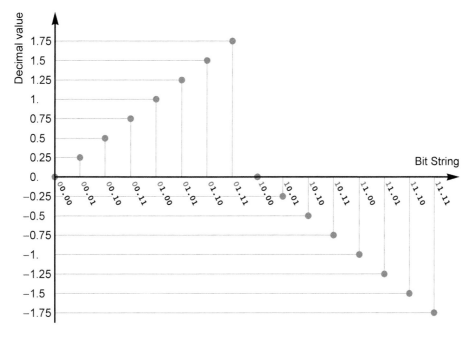

Here is the format for a signed number with nine bits total, and in this case only one bit to the left of the binary point:

s	b_7	b_6	b_5	b_4	b_3	b_2	b_1	b_0
$\pm$	$2^0 \cdot$	2^{-1}	2^{-2}	2^{-3}	2^{-4}	2^{-5}	2^{-6}	2^{-7}

For example, if the nine bits in the string are `100000001`, the string represents $-2^{-7} = -\frac{1}{128} = -0.0078125$. The first "1" means the number is negative, and the rest of the digits mean `0.0000001` in binary.

In general, with n bits and a fixed binary point e bits from the right, the bit string $s\, b_{n-2}\, b_{n-1} \ldots b_0$ means

$$(-1)^s \left(\frac{b_{n-2}\, 2^{n-2} + \ldots + b_2\, 2^2 + b_1\, 2^1 + b_0\, 2^0}{2^e} \right) =$$
$$(-1)^s \left(b_{n-2}\, 2^{n-2-e} + \ldots + b_2\, 2^{2-e} + b_1\, 2^{1-e} + b_0\, 2^{-e} \right).$$

The math a computer can perform with such a rigid format is quite limited, but it was popular in the early days of computing because it was relatively easy to build computers that used such a format. Arithmetic based on the fixed point format does not take many logic gates, and it reduces the number of cases where multiplication and division underflow.

Fixed point works well for financial calculations, where most of the math is adding and subtracting. Using fixed point for scientific calculations is a lot of work, because it requires constantly having to watch out for running out of digits to the right or the left of the binary point when multiplying or dividing. The closure plots are for a five-bit sign-magnitude number with two magnitude bits to the left and two bits to the right of the binary point:

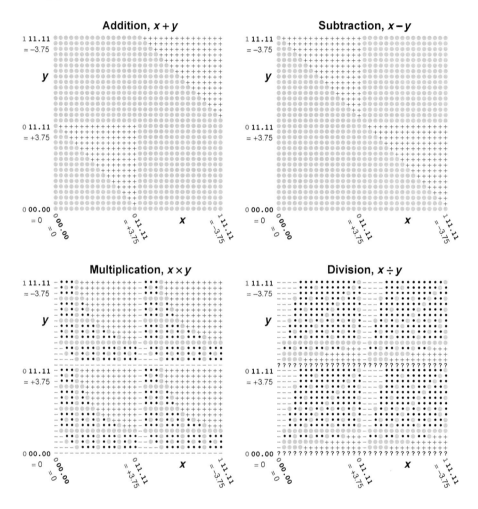

Addition and subtraction are unchanged by the scaling, but multiplication and division are beginning to resemble quilting patterns. For multiplication, the range of valid numbers is now balanced between underflow and overflow regions, but at a cost. The cost is that many products of *x* and *y* require more than four contiguous binary digits, so the L-shaped regions are now permeated with values that *cannot be represented in the format*. For example, **11.01** times **00.11** in binary (3.25 times .75 in decimal) could be represented as **10.0111** if we had six bits after the sign bit. The number **10.10** is a close approximation, but it represents a **different number**. In decimal, those values are 2.4375 and 2.5. Should we accept 2.5 as the "rounded" value of 2.4375?

> Accepting results that are incorrect but "close" also would allow filling in many of the gaps in the Division closure plot. There lies a slippery slope.

The next step is to create a *new field* within the bit string, one that indicates *where* the binary point is instead of keeping it at a fixed location. We use bits to represent the sign and magnitude, but also a variable *scale factor* as part of each number.

2.4 Floating point format, almost

The ancestor of floating point is scientific notation, as mentioned in the beginning. Using scientific notation with a sign, three significant decimal digits, and a signed exponent with two decimal digits, for example, we could represent

$$-9.99 \times 10^{+99} \text{ to } -1.00 \times 10^{-99}, 0, 1.00 \times 10^{-99} \text{ to } 9.99 \times 10^{+99}$$

Representing zero is a special case. For every other number, the exponent is adjusted so that there is a digit 1 through 9 just before the decimal point. You would not write 0.06×10^{17} because that would be like a spelling error; the format rules say that you *normalize* the number to 6.00×10^{15}. The one exception is allowing 0.00 for the significance digits, and then not caring what the exponent or sign is. Making an exception for 0.00 starts to look inelegant.

An alternative: *If* the exponent is the minimum value, –99, allow the fraction part to go below 1.00, so that the numbers

$$0.01 \times 10^{-99}, 0.02 \times 10^{-99}, \dots, 0.99 \times 10^{-99}$$

evenly fill in the gap between zero and the smallest positive number, 1.00×10^{-99}. (And similarly for the negative versions of those numbers.) This is called *gradual underflow*, and the numbers are called *subnormal* because the leading digit is less than one. This seems to some people like a lot of work to go through just to support unusually small numbers. It turns out that when the representation is binary, and the length of the bit string is small, it is *absolutely essential* to use this idea. The value plots will bear this out.

The binary version of scientific notation says that the significant digits, or *fraction*, represent a value between **1.000...0** and **1.111...1**, to however many bits have been allocated to store the fraction. Konrad Zuse was the first to notice that the first bit is always a **1**, so there is no reason to store that first **1** in the format. (That trick does not have an obvious equivalent with decimal fractions, where the first digit could be anything from **1** through 9 and has to be stored.) So the implied **1** bit to the left of the binary point is called the *hidden bit*. If we use some of the bits to show the scale factor, where in the string should those scale bits go? To the left of the sign bit? To the right of the fraction bits? And how should both positive and negative values for the scale factor be stored, since there are many ways to represent negative numbers? Does the scale have to represent a power of two, or could it be a power of a different base, like 16? Should there be a way to represent infinity?

With so many choices, it is no surprise that every computer company came up with a different float format in the early days of electronic computing. If you had data archives containing floats, they would only make sense on a particular computer brand. If you wrote a program to *convert* from one format to another using the same number of bits for the number, you typically would lose either precision or dynamic range because fixed-size conversions are limited to *whichever bit field is the smallest* of the two formats. Also, calculations with floats almost always gave different answers on different computer systems.

There is, however, a format that seems the simplest and most elegant mathematically, and that is consistent with (but not identical to) the current IEEE Standard. The exponent bits are colored blue for clarity, here and throughout this book:

| s | e | e | e | e | f | f | f | f | f |

$$\pm \quad 2^{exponent-bias} \times 1.+ \quad \text{fraction bits}$$

The exponent bits should be *between* the sign bit and the fraction bits. And they should use a *bias* to indicate negative exponent values, like the first example in Section 1.2. If there are *es* bits in the exponent, then the bias is $2^{es-1} - 1$. That definition still does not provide a way to represent zero, however.

Here is the value graph for a five-bit representation with a sign bit, two-bit exponent (biased by 1, so it represents −1, 0, 1, or 2), and two-bit fraction. This is not yet an IEEE float, but it is getting closer:

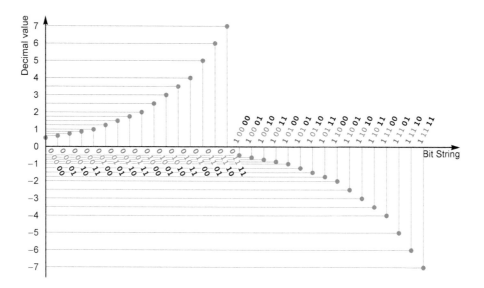

When the exponent increases by one, the slope doubles. Notice, however, that there is no way to represent *zero*.

In the early days of floating point, hardware designers decided to use the following exception: "If the exponent bits are all **0**, then the number is zero." That gives you something that looks like the plot on the next page, where several bit strings redundantly represent zero (marked with hollow circles):

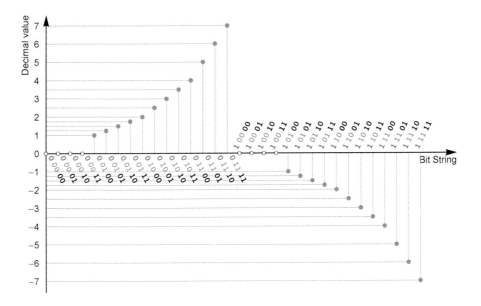

Ignoring the fraction bits wastes a lot of bit strings that could have unique and useful meanings. A far better solution is to double the slope of the very first ramp and start it at zero. The rule that does so is this: "If all exponent bits are **0**, then *add one to the exponent and use **0** for the hidden bit.*" Why this works is not obvious, but it makes the points near zero ramp *exactly like fixed point representation*:

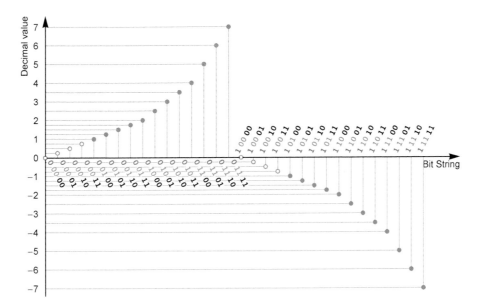

This is the binary version of "gradual underflow," and some hardware designers still object to it, claiming that it is harder to design, or takes more transistors, or runs slower. Clipping to zero and gradual underflow both require doing something exceptional if all the bits of the exponent are 0, so the number of transistors and execution times are very similar either way.

As with scientific notation, the numbers represented with the smallest possible exponent and fractional part less than 1 are called *subnormal* numbers. A normal float is scaled so that the fraction is between 1 and 2, but a subnormal has a fraction between 0 and 1. The elegant thing about this approach is this: If there is only *one* exponent bit, then the exponent bit works exactly as if it were part of the fraction! For example, with three bits in the fraction field and just a single-bit exponent, it looks just like a four-bit fixed point representation:

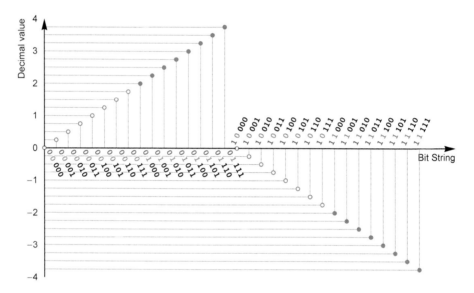

The first half of each ramp in the last two value plots consists of subnormal numbers (shown with hollow circles). If what we seek is a unified approach to representing numbers, then we clearly need "subnormal" values, one of the best ideas of the IEEE Standard, since it unifies fixed point and floating point representations. (Unfortunately, the IEEE Standard made support for subnormals *optional*, thereby creating another source of inexplicably different results for identical programs run on different computer systems.)

Here is the formula for the value of a number with three fields: One bit for the sign, *es* bits for the exponent, and *fs* bits for the fraction, where **Boole[***test***]** is 0 if the *test* is false, and 1 if *test* is true:

$$x = (-1)^s \, 2^{e-(2^{es-1}-1)+(1-\text{\textbf{Boole}}[e>0])} \times \left(\text{\textbf{Boole}}[e>0] + \frac{f}{2^{fs}} \right)$$

The following may be easier to read, and it means the same thing:

$$
x = \begin{cases} (-1)^S \, 2^{e-2^{es-1}} \times \left(\dfrac{f}{2^{fs}}\right) & \text{if } e = 0, \\[2ex] (-1)^S \, 2^{e-2^{es-1}-1} \times \left(1 + \dfrac{f}{2^{fs}}\right) & \text{if } e > 0. \end{cases}
$$

This is almost an IEEE-style float; the only difference is that we have not yet dealt with exceptions for infinity, negative infinity, and Not-a-Number (NaN). For now, we want to know if the closure plots are improved with all the extra scaling machinery in the representation. Try it with just *five bits*, say, a sign, two-bit exponent, and a two-bit fraction. That is, e and f can be any of {0, 1, 2, 3} and the value represented is

$$
x = (-1)^S \, 2^{e - \texttt{Boole}[e > 0]} \times \left(\texttt{Boole}[e > 0] + \frac{f}{4} \right)
$$

Here are the "almost IEEE" float values generated by all 32 of the five-bit strings using the formula above, from `0 00 00` to `1 11 11`:

$$
0 \quad \frac{1}{4} \quad \frac{1}{2} \quad \frac{3}{4} \quad 1 \quad \frac{5}{4} \quad \frac{3}{2} \quad \frac{7}{4} \quad 2 \quad \frac{5}{2} \quad 3 \quad \frac{7}{2} \quad 4 \quad 5 \quad 6 \quad 7
$$

$$
0 \quad -\frac{1}{4} \quad -\frac{1}{2} \quad -\frac{3}{4} \quad -1 \quad -\frac{5}{4} \quad -\frac{3}{2} \quad -\frac{7}{4} \quad -2 \quad -\frac{5}{2} \quad -3 \quad -\frac{7}{2} \quad -4 \quad -5 \quad -6 \quad -7
$$

Five is an awfully small number of bits for a floating point representation, but we can still use closure plots to see when the result of a plus-minus-times-divide operation is *exactly* representable in the same format. The generated patterns are striking:

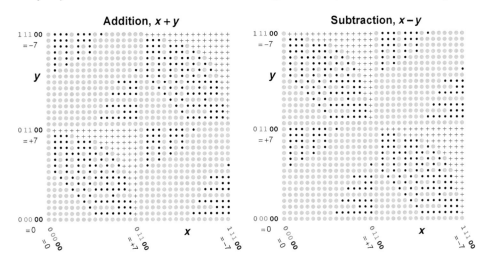

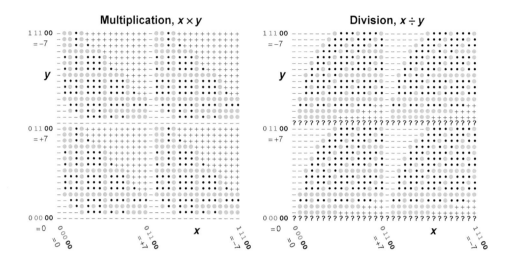

Besides the overflow cases, the Addition and Subtraction closure plots now show intermingled cases where the result cannot be represented. For example, if we add $\frac{1}{4}$ and 5, the result $\frac{21}{4}$ cannot be represented with only five bits in this format. Multiplication and Division plots also have unrepresentable results, as they did before, but they produce exact answers for a more robust range of values than with fixed point. The float format asks us to accept several *departures from the truth*:

- Accept ∞ (or $-\infty$) as a substitute for large-magnitude finite numbers (the "+" symbols).
- Accept 0 as a substitute for small-magnitude nonzero numbers (the "−" signs).
- Accept incorrect exact floats for nearby, but distinct, exact real numbers (the ·" dots).

The "?" signs at the bottom of the Division closure plot above do not indicate a departure from the truth, but merely a warning that division by zero produces an indeterminate result.

2.5 What about infinity and NaN? Improving on IEEE rules

Even the earliest floating point hardware made an exception for representing infinity, both positive and negative. This is where we first deviate from the IEEE 754 Standard, but we do it in a way that *preserves more information* than the IEEE Standard does.

The IEEE Standard tests if all exponent bits are **1**, and uses that case for all of the exception values. If the fraction bits are all **0**, the value is ∞ or $-\infty$ (depending on the sign bit). If the fraction bits are *anything else*, then the string represents Not-a-Number, or "NaN."

So with our five-bit float example, the value plot with IEEE rules looks like this:

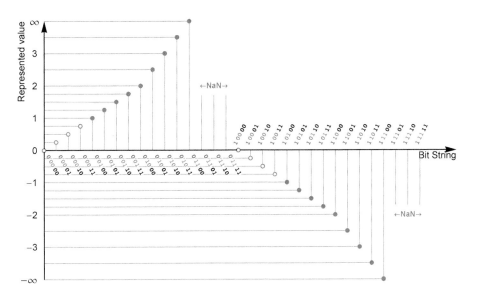

Do we really need *six* different ways to say something is a NaN, that it has an indeterminate value? The IEEE Standard specifies that there be "quiet" NaNs that continue computing (but of course any operation with a NaN input produces a NaN as output) and "signaling" NaNs that halt the computation and alert the computer user. That seems reasonable, so there should be *two* kinds of NaNs.

The situation with IEEE floats is actually much more wasteful than shown above. In single precision, there are over *sixteen million* ways to say that a result is indeterminate, and in double precision, there are over *nine quadrillion* (9×10^{15}) ways to say it. How many do we really need? *Two.* A bit string for a quiet NaN, and another bit string for a signaling NaN. The IEEE designers may have envisioned, in the 1980s, that a plethora of useful categories would eventually consume the vast set of ways to say NaN, like one that says the answer is an imaginary number like $\sqrt{-1}$, and one that says the answer is infinite but of unknown sign, like $1 \div 0$, and so on.

It didn't happen.

The IEEE designers meant well, but when you discover you are computing garbage, there is really very little need to carefully sort and classify what kind of garbage it is. So instead we will define the case where all bits in the exponent and fraction are 1 to represent infinity (using the sign bit to determine $+\infty$ or $-\infty$), and use all the other bit strings where just the exponent is all 1 bits to mean an actual finite value. That *doubles* the range of values we can represent. Later we will show a better way to represent the two kinds of NaNs that makes mathematical sense.

The IEEE folks figured out gradual underflow, but it is unfortunate that they did not notice that they also could have supported "gradual overflow" for bit patterns that have a maximum exponent. Instead, their Standard throws big numbers upwards to infinity or NaN, the mirror image of what some hardware designers do when they clip subnormals to zero.

For the five-bit float example, the value plot for our improved format looks like this:

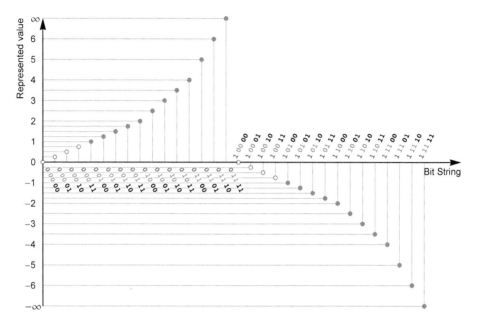

The formula for the meaning of a bit pattern with *es* exponent bits and *fs* fraction bits is as follows:

$$x = (-1)^s \times \begin{cases} 2^{e-2^{es-1}} \times \left(\dfrac{f}{2^{fs}}\right) & \text{if } e = \text{all 0 bits,} \\ \infty & \text{if } e \text{ and } f \text{ are all 1 bits,} \\ 2^{e-2^{es-1}-1} \times \left(1 + \dfrac{f}{2^{fs}}\right) & \text{otherwise.} \end{cases}$$

It is always dangerous to leave a pile of unused bit representations lying around, because someone inevitably discovers them and creates a hack. So it is with the trillions of NaN bit strings. There is now a technique in JavaScript that uses the 2^{52} different bit strings for NaN for something called "NaN boxing." It breaks just about every rule there is about creating maintainable standards.

There was discussion in the IEEE Standard committee meetings that multiple NaNs could be used to encode the location of where in the program a NaN error occurred, which sounds like the beginning of a good idea. But how would you ever be able to make that idea work uniformly for half, single, double, and quad precision floats?

3 The "original sin" of computer arithmetic

3.1 The acceptance of incorrect answers

Binary floats can express very large and very small numbers, with a wide range of precisions. But they cannot express all real numbers. They can only represent rational numbers where the denominator is some power of 2. Some readers might be taken aback by this statement, since they have long ago gotten used to having floats represent square roots, π, decimals, and so on, and were even taught that floats were designed to represent real numbers. What is the problem?

The "original sin" of computer arithmetic is committed when a user accepts the following proposition:

*The computer cannot give you the exact value, sorry. Use **this** value instead. It's close.*

And if anyone protests, they are typically told, "The only way to represent every real number is by using an infinite number of bits. Computers have only finite precision since they have finite memory. Accept it."

There is no reason to accept it. The reasoning is flawed. Humans have had a simple, finite way of expressing the vast set of real numbers for hundreds of years. If asked, "What is the value of π?" an intelligent and correct reply is "3.14···" where the "···" indicates *there are more digits after the last one*. If instead asked "What is seven divided by two?" then the answer can be expressed as "3.5" and the only thing after the last digit is a space. We sometimes need something more explicit than a space to show exactness.

> **Definition**: The "↓" symbol, when placed at the end of a set of displayed fraction digits, emphasizes that all digits to the right are zero; the fraction shown is *exact*.

Just as the sign bit positive (blank) or − is written to the left of the leftmost digit as a two-state indicator of the sign, exactness is a two-state indicator where the bit indicating exact (blank) or ⋯ is written just to the right of the rightmost digit. When we want to be more clear, we explicitly put a "+" in front of a positive number instead of a blank. Similarly, we can put the symbol "↓" after the digit sequence of an exact number, to emphasize "The digits stop **here** ↓." Or put another way, all the bits after the last one shown are **0**. It only takes one bit to store the sign of a number, and it only takes one bit, the *ubit*, to store whether it does or does not have more bits than the format is able to store.

> **Definition**: The *ubit* (uncertainty bit) is a bit at the end of a fraction that is **0** if the fraction is exact, **1** if there are more nonzero bits in the fraction but no space to store them in the particular format settings of the unum.

If the bits in the fraction to the right of the last one shown are all **0**, it is an exact value. If there is an infinite string of **1** bits, it is again exact and equal to the fraction that is one bit greater in the last place, the same way 5.999999… is exactly the same as 6 if the number of 9s is infinite). If the ubit is set, then the quantity expressed is *all real numbers* between the exact value of an infinite string of **0**s and the exact value of an infinite string of **1**s. The closest thing to a historical precedent for the ubit is a "rounding to sticky bit" described by Pat Sterbenz in his long out-of-print book, *Floating-Point Computation* (1974). He used the bit to indicate "it is possible to calculate more bits"; he, too, put it at the end of the fraction.

There is a jargon term that numerical analysts use for the difference between a number and the number represented by the next larger fraction value: An *ULP*.

> **Definition**: An *ULP* (rhymes with gulp) is the difference between exact values represented by bit strings that differ by one Unit in the Last Place, the last bit of the fraction.

(Some texts use "Unit of Least Precision" as the abbreviated phrase.) For example, if we had the fixed-point binary number **1.0010**, the ULP is the last 0, and a number that is one ULP larger would be **1.0011**.

So we write **1.0010↓** if the number is exact, and **1.0010···** to mean *all values between* **1.0010** *and* **1.0011**, *not including the endpoints*. That is, it is an open interval, which is typically written with two numbers in parentheses like (**1.0010, 1.0011**). A *closed* interval includes the endpoints and is written with brackets, [**1.0010, 1.0011**], but that is not what we want here. If the ubit is set, it means *an open interval one ULP wide*. Here are the annotated bit fields, where the ubit is shown in magenta as a convention throughout this book:

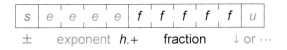

The "*h*" represents the hidden bit: **0** if the exponent is all zero bits, and **1** otherwise. Suppose we again have only five bits for a float, but now use one bit for the sign, two bits for the exponent, one bit for the fraction (not counting the hidden bit), and use the last bit as the ubit. What values can we represent? Some examples:

- **0 00 1 0** means exactly $\frac{1}{2}$. (An all-zero exponent **00** means the hidden bit is zero.)

- **0 00 1 1** means the open interval $\left(\frac{1}{2}, 1\right)$. All real values *x* that satisfy $\frac{1}{2} < x < 1$.

- **0 01 0 0** means the exact number 1.

The meanings increase monotonically, and neatly "tile" the real number line by alternating between exact numbers and the ranges between exact numbers. We no longer have to commit the "original sin" of substituting an incorrect exact number for the correct one. If you compute 2 ÷ 3, the answer with this five-bit format is "the open interval $\left(\frac{1}{2}, 1\right)$" which in this format is **0 00 1 1**. That may be a very low-precision statement about the value $\frac{2}{3}$, but at least it is the truth. If you instead use a float, you have to choose between two lies: $\frac{2}{3} = \frac{1}{2}$ or $\frac{2}{3} = 1$.

The designers of hardware for floats and their math libraries have to jump through hoops to do "correct rounding," by which they mean the requirement that rounding error never be more than 0.5 ULP. Imagine how many extra bits you have to compute, to figure out which way to round. For many common functions like e^x, there is no way to predict in advance how many digits have to be computed to round it correctly! At least the ubit eliminates the extra arithmetic needed for rounding. As soon as a computer can determine that there are nonzero bits beyond the rightmost bit in the fraction, it simply sets the ubit to **1** and is done.

The next section shows an unexpected payoff from the ubit: It provides an elegant way to represent the two kinds of NaNs.

3.2 "Almost infinite" and "beyond infinity"

This is a frame from the "JUPITER AND BEYOND THE INFINITE" final sequence of the motion picture *2001: A Space Odyssey*. The title of that sequence was probably the inspiration for the catchphrase of Pixar's Buzz Lightyear character, "To Infinity... and Beyond!" Going "beyond infinity" is exactly the way unums express NaN quantities.

Recall that we used the largest magnitude bit strings (all bits set to **1** in both the exponent and the fraction), with the sign bit, to mean positive and negative infinity. The largest magnitude finite numbers have a fraction just one ULP smaller, so their rightmost fraction bit is a **0**. What does appending a ubit mean for those cases?

It may seem peculiar, but $+\infty$ and $-\infty$ are "exact" numbers, for our purposes. Sometimes people will say that some quantity, like the number of stars in the sky, is "almost infinite," and the grammarian in the crowd will point out that "almost infinite" is contradictory, since it means "finite." Actually, "almost infinite" is a useful computational concept, because it means *a finite number that is too big to express.*

Definition: The value *maxreal* is the largest finite value that a format can express. It is one ULP less than the representation of infinity.

With the ubit notation, it is possible to express the open intervals (*maxreal*, ∞) and $(-\infty, -maxreal)$, which mean "almost infinite." That means every real number that is finite but too large in magnitude to express exactly with the available bits. This is a very different approach from *extended-precision arithmetic*, which uses software routines to express and perform arithmetic with oversize numbers by tacking on additional bits until memory is exhausted.

In terms of the five-bit examples of Section 2.5, where we presented a non-IEEE way to represent positive and negative infinity,

0 11 1 0   is +∞ (sign bit = 0, so value is positive; all exponent and fraction bits are 1),
1 11 1 0   is −∞ (sign bit = 1, so the value is negative).

The 0 ubit that ends the string marks the infinity values as exact. The largest magnitude *finite* values we can represent are one bit smaller in the fraction bits:

0 11 0 0   represents the number 4 by the rules for floats (hidden bit = 1), and
1 11 0 0   is −4. The *maxreal* and *−maxreal* values, one ULP less than ∞ and −∞.

When we turn the ubit on for the biggest finite numbers, it means "almost infinite." That is, bigger than we can express, but still finite:

0 11 0 1   is the open interval (4, ∞), and
1 11 0 1   is the open interval (−∞, −4).

If setting the ubit means "a one ULP open interval beyond the number shown," *what does it mean to set the ubit for infinity?* "Beyond infinity" makes us smile, because the phrase is nonsense. But we *need* a way to represent nonsense: NaN! So we can define "beyond positive infinity" as quiet NaN, and "beyond negative infinity" as signaling NaN. Sometimes we use red type to indicate signaling NaN, since it means "Stop computing!" In the five-bit case,

0 11 1 1   is quiet NaN, and
1 11 1 1   is signaling NaN.

There is an equivalent to *maxreal* at the opposite end of the dynamic range, which we call *smallsubnormal*.

> **Definition**: The value *smallsubnormal* is the smallest value greater than zero that a format can express. It is one ULP larger than zero.

Sometimes people use the expression "almost nothing," and the same grammarians again might say, "That means 'something', so 'almost nothing' is an oxymoron, like 'jumbo shrimp'." However, the phrase "almost nothing" is just as useful as "almost infinite." It means "not zero, but *a smaller number than we can express*."

The ubit elegantly lets us represent that concept as well: Beyond *zero*. And it gives meaning to "negative zero" since that tells which direction to point the ULP interval that has zero as one end.

0 00 0 0   is exactly zero, and so is
1 00 0 0   because we do not care what the sign bit is for zero. *But:*

0 00 0 1   is the open interval (0, *smallsubnormal*) and
1 00 0 1   is the open interval (−*smallsubnormal*, 0).

Incidentally, meteorologists already use this idea when they write that the temperature is negative zero, −0°. That means a temperature between −0.5° and 0°, whereas writing 0° means a temperature between 0° and 0.5°. For weathermen, it is obviously crucial to convey whether the temperature is *above* or *below* the freezing point of water (0° Celsius).

The smallest representable non-zero number has a 1 in the last bit of the fraction, and all other bits are 0. Because having all 0 bits in the exponent makes it a subnormal number, we call it *smallsubnormal* no matter how many exponent and fraction bits we are using. In this low-precision case of only five bits, that smallest nonzero number is $\frac{1}{2}$, but of course it will usually be a much tinier quantity once we start using larger bit strings.

3.3 No overflow, no underflow, and no rounding

If a number becomes too large to express in a particular floating point precision (overflow), what should we do? The IEEE Standard says that when a calculation overflows, the value ∞ should be used for further calculations. However, setting the finite overflow result to exactly ∞ is *infinitely* wrong. Similarly, when a number becomes too small to express, the Standard says to use 0 instead. Both substitutions are potentially catastrophic things to do to a calculation, depending on how the results are used. The Standard also says that a flag bit should be set in a *processor register* to indicate if an overflow occurred. There is a different flag for underflow, and even one for "rounded."

No one ever looks at these flags.

Computer languages provide no way to view the flags, and if a computing environment starts reporting intermediate underflow or overflow warnings even when the results look just fine, most users find out how to disable that reporting, which often improves the speed of the calculation. Chip designers are burdened with the requirement to put flags into the float arithmetic unit despite the fact that computer users find them useless at best, and annoying at worst.

A fundamental mistake in the IEEE design is putting the "inexact" description in three processor flags that are very difficult for a computer user to access. The right place is *in the number itself*, with a ubit at the end of the fraction. Putting just that single bit in the number *eliminates the need for overflow, underflow, and rounding flags*.

We do not need overflow because instead we have "almost infinite" values (*maxreal*, ∞) and (−∞, −*maxreal*). We do not need underflow because we have the "almost nothing" values (−*smallsubnormal*, 0) and (0, *smallsubnormal*). A computation need never erroneously tell you, say, that $10^{-100000}$ is "equal to zero" but with a hidden (or intentionally disabled) underflow error. Instead, the result is marked *strictly greater than zero but strictly less than the smallest representable number*. Similarly, if you try to compute something like a billion to the billionth power, there is no need to incorrectly substitute infinity, and no need to set off an overflow alarm.

If you ask for the value of π, with five bits of fraction and a ubit, what you get is that the value lies in the open interval (3.125, 3.25), which is a true statement. There is no need to set a "rounded" flag in the processor, which is a "sticky" bit in that it stays set until something in the operating system resets it. It could happen, say, that a result like (3.125, 3.25) gets multiplied by zero and becomes exact again. The "rounded" flag is like a car with a "CHECK ENGINE" light that is always on, yet still seems to run fine.

One last comment about recording the exact-inexact description in the number format itself. If you enter a big positive number on a calculator and press the square root button over and over, eventually the display will show 1.0000000 *exactly*. Which is incorrect, of course. The error is very much like setting an overflow number to infinity or an underflow number to zero. With the ubit after the fraction, the answer can be "1.00000···" with a "···" that *never goes away*. The number is always marked as "greater than 1, but by an amount too small to express." Traditional interval arithmetic cannot do this and makes the same mistake that floats do, accepting exact 1 as a possible result.

Here is some code that iterates until the square root is exactly 1, so mathematically, it should be an *infinite loop*:

```
Module[{x = 1.000001}, While[x ≠ 1, x = √x ; Print[NumberForm[x, 15]]]]
```

1.00000049999988

1.00000024999991

1.00000012499995

1.00000006249997

1.00000003124999

1.00000001562499

1.0000000078125

1.00000000390625

1.00000000195312

1.00000000097656

1.00000000048828

1.00000000024414

1.00000000012207

1.00000000006104

1.00000000003052

1.00000000001526

1.00000000000763

1.00000000000382

1.00000000000191

1.00000000000095

1.00000000000048

1.00000000000024

1.00000000000012

1.00000000000006

1.00000000000003

1.00000000000002

1.00000000000001

It stopped iterating. Double precision floats erroneously declared the last result *exactly* equal to 1. Many scientific calculations involve iteration and testing convergence, and the "stopping criterion" is a source of bugs and extra programming effort when using floating point representation. The ubit protects us from mistaking an exact number for a value that is *asymptotically approaching it*.

3.4 Visualizing ubit-enabled numbers

Floats were designed to allow calculations with (approximate) real numbers. For all formats discussed so far, bit strings could only represent discrete, exact values, so the value plots consisted of a series of points. With the ubit extension, we can now cover the *entire real number line*, as well as −∞ and ∞, and the two NaN types.

For the five-bit case used previously, the represented values look like the value plot below, with dots for exact numbers and rectangles that have rounded corners to remind us that they are the *open* intervals between exact numbers:

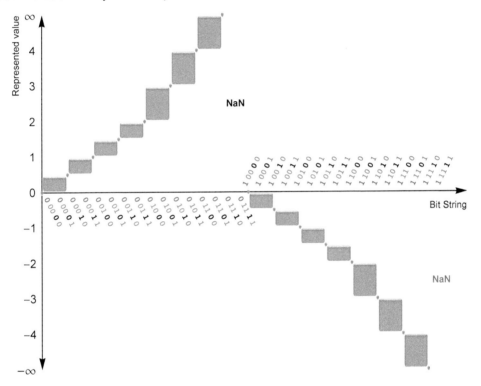

There are two subtle changes from previous graphs, other than the inclusion of open regions instead of just point values: The *y* coordinates where you might expect the numbers −5 and 5 instead have −∞ and ∞ as exact values, and the arrows on the *y* axis point both directions to remind us that the *entire real number line* is now repre-sented by the bit strings. We change the label for the *y* axis from "Decimal value" to "Represented value."

The next question is whether things are better in terms of *closure*, when doing plus-minus-times-divide arithmetic. They are, but they are still not perfect. The reason is that we can only represent ranges that are one ULP wide, and many operations produce results that are wider than one ULP. For example, $\left(0, \frac{1}{2}\right) + \left(0, \frac{1}{2}\right) = (0, 1)$, but there is no way to express the interval (0, 1) in this format.

The closure plots with the ubit enabled look like the following, for a five-bit string:

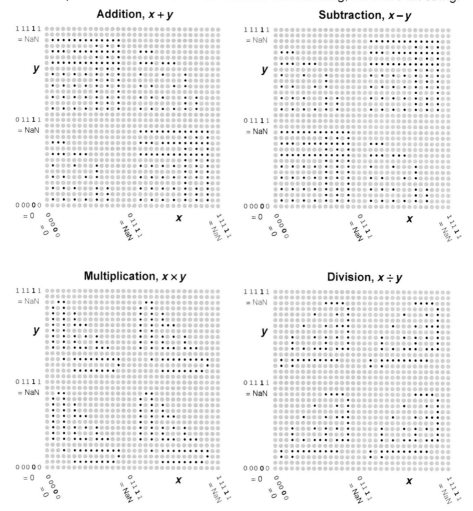

Overflow and underflow cases have been eliminated. We are tantalizingly close to having a *closed system*, where the results of arithmetic are always representable without rounding or otherwise compromising correctness. Whenever we add something to accommodate a new type of value, we risk creating yet *more* cases that do not fit the format. Is this an endless process, or is it possible to create a closed system for honest arithmetic on real numbers with a limited number of bits?

Exercise for the reader: It is possible to have *no* bits for the fraction and still get a number set that covers the real line, using the ubit. What are the distinct values representable by a sign bit, two exponent bits, no fraction bits, and the ubit, following the format definitions described up to this point for exponent bias, hidden bit, and so forth? How many different values can be represented if IEEE rules are used for infinity and NaN values, instead of the approach described here?

4 The complete unum format

4.1 Overcoming the tyranny of fixed storage size

Ivan Sutherland talks about "the tyranny of the clock," referring to the way traditional computer design requires operations to fit into clock cycles instead of letting operations go as fast as possible. There is a similar tyranny in the way we are currently forced to use a fixed storage size for numbers, instead of letting them fit in as small a space as they can. The two concepts go hand-in-hand; smaller numbers should be allowed to do arithmetic as quickly as possible, and larger numbers should be allowed more time to complete their calculations.

In doing arithmetic using pencil-and-paper, it is natural to increase and decrease the number of digits as needed. Computer designers find it more convenient to have fixed sizes, usually a number of bits that is a power of two. They get very uncomfortable with the idea that a numeric value could have a variable number of bits. They accept that a program might declare a *type*, which means supporting different instructions for different types, like `byte`, `short`, `real`, `double`, etc. Yet, *most* of the computer data that processors work with has variable size: Strings, data files, programs, network messages, graphics commands, and so on.

Strings are some number of bytes (characters) long, and files in mass storage are some number of "blocks" long, but the techniques for managing variable-size data are independent of the size of the units that the data is made out of. There are many libraries for extended-precision arithmetic that use software to build operations out of lists of fixed-precision numbers. Why not go all the way down to a *single bit* as the unit building block for numerical storage?

Suppose written English used the fixed-size tyranny we now have with numbers and required, say, that every word fit into a sixteen-letter allocated space.

It · · · · · · · · · · · ·	would · · · · · · · · · ·	read · · · · · · · · · · ·	something · · · · · · ·
like · · · · · · · · · · ·	this · · · · · · · · · · ·	and · · · · · · · · · · · ·	words · · · · · · · · · ·
longer · · · · · · · · · ·	than · · · · · · · · · · ·	sixteen · · · · · · · · ·	characters · · · · · ·
would · · · · · · · · · ·	not · · · · · · · · · · · ·	be · · · · · · · · · · · · ·	permitted · · · · · · ·

Perhaps someone designing word processors would argue that it is *much* easier and faster to do typesetting if every word is always sixteen letters long. Since the *average* word length is much less than sixteen letters, this wastes a lot of space. Also, since some English words are more than sixteen letters long, writers have to watch out for "overflow" and use a thesaurus to substitute shorter words.

Morse code is a binary form of communication that is flexible down to the single bit level. It was designed at a time when every bit counted, since sending messages across the country in 1836 was no small feat. Morse estimated the ranking of most common letters and assigned those the shortest dot-dash sequences. In contrast, the ASCII character set makes every character occupy eight bits, and actually uses more bits of information to send English text than does Morse code.

> *Fixed size numerical storage, like a synchronous clock, is a convenience for computer **hardware designers** and not for computer **users**.*

We will later discuss in more detail how to actually build a computer that uses a variable-length format, and that unpacks variable-length numbers into fixed-length storage that is easier and faster to work with. But for now, consider one last point: Hardware designers have to create instructions for half, single, and double precision floats, so they are *already* coping with variable data sizes.

4.2 The IEEE Standard float formats

We have been using such tiny bit strings to represent numbers, it is almost a shock to see the sizes used for modern floating point math. The IEEE Standard defines four standard sizes for binary floats: 16, 32, 64, and 128 bits long. It also defines the number of exponent and fraction bits for each size:

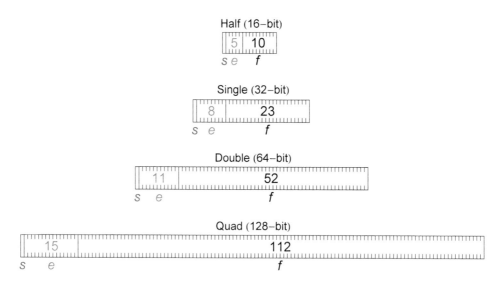

The first thing you might wonder is, why are those the "right" number of bits for the exponent and fraction? Is there a mathematical basis for the number of exponent bits? *There is not.* They were chosen by committee. One of the influences on the committee was that it is much easier to build exponent hardware than fraction hardware, so the exponents allow dynamic ranges *vastly* larger than usually required in real-world computations.

One can only imagine the long and contentious arguments in the 1980s about whether a particular size float should have a particular number of bits in its exponent. The sub-field sizes were artistically selected, not mathematically determined. Another argument for large exponents might have been that underflow and overflow are usually bigger disasters than rounding error, so better to use plenty of exponent bits and fewer fraction bits.

The scale factor of a number (like a fixed point number) was originally kept separate from the digits of the number and applied to all the numbers in a computation. Eventually designers realized that each number needed its own *individual* scale factor, so they attached an exponent field to a fraction field and called it a floating point number. We have the same situation today at the next level up: A block of numbers is specified as all being a particular precision, even though each number needs its own individual exponent size and fraction size. The next logical step, then, is similar to what was done with the scale factor: What if a number were made *self-descriptive* about its exponent size and fraction size, with that information *as part of its format*?

Before exploring that idea further, here is a general description of the state of the four IEEE Standard sizes, at the time of this writing:

- **Half precision**, 16 bits, is relatively new as a standard size. It was promoted starting around 2002 by graphics and motion picture companies as a format that could store visual quantities like light intensity, with a reasonable coverage of the accuracy and dynamic range of the human eye. It can represent all the integers from −2048 to +2048 exactly. The largest magnitude numbers it can represent are ±65,520, and the smallest magnitude numbers it can represent are about $\pm 6 \times 10^{-8}$. It has about three significant decimals of accuracy. Only recently have chip designers started to support the format as a way to store numbers, and existing processors do not perform native arithmetic in half precision; they promote to single precision internally and then demote the final result.

- **Single precision** is 32 bits total. With an 8-bit exponent, the dynamic range is about 10^{-45} to 10^{-38} (subnormal) and 10^{-38} to 10^{38} (normal) . To put this dynamic range in perspective, the ratio of the size of the known universe to the size of a proton is about 10^{40}, so single precision covers this range with about 43 orders of magnitude to spare. The 23-bit fraction provides about seven decimals of accuracy. Single precision floats are often used in imaging and audio applications (medical scans, seismic measurements, video games) where their accuracy and range is more than sufficient.

- **Double precision** is 64 bits total, and since the late 1970s has become the de facto size for serious computing involving physics simulations. It has about fifteen decimals of accuracy and a vast dynamic range: About 10^{-324} to 10^{-308} (subnormal) and about 10^{-308} to 10^{+308} (normal). Double precision and single precision are the most common sizes built into processor chips as fast, hard-wired data types.

- **Quad precision**, 128 bits, is mainly available through software libraries at the time of this writing, with no mainstream commercial processor chips designed to do native arithmetic on such large bit strings. The most common demand for quad precision is when a programmer has discovered that the double precision result is very different from the single precision result, and wants assurance that the double precision result is adequate. A quad precision number has 34 decimals in its fraction and a dynamic range of almost 10^{-5000} to 10^{+5000}. Because quad precision arithmetic is typically executed with software instead of native hardware, it is about twenty times slower than double precision. Despite its impressive accuracy and dynamic range, quad precision is every bit as capable of making disastrous mathematical errors as its lower-precision kin. And, of course, it is even more wasteful of memory, bandwidth, energy, and power than double precision.

> With four different binary float sizes, the programmer must choose the best one, and the computer system must support an instruction set for all four types. Because few programmers have the training or patience to right-size the precision of each operation, most resort to overkill. Even though *input* values might only be known to three or four decimal places, and the *results* are only needed to about three or four places, common practice is to make *every* float double precision with the vague hope that using **fifteen decimals** for all intermediate work will take care of rounding errors. Years ago, the only downside of demanding double precision everywhere was that the program consumed more memory and ran slightly slower. Now, such excess insurance aggravates the problems of energy and power consumption, since computers run up against the limits of their power supply whether it is the battery in a cell phone or the utility substation for a data center. Also, programs now run *much* slower with high precision, because performance is usually limited by system bandwidth.

4.3 Unum format: Flexible range and precision

The final step to achieving a "universal number" format is to attach two more self-descriptive fields: The *exponent size* (*es*) field and the *fraction size* (*fs*) field. They go to the right of the ubit, and we color them green and gray for clarity.

The Universal Number (Unum) Format

←—*es* bits—→		←————————*fs* bits————————→				
s	e	f	u	es−1	fs−1
sign	exponent	fraction	ubit	exponent size	fraction size

Sometimes such fields are called "metadata," that is, data describing data. How big should the new fields be? That is, how many bits are needed to specify the exponent size and the fraction size?

Suppose we store, in binary, the number of exponent bits in each IEEE float type. The exponent sizes 5, 8, 11, and 15 become 101, 1000, 1011, and 1111 in binary, so four bits suffice to cover all the IEEE cases for the exponent size. But we can be more clever by noticing that there is always *at least one* exponent bit, so we bias the binary string (treated as if it were a counting number) upward by one. The four exponent sizes can be represented with 100, 111, 1010, and 1110 in binary. Four bits would be the most we would need to specify every exponent size from a single bit (representations that act exactly like fixed point format) to 16 (which is even more exponent bits than used in IEEE quad precision). That is why the exponent size field is marked as storing the value *es* – 1 instead of *es* in the above diagram.

Similarly, the four IEEE *fraction* sizes 10, 23, 52, and 112 become 1010, 10111, 110100, and 1110000 in binary, so seven bits suffice to cover all the IEEE cases. But we similarly can require at least one fraction bit, so the binary counting number stored in the *fs* field is actually one less than the number of bits in the fraction field. With a seven-bit *fs* field, we can have fraction sizes that range anywhere from one bit to $2^7 = 128$ bits. The *es* and *fs* values can thus be customized based on the needs of the application and the user, not set by committee.

> **Definition**: The *esizesize* is the number of bits allocated to store the maximum number of bits in the **exponent** field of a unum. The number of exponent bits, *es*, ranges from 1 to $2^{esizesize}$.

> **Definition**: The *fsizesize* is the number of bits allocated to store the maximum number of bits in the **fraction** field of a unum. The number of fraction bits, *fs*, ranges from 1 to $2^{fsizesize}$.

It is possible to have *esizesize* be as small as zero, which means there is *no* bit field, and the number of exponent bits is exactly $2^0 = 1$. Similarly for *fsizesize*. The reason for putting them to the right of the ubit is so that all of the new fields are together and can be stripped away easily if you just want the float part of the number. The reason the ubit is where it is was shown previously; it is an extension of the fraction by one more bit, and creates a monotone sequence of values just like the positional bits in the fraction.

It looks odd to see "*size*" twice in *esizesize* and *fsizesize*, but they are the size of a size, so that is the logical name for them. It is like taking the logarithm of a logarithm, which is why *esizesize* and *fsizesize* are usually small, single-digit numbers. For example, suppose we have

Fraction value = 110101011 in binary; then
Fraction size = 9 bits, which is 1001 in binary, and
Fraction size size = 4 bits (number of bits needed to express a fraction size of 9).

> **Definition**: The *utag* is the set of three self-descriptive fields that distinguish a unum from a float: The ubit, the exponent size bits, and the fraction size bits.

The utag is the "tax" we pay for flexibility, compactness, and exactness information, just as having exponents embedded in a number is the "tax" that floating point numbers pay for describing their individual scale factor. The additional self-descriptive information is why **unums are to floating point numbers what floating point numbers are to integers.**

4.4 How can appending extra bits *save* storage?

The size of the "tax" is important enough that it deserves a name: *utagsize*.

> **Definition**: The *utagsize* is the length (in number of bits) of the utag. Its value is 1 + *esizesize* + *fsizesize*.

If we want to create a superset of all the IEEE floats, *utagsize* will be $1 + 4 + 7 = 12$ bits. Besides the utag bits, there is always a sign bit, at least one exponent bit, and at least one fraction bit, so the minimum number of bits a unum can occupy is *utagsize* + 3. The maximum possible number of bits also has a name: *maxubits*.

> **Definition**: *maxubits* is the maximum number of bits a unum can have. Its value is $2 + esizesize + fsizesize + 2^{esizesize} + 2^{fsizesize}$.

You might wonder: How is the unum approach going to help reduce memory and energy demands compared to floats, if it adds *more* bits to the format?

The principal reason is that it frequently allows us to use far *fewer* bits for the exponent and fraction than a "one-size-fits-all" precision choice. Experiments that show this are in chapters yet to come, but for now it might seem plausible that most calculations use far fewer exponent bits than the 8 and 11 bits allocated for single and double precision floats (dynamic ranges of about 83 and 632 orders of magnitude, not exactly your typical engineering problem).

In the example where all English words had to fill 16-character spaces, the paragraph took up far more space than if every word was simply as long as it needed to be. Similarly, the fraction and the exponent size of a unum increases and decreases as needed, and the *average* size is so much less than the *worst case* size of an IEEE float that the savings is more than enough to pay for the utag. Remember the case of Avogadro's number in the very beginning? Annotating a number with scaling information saved a lot of space compared to storing it as a gigantic integer. The next question might be, does the programmer then have to manage the variable fraction and exponent sizes?

No. That can be done automatically by the computer. Automatic range and precision adjustment is inherent in the unum approach, the same way float arithmetic automatically adjusts the exponent. The key is the ubit, since it tells us what the level of certainty or uncertainty is in a value. The exponent size and fraction size change the meaning of an ULP, so the three fields together provide the computer with what it needs, at last, to automate the control of accuracy loss. If you add only one of any of the three fields in the utag, you will not get a very satisfactory number system.

There is something called "significance arithmetic," for example, that annotates each value with the number of significant figures. That approach is like a utag that only has a field for the number of bits in the fraction and that assumes every answer is inexact. After only a few operations, a significance arithmetic scheme typically gives even more pessimistic assessments than interval arithmetic, and incorrectly declares that all significance has been lost.

> It is essential to have all three sub-fields in the utag of the unum format to avoid the disadvantages of significance arithmetic and interval arithmetic.

Imagine never having to specify what type of floating point number to use for variables, but instead simply say that a variable is a real number, and let the computer do the rest. There are symbolic programming environments that can do this (*Mathematica* and Maple, for example), but to do so they use very large data structures that use far more bits than IEEE floats. We will demonstrate that unums get better answers than floats but do it with *fewer* bits than IEEE floats, even after including the cost of having a utag on every value.

4.5 Ludicrous precision? The vast range of unums

If we use *esizesize* = 4 and *fsizesize* = 7, we call that "a {4, 7} environment" for short, or say that a unum uses "a {4, 7} utag." While a computer can manage its own environment sizes by detecting unsatisfactory results and recalculating with a bigger utag, the programmer might want some control over this for improved efficiency. It is usually easy for a programmer to make a good estimate of the *overall* accuracy needs of an application, which still leaves the heavy lifting (dynamically figuring out the precision requirements of each operation) to the computer.

For example, the {3, 4} unum environment looks appropriate for graphics and other low-precision demands; it provides for an exponent up to $2^3 = 8$ bits long and a fraction up to $2^4 = 16$ bits long. Therefore, its maximum dynamic range matches that of a 32-bit float and its fraction has more than five decimals of accuracy, yet it cannot require more than 33 bits of total storage (and usually takes far less than that). Using a {3, 4} environment instead of a {4, 7} environment reduces the utag length to 8 instead of 12, a small but welcome savings of storage.

A surprising amount of very useful computation (for seismic signal processing, say) can be done with a {2, 2} environment, because of the absence of rounding, overflow, and underflow errors. There is such a thing as a {0, 0} environment, and it is so useful it will receive its own discussion in Chapter 7. The interesting thing about {0, 0} unums is that they are all the same size: Four bits.

At the other extreme, unums can easily go well beyond the limits of human compre-hension for dynamic range and precision. The number represented in decimal by one followed by a hundred zeros is called a "googol": 10^{100}. The number represented by one followed by a *googol* zeros is called a "googolplex," or $10^{10^{100}}$. How big a utag would we need to represent all the integers up to a googol, scaled by a dynamic range of a googolplex?

It works out to only *9 bits* each for *fsizesize* and *esizesize*, a total utag size of 19 bits! With just a handful of bits, we can create a unum with ludicrous precision. While it seems very unlikely that anyone can justify the use of such precision for a real-world engineering problem, the unum format is certainly up to the task if anyone demands it. Furthermore, such numbers are well within the computing abilities of even the most modest computer. The maximum size of a unum with such a utag is only 1044 bits, about the same amount of storage as 16 double-precision floats.

There *are* practical problems that demand ultrahigh precision, like e-commerce encryption security that currently requires integers that have hundreds of digits. Even a cell phone uses very large integers for encryption and decryption of data. The need for large integer multiplication for public-key encryption means, incidentally, that hardware designers have already been unwittingly working toward building the type of circuitry needed for fast unum calculations. Computational scientist David H. Bailey has collected many examples where very high precision is needed either to solve a problem, or to figure out that it is unsolvable.

4.6 Changing environment settings within a computing task

The environment can be changed in the middle of a computation, and this actually is a very useful thing to do. For example, the computer might decide higher precision is needed for some task, and temporarily increase *fsizesize*, resetting it to where it was when finished. In that case, there is no need for a programmer to do anything to the values that have already been calculated.

It can also happen that previously computed unums will be needed in an environment that has different settings. In that case, The program has to know to promote or demote the previously computed unums to the new size, just as one would promote single-precision floats to double precision. It is good practice to annotate a block of unums with their {*esizesize, fsizesize*} value if they are archived or communicated to another program. Doing so should take only one 8-bit byte, since it is hard to imagine either *esizesize* or *fsizesize* taking up more than four bits.

Exercise for the reader: If the bit lengths for *esizesize* and *fsizesize* are both limited to the range 0 to 15 so that four bits suffice to express each value, what is the maximum number of bits in the unum for *maxreal*, the largest positive real number in a {15, 15} environment? If the byte is instead divided into three bits for *esizesize* and five bits for *fsizesize*, what is the maximum number of decimal digits accuracy possible in a {7, 31} environment?

With fixed-size floats, you tend to think of loading or storing all their bits at once. With unums, you have a two-step process like you would have reading character strings of variable length. Based on the environment settings, the computer loads the bits of the *es* and *fs* fields, which point to where the sign bit, exponent, and fraction must be to the left of the utag. It then loads the rest of the unum and points to the next unum. This is why *esizesize* and *fsizesize* are like processor control values; they tell the processor how to interpret bit strings. A collection of unums are packed together exactly the way the words in this paragraph are packed together. As long as a set of unums is of substantial length, there is little waste in using conventional power-of-two sizes to move the unum data in blocks, such as cache lines or pages of memory. Only the last block is "ragged" in general, that is, only partly full. Blocking the set of unums can also make it easier to do random access into the set.

These are all classic data management problems, no different than the problem of organizing records on a disk drive. To use the vocabulary of data structures, packed unums form a *singly linked list*. Each value contains a pointer to where the next value begins, since the utag specifies the total length of the bit string. Chapter 7 will show how unums can be unpacked into a fixed-size form that is actually faster to work with than floats. That restores the ability to use *indexed arrays*, the way floats are often stored.

In summary, there is no difficulty dealing with multiple environment settings within a calculation, or with multiple unum lengths within a set of unums that are loaded or stored at any particular setting. These are problems that have been solved for many decades, and the solutions are part of every operating system. They simply need to be applied to unums as they have been to many other kinds of variable-sized data.

4.7 The reference prototype

The text you have been reading was not created with a word processor or conventional text editor. **Everything you have been reading is the input, output, or comments of a computer program.**

The graphics shown here are computed output, as are the numerical examples. This book is a "notebook" written in *Mathematica*. This has the advantage of eliminating many sources of errors, and making it easier to make changes and additions. The code contained here is the *reference prototype* of the unum computing environment.

While much of the discussion so far has been conceptual, the unum concept *has been reduced to practice*, and this book is that reduction. The prototype contains a collection of tools for interactively working with unums and many other ideas. In explaining how the tools work, there are two challenges: One is that we risk sounding like some user's manual instead of a book about a completely new approach to numerical computing. The other is that it is difficult to shield the reader from seeing *some* computer code. The example of Section 3.3 had a code snippet that took the square root of a starting value until it "converged" to 1. That resulting series of values was actually computed, not typed in.

Sections of code have a light green background to make it easier to skip them if you (like most people) do not enjoy reading computer code written by others. Most of them are short, like the following command to set the computing environment:

```
setenv[{3,4}]
```

The **setenv** function sets up the environment for specified settings of *esizesize* and *fsizesize*. It takes the desired {*esizesize, fsizesize*} pair as its argument. It computes all the useful pointers and bit combinations that are used to pull apart and interpret any given unum as a value. The full description of what **setenv** does is shown in a code listing in Appendix C.1. The line of code shown above just set the unum environment to a maximum exponent size of 2^3 bits and a maximum fraction size of 2^4 bits, and also created all the useful constants that go along with the environment, like *maxreal* and *utagsize*. Just as we use **this typeface** to indicate computerese like binary numbers, functions that are displayed like **setenv** also indicate that a function exists in the prototype. A value like *maxreal* becomes **maxreal** when it actually has to participate in a calculation. Whereas defined concepts like *maxreal* are listed in the Glossary, defined computer variables and functions are listed in the Appendices for those interested in that level of detail.

One of the things **setenv** computes is *bit masks*. When you need to extract a set of bits from a number, you "AND" each bit with a string that has **1** values in just those bits. For instance, if you "AND" a number with **000110** in binary, you will get just the values of the bits in the fourth and fifth position, and all other bits will always be zero. Here are some bit masks that will be useful in writing software for unum arithmetic: **ubitmask** picks out the ubit, **fsizemask** selects the bits that indicate the number of fraction bits, and **esizemask** selects the bits that indicate the number of exponent bits. We also will find combinations handy: **efsizemask** for both exponent and fraction size fields, and **utagmask** selects the entire utag. For clarity, we show the three bit fields of the utag, with annotations under each field as a reminder for what the contents of each field represent. The first field, the ubit, is annotated with "···" if there are more bits after the last fraction bit, but "↓" if the fraction is exact. The second and third fields hold binary counting numbers that are one less than the fraction and exponent sizes *es* and *fs*.

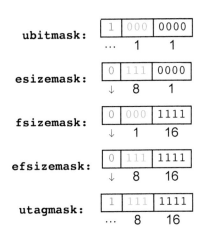

With floats, the ULP bit is the rightmost bit of the number. With unums, the ULP bit is *the bit just to the left of the utag*. A unum value can have a range of ULP sizes, since the number of bits in the fraction is allowed to vary. The ULP bit mask in a unum is easily computed by shifting a `1` bit left by the number of bits in the utag, *utagsize*. The value the ULP bit represents depends on the utag bits. The represented real value is an ULP, but the unum mask for it we call `ulpu`. As a general rule, the names of variables that are unums and functions that produce unums will end in the letter `u`. Since we currently are in a {3, 5} environment, the utag is nine bits long and `ulpu` will be a `1` bit followed by the nine empty bits of the utag, or `1` `0 000 00000` in binary. Besides color-coding the bits for clarity, notice that we use a convention that the sign, exponent, and fraction bits are in **boldface** but the utag bits are not.

Here is a visual way to compare unums and float formats: A grid of dots that shows the possible exponent bits (horizontal) and fraction bits (vertical), color-coded for size comparison with IEEE floats. This diagram shows when a unum has greater fraction accuracy than a standard float type (above a *horizontal* line), or when a unum has greater dynamic range (to the right of a *vertical* line).

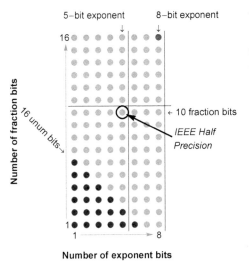

Number of exponent bits

This is the diagram for a {3, 4} environment. As mentioned above, it looks robust enough for graphics and other low-precision demands. The size of the utag is just 8 bits, and it includes half-precision IEEE floats but also has the dynamic range of a 32-bit float and a couple more decimals of precision than 16-bit floats.

The blue dots indicate unums that take more space than a half precision float, but no more than required for single precision. Dark red dots show unums that are as small as a half precision float or smaller. The dark orange dot is the only unum size in a {3, 4} environment that takes more space than IEEE single precision; the largest {3, 4} unum takes 33 bits. The smallest one takes only 11 bits.

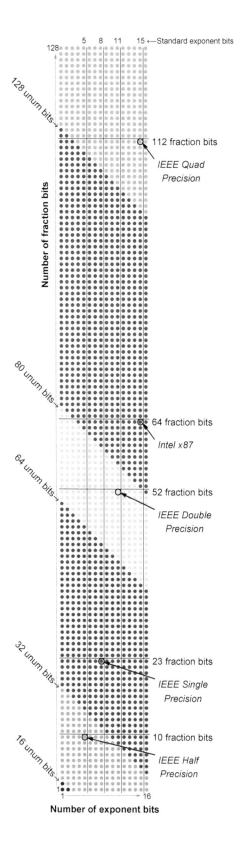

5   8   11   15 ←—Standard exponent bits

128

128 unum bits

Number of fraction bits

112 fraction bits

IEEE Quad Precision

80 unum bits

64 fraction bits

Intel x87

64 unum bits

52 fraction bits

IEEE Double Precision

32 unum bits

23 fraction bits

IEEE Single Precision

16 unum bits

10 fraction bits

IEEE Half Precision

1

1   16

Number of exponent bits

The diagram for a {4, 7} environment is so tall that it barely fits on a page. At the top right corner, the largest unum has a larger dynamic range and precision than IEEE quad format. There are

$$2^4 \cdot 2^7 = 2048$$

possible combinations for the sizes of the exponent and fraction, four of which match IEEE Standard binary floats, and one of which matches the format used in the original Intel math coprocessors (sometimes called "extended double"). In the 1980s, Intel introduced a coprocessor called the i8087 that *internally* used 64 bits of fraction and 15 bits of exponent, for a total size of 80 bits. It had successor versions with product numbers ending in 87, so here that format is called "Intel x87." The 80-bit float was initially one of the IEEE Standard sizes, but exposed an interesting technical-social issue: Most users wanted *consistent* results from all computers, not results that were *more accurate* on one vendor's computers! Using the extra size of the x87 for scratch results reduced rounding error and occurrences of underflow/overflow, but the results then differed mysteriously from those using double precision floats. We will show a general solution for this problem in the next chapter.

There has been much hand-wringing by floating point hardware designers over how many exponent bits and how many fraction bits to allocate. Imagine instead always being able to select the right size *as needed*.

Note: Be careful when requesting the exponent field size. Even casually asking for an *esizesize* of 6 can launch you into the land of the outrageous. The prototype environment in particular is not designed to handle such huge numbers.

A unum with this {4, 7} utag can take as few as 15 bits, which is even more compact than a half-precision float. In the bottom left corner is the minimal unum format, with just three bits of float to the left of the utag.

4.8 Special values in a flexible precision environment

For floats with just a ubit appended, we can use the maximum exponent and fraction fields (all bits set to 1) to represent infinity (ubit = 0) or NaN (ubit = 1). With *flexible precision* unums, the largest magnitude number happens *only when we are using the largest possible exponent field and the largest possible fraction field*. There is no need to waste bit patterns with multiple ways to represent infinities and NaNs for each possible combination of *es* and *fs*. Unum bit strings represent real numbers (or ULP-wide open intervals) unless the *es* and *fs* are at maximum, corresponding to the extreme top right-hand corner dot of the two preceding diagrams, which means they represent infinities or NaNs.

Our convention is that unum variable names end in "u" so we do not confuse bit strings with the real values they represent. For instance, *maxreal* is the largest real value, but the unum that represents it in a particular environment is `maxrealu`.

The function `big[`*u*`]` returns the biggest real value representable by a unum with an *es* and *fs* matching that of a unum *u*. The function `bigu[`*u*`]` returns the unum bit string that represents that value; unless the *es* and *fs* fields are at their maximum, `bigu[`*u*`]` is the unum with all 1s in the exponent and fraction fields. If the *es* and *fs* fields are at their maximum (all 1 bits), then we have to back off by one ULP, since *that* bit pattern is reserved for ±∞. In that case, `big[`*u*`]` is the same as *maxreal*. Think of *maxreal* as "biggest of the `big`." (It was tempting to call the `big` function `mightybig` because of the way `mightybig[`*u*`]` would be pronounced.)

A set of values is calculated whenever `setenv` is called. The following table shows the names of special unums and approximate decimal versions of the *maxreal* and *smallsubnormal* real values computed when the environment is set to {3, 2}. The unum for *smallsubnormal*, like *maxreal*, occurs when *es* and *fs* are at their maximum:

Name	Meaning	Value in a {3,2} environment
esizesize	Size of size of exponent	3
fsizesize	Size of size of fraction	2
utagsize	Number of bits in the utag	6
maxubits	Maximum bits in a unum	19
`posinfu`	The unum for +∞	0 11111111 1111 0 111 11
`neginfu`	The unum for −∞	1 11111111 1111 0 111 11
`qNaNu`	The unum for quiet NaN	0 11111111 1111 1 111 11
`sNaNu`	The unum for signaling NaN	1 11111111 1111 1 111 11
`maxrealu`	Finite unum closest to +∞	0 11111111 1110 0 111 11
`negbigu`	The largest magnitude negative unum	1 11111111 1110 0 111 11
maxreal	The largest representable real	$\sim 6.38 \times 10^{38}$
`smallsubnormalu`	Unum for the smallest real	0 00000000 0001 0 111 11
smallsubnormal	Smallest representable real > 0	$\sim 7.35 \times 10^{-40}$

Larger exponents extend expressible magnitudes in both directions, so the largest exponent field expresses the smallest magnitude number. The representation of *smallsubnormal* has just one bit set to 1 in the fraction, the ULP bit in the last place; all other bits to the left are 0. The tabled values, and others computed by `setenv`, are mainly of interest if you want to write low-level programs with unums. If there is ever any question about what the current environment is, simply look at the values of *esizesize* and *fsizesize*. In the prototype, here is one way to show those values:

`{esizesize,fsizesize}`

`{3, 2}`

In this book, most computed output is in `this typeface` (Courier plain, not bold).

4.9 Converting exact unums to real numbers

We need a way to convert the floating point part of a unum into its mathematical value. First we use the self-descriptive bits in the utag to determine the fraction and exponent sizes. Then we can extract the sign, exponent, and fraction bits using bit masks. From those values we build a function that converts the part of unum number left of the utag into a real number using IEEE binary float rules. When the ubit is 0, the formula for a unum value is exact.

◄—*es* bits—►	◄———————*fs* bits———————►			
s	*e*	*f*	0	*es*−1	*fs*−1
sign	exponent	fraction	ubit	exponent size	fraction size

The formula looks very much like the formula for a float (not the strict IEEE float, but the improved one that does not waste huge numbers of bit patterns on NaN).

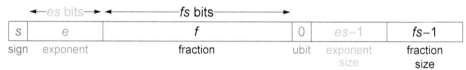

$$x = (-1)^s \times \begin{cases} 2^{e-2^{es-1}} \times \left(\dfrac{f}{2^{fs}}\right) & \text{if } e = \text{all } \mathbf{0} \text{ bits,} \\ \infty & \text{if } e, f, es, \text{ and } fs \text{ have all their bits set to 1,} \\ 2^{e-2^{es-1}-1} \times \left(1 + \dfrac{f}{2^{fs}}\right) & \text{otherwise.} \end{cases}$$

The code version of this is `u2f[u]` ("unum to float") where *u* is a unum. The code is in Appendix C.3.

To reinforce the connection between unums and floats, here is a utag for IEEE single precision (8 bits of exponent, 23 bits of fraction) in a {3, 5} environment:

0	111	10110
↓	8	23

If you put 32 bits to the left of that utag, it will express the same values as an IEEE single precision float, except that it avoids clipping the maximum exponent to NaN.

Similarly, here is a utag for IEEE half precision in a {3, 4} environment:

> **Exercise for the reader**: What is the smallest environment capable of storing Intel's x87 format, and what is the binary string for its utag?

The `utagview[`*ut*`]` function displays the utags as shown above, where *ut* is a utag. It can be used even if the environment is {0, 0}, with *no* bits for *es* and *fs*. It does not have a way of showing a nonexistent bit so it draws the last two squares anyway, empty; *es* and *fs* are always zero. The only thing left in the utag is the ubit:

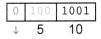

It is easy to convert an *exact* unum to an IEEE Standard float (at least for unums with exponent and fraction sizes no larger than the largest IEEE type, quad precision). To do that, you find the smallest size IEEE float with fields large enough for the unum exponent and fraction, and pad the unum with extra bits to match the worst-case size allotment in the float. If the exponent is all **1** bits and represents a finite value, you *discard* all the fraction information and replace it with **0** bits, to represent infinity in the style of the IEEE Standard. There are no NaN values to convert, since a unum NaN is not exact; its ubit is set to **1**. And, of course, you strip off the utag. For example, with a unum tag for IEEE half precision,

the largest expressible real number is 131 008. (We can write the binary string for that unum as **0 11111 1111111111** 0 100 1001. Using boldface for the float part but not for the utag helps make such long bit strings more readable.) The half precision IEEE Standard only allows values half as large as this: 65 504. Every bit pattern larger than that means either infinity or NaN. The decision to have a vast number of NaN types looks particularly dubious when the precision or dynamic range is small.

You might think, "Wait, all the bits are set to **1** in that unum. Doesn't that mean it has to express infinity, since we reserve that largest number for infinity?" No, because it is *not* the largest number possible in the unum environment. With the same size utag, we could also have *es* and *fs* values as large as 8 and 16, which permits expression of a much larger dynamic range and precision:

allows 680 554 349 248 159 857 271 492 153 870 877 982 720 as the largest finite number! (About 6.8×10^{38}, twice as large as *maxreal* in IEEE single precision since we do not waste bit patterns on a plethora of NaNs.)

4.10 A complete exact unum set for a small utag

If we set the environment to a small utag, say {2, 2}, the set of possible exact unum values is so small that we can list *all* of them. Here are the non-negative values:

0 $\frac{1}{1024}$ $\frac{1}{512}$ $\frac{3}{1024}$ $\frac{1}{256}$ $\frac{5}{1024}$ $\frac{3}{512}$ $\frac{7}{1024}$ $\frac{1}{128}$ $\frac{9}{1024}$ $\frac{5}{512}$ $\frac{11}{1024}$ $\frac{3}{256}$ $\frac{13}{1024}$ $\frac{7}{512}$ $\frac{15}{1024}$ $\frac{1}{64}$ $\frac{17}{1024}$ $\frac{9}{512}$ $\frac{19}{1024}$ $\frac{5}{256}$

$\frac{21}{1024}$ $\frac{11}{512}$ $\frac{23}{1024}$ $\frac{3}{128}$ $\frac{25}{1024}$ $\frac{13}{512}$ $\frac{27}{1024}$ $\frac{7}{256}$ $\frac{29}{1024}$ $\frac{15}{512}$ $\frac{31}{1024}$ $\frac{1}{32}$ $\frac{17}{512}$ $\frac{9}{256}$ $\frac{19}{512}$ $\frac{5}{128}$ $\frac{21}{512}$ $\frac{11}{256}$ $\frac{23}{512}$ $\frac{3}{64}$ $\frac{25}{512}$

$\frac{13}{256}$ $\frac{27}{512}$ $\frac{7}{128}$ $\frac{29}{512}$ $\frac{15}{256}$ $\frac{31}{512}$ $\frac{1}{16}$ $\frac{17}{256}$ $\frac{9}{128}$ $\frac{19}{256}$ $\frac{5}{64}$ $\frac{21}{256}$ $\frac{11}{128}$ $\frac{23}{256}$ $\frac{3}{32}$ $\frac{25}{256}$ $\frac{13}{128}$ $\frac{27}{256}$ $\frac{7}{64}$ $\frac{29}{256}$ $\frac{15}{128}$

$\frac{31}{256}$ $\frac{1}{8}$ $\frac{17}{128}$ $\frac{9}{64}$ $\frac{19}{128}$ $\frac{5}{32}$ $\frac{21}{128}$ $\frac{11}{64}$ $\frac{23}{128}$ $\frac{3}{16}$ $\frac{25}{128}$ $\frac{13}{64}$ $\frac{27}{128}$ $\frac{7}{32}$ $\frac{29}{128}$ $\frac{15}{64}$ $\frac{31}{128}$ $\frac{1}{4}$ $\frac{17}{64}$ $\frac{9}{32}$ $\frac{19}{64}$

$\frac{5}{16}$ $\frac{21}{64}$ $\frac{11}{32}$ $\frac{23}{64}$ $\frac{3}{8}$ $\frac{25}{64}$ $\frac{13}{32}$ $\frac{27}{64}$ $\frac{7}{16}$ $\frac{29}{64}$ $\frac{15}{32}$ $\frac{31}{64}$ $\frac{1}{2}$ $\frac{17}{32}$ $\frac{9}{16}$ $\frac{19}{32}$ $\frac{5}{8}$ $\frac{21}{32}$ $\frac{11}{16}$ $\frac{23}{32}$ $\frac{3}{4}$ $\frac{25}{32}$ $\frac{13}{16}$ $\frac{27}{32}$ $\frac{7}{8}$ $\frac{29}{32}$ $\frac{15}{16}$ $\frac{31}{32}$

1 $\frac{17}{16}$ $\frac{9}{8}$ $\frac{19}{16}$ $\frac{5}{4}$ $\frac{21}{16}$ $\frac{11}{8}$ $\frac{23}{16}$ $\frac{3}{2}$ $\frac{25}{16}$ $\frac{13}{8}$ $\frac{27}{16}$ $\frac{7}{4}$ $\frac{29}{16}$ $\frac{15}{8}$ $\frac{31}{16}$ 2 $\frac{17}{8}$ $\frac{9}{4}$ $\frac{19}{8}$ $\frac{5}{2}$ $\frac{21}{8}$ $\frac{11}{4}$ $\frac{23}{8}$

3 $\frac{25}{8}$ $\frac{13}{4}$ $\frac{27}{8}$ $\frac{7}{2}$ $\frac{29}{8}$ $\frac{15}{4}$ $\frac{31}{8}$ 4 $\frac{17}{4}$ $\frac{9}{2}$ $\frac{19}{4}$ 5 $\frac{21}{4}$ $\frac{11}{2}$ $\frac{23}{4}$ 6 $\frac{25}{4}$ $\frac{13}{2}$ $\frac{27}{4}$ 7

$\frac{29}{4}$ $\frac{15}{2}$ $\frac{31}{4}$ 8 $\frac{17}{2}$ 9 $\frac{19}{2}$ 10 $\frac{21}{2}$ 11 $\frac{23}{2}$ 12 $\frac{25}{2}$ 13 $\frac{27}{2}$ 14 $\frac{29}{2}$ 15 $\frac{31}{2}$ 16 17

18 19 20 21 22 23 24 25 26 27 28 29 30 31 32 34 36 38 40 42 44

46 48 50 52 54 56 58 60 62 64 68 72 76 80 84 88 92 96 100 104 108

112 116 120 124 128 136 144 152 160 168 176 184 192 200 208 216 224 232 240 248 256

272 288 304 320 336 352 368 384 400 416 432 448 464 480 ∞

The unum bit strings for these numbers require from eight to thirteen bits, total. Unums have more than one way to represent exact numbers, so we use one that requires the fewest bits. Even that does not make a unum unique; sometimes there is more than one way to represent an exact number with the minimum number of bits.

Exercise for the reader: In a {2, 2} environment, what number is expressed by the unum bit string `0 00 1 0 0 01 00` ? Can you find another bit string the same length that expresses the same value?

Instead of fixing the utag contents, we can also limit the total number of *exponent and fraction bits*, since that determines the total number of bits of storage. The total additional bits cannot be less than two, since we always have at least one exponent bit and one fraction bit. There are 66 non-negative exact numbers expressible with just *two to five bits* in addition to the five-bit utag and the sign bit:

0 $\frac{1}{128}$ $\frac{1}{64}$ $\frac{3}{128}$ $\frac{1}{32}$ $\frac{3}{64}$ $\frac{1}{16}$ $\frac{3}{32}$ $\frac{1}{8}$ $\frac{3}{16}$ $\frac{1}{4}$ $\frac{5}{16}$ $\frac{3}{8}$ $\frac{7}{16}$ $\frac{1}{2}$ $\frac{5}{8}$ $\frac{3}{4}$ $\frac{7}{8}$ 1 $\frac{9}{8}$ $\frac{5}{4}$ $\frac{11}{8}$ $\frac{3}{2}$ $\frac{13}{8}$ $\frac{7}{4}$ $\frac{15}{8}$ 2 $\frac{17}{8}$ $\frac{9}{4}$ $\frac{19}{8}$ $\frac{5}{2}$ $\frac{21}{8}$ $\frac{11}{4}$

$\frac{23}{8}$ 3 $\frac{25}{8}$ $\frac{13}{4}$ $\frac{27}{8}$ $\frac{7}{2}$ $\frac{29}{8}$ $\frac{15}{4}$ $\frac{31}{8}$ 4 $\frac{9}{2}$ 5 $\frac{11}{2}$ 6 $\frac{13}{2}$ 7 $\frac{15}{2}$ 8 10 12 14 16 20 24 28 32 48 64 96 128 192 256 384

Some of those 66 numbers have more than one unum representation. Many of the duplications are like fractions that have not been reduced to lowest terms; there is no need for exact representations like $\frac{12}{8}$ and $\frac{6}{4}$ when $\frac{3}{2}$ means the same thing and requires fewer bits to store. This is one reason unums are more concise than floats. There are 34 values in the preceding set that have more than one representation:

0 $\frac{1}{16}$ $\frac{1}{8}$ $\frac{3}{16}$ $\frac{1}{4}$ $\frac{3}{8}$ $\frac{1}{2}$ $\frac{3}{4}$ 1 $\frac{9}{8}$ $\frac{5}{4}$ $\frac{11}{8}$ $\frac{3}{2}$ $\frac{13}{8}$ $\frac{7}{4}$ $\frac{15}{8}$ 2 $\frac{9}{4}$ $\frac{5}{2}$ $\frac{11}{4}$ 3 $\frac{13}{4}$ $\frac{7}{2}$ $\frac{15}{4}$ 4 5 6 7 8 12 16 24

The distribution of the logarithm of possible numbers shows that for a given bit-length budget, *the closer the numbers are to 1 in magnitude, the higher the precision.* This seems like a practical choice that saves storage. This differs from floats in that it dedicates more bits for precision for numbers near 1 and more bits for dynamic range for the largest and smallest numbers, whereas floats give equal precision to all values. If we sort the values generated above and plot the logarithm of their value, the curve is less steep in the middle, indicating more precision for numbers that are neither huge nor tiny. This allows us to cover almost five orders of magnitude with only a handful of bits, yet have about one decimal of accuracy for numbers near unity.

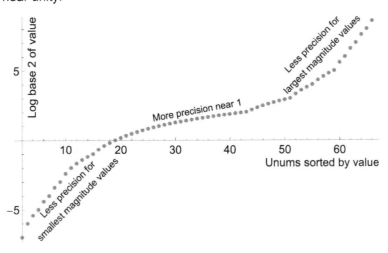

When people write application programs, they select units of measurement that make the numbers easy to grasp, which means not too many orders of magnitude away from small counting numbers. Pharmacists uses milligrams, cosmologists use light years, chip designers use microns, and chemists measure reaction times in femtoseconds $\left(10^{-15}\right.$ second). Floats are designed to treat, say, the numerical range 10^{306} to 10^{307} as if it were just as deserving of precision bits as the numbers one to ten, which ignores the human factor. As with Morse code, unums make the more common data expressible with fewer bits.

4.11 Inexact unums

A unum that has its ubit set to 1 indicates a range of values strictly between two exact values, that is, an open bound. An "inexact" unum is *not* the same thing as a rounded floating point number. In fact, it is the opposite, since IEEE floats return inexact calculations as exact (incorrect) numbers.

> **An inexact unum represents the set of all real numbers in the *open* interval between the floating point part of the unum and the floating point number one ULP farther from zero.**

The formula for the value should be simple, but we do need to be careful of the values that express infinity and *maxreal*, and remember that if we go "beyond infinity" by setting the ubit, we get a NaN. Also, if the number is an inexact zero, then the direction "farther from zero" is determined by checking the sign bit.

Recall that in the prototype, `u2f` (unum to float) is the function that converts a unum into the real number represented by its float parts, and `ulpu` is the unum bit string with a `1` bit in the last bit of the fraction and zero for its other bits. Occasionally we want to change a unum to the one representing the exact value closer to zero; `exact[`*u*`]` does just that. If *u* is already exact, the function leaves it as is. Also, `inexQ[`*u*`]` is a Boolean test of whether a unum *u* is inexact; it returns `True` if it is and `False` if it is exact. Just for completeness, `exQ[`*u*`]` does the reverse, to reduce the number of "not" symbols in the program.

Here is the kind of picture you should keep in your head when thinking about inexact unums, where rounded shapes help remind us that the endpoint is not included in the bound. For positive unums:

Open Interval Represented by a Positive Inexact Unum

The way to test if a unum is positive or negative in the prototype is `sign[`*u*`]`, which returns the sign bit: `0` if positive, `1` if negative. Adding an ULP always makes the number farther from zero, so for negative unums, adding an ULP makes the represented value *more negative*:

Open Interval Represented by a Negative Inexact Unum

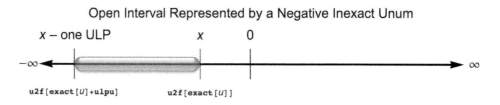

So `u2f[`*u*`]` handles the case of an exact unum *u*, and the above figures show how to handle inexact cases. We also need to convert unum strings for signaling and quiet NaN values, `sNaNu` and `qNaNu`. In the prototype, we use `exact[`*u*`]`, `big[`*u*`]`, `bigu[`*u*`]`, `signmask[`*u*`]`, and `sign[`*u*`]` to express the general conversion function (where `signmask[`*u*`]` is a `1` bit where the sign bit of *u* is and all `0` bits elsewhere):

$$x = \begin{cases} \texttt{u2f}[u] & \text{if } \texttt{exQ}[u] \text{ (that is, the ubit of } u \text{ is 0), else} \\ \text{NaN} & \text{if } u = \texttt{sNaNu or } u = \texttt{qNaNu, else} \\ (\texttt{big}[u], \infty) & \text{if } \texttt{exact}[u] = \texttt{bigu}[u], \text{ else} \\ (-\infty, -\texttt{big}[u]) & \text{if } \texttt{exact}[u] = \texttt{bigu}[u] + \texttt{signmask}[u], \text{ else} \\ (\texttt{u2f}[\texttt{exact}[u]], \texttt{u2f}[\texttt{exact}[u] + \texttt{ulpu}]) & \text{if } \texttt{sign}[u] = 0, \text{ else} \\ (\texttt{u2f}[\texttt{exact}[u] + \texttt{ulpu}], \texttt{u2f}[\texttt{exact}[u]]) & \texttt{sign}[u] = 1, \text{ which covers all other cases.} \end{cases}$$

Computer logic might first test if all the *e*, *f*, *es*, and *fs* bits are set to 1, and if they are, sort out the four possible special cases of signaling NaN, quiet NaN, $-\infty$, and $+\infty$. The next thing to test is for $\pm\texttt{bigu}[u]$ because those are the only cases where adding `ulpu` would cause integer overflow of the combined exponent and fraction fields, treated as a binary counting number. With all those cases taken care of, we can find the open interval between floats as shown in the two preceding illustrations of the real number line. Notice that the last two lines in the above function take care to express the open interval the way mathematicians usually write open intervals: (*a*, *b*), sorted so *a* < *b*. Also notice that the first two lines express floats, but the last four are open intervals that will require a new internal data structure; that is the subject of Chapter 5.

4.12 A visualizer for unum strings

The reader is *not* expected to learn to interpret unum bit strings! That is what computers are for. Understanding the bit strings that represent floats is already difficult enough. Recall that we can view the utag using the **utagview[**u**]** function, like this one (which describes an IEEE quad-precision float):

0	1110	1101111
↓	15	112

The **unumview[**u**]** function similarly annotates each field of a unum *u*, when we want to see all the machinery in operation. It shows the exponent used in scaling (accounting for the bias) and the value of the hidden bit. It also shows the represented value as both a fraction and a decimal, whenever the two forms are different. Sometimes decimals are easier to read, and sometimes the fraction is easier. The decimals shown are always *exact*, since a fraction with a power of 2 in the denominator always can be expressed with a finite decimal.

For example, here is how unum format represents π in a {1, 4} environment:

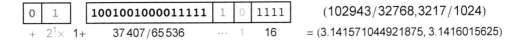

0	1		1001001000011111	1	0	1111	(102943 / 32768, 3217 / 1024)
+	$2^1 \times$	1+	37 407 / 65 536	...	1	16	= (3.141571044921875, 3.1416015625)

You expect irrational numbers like π to use the maximum number of fraction bits, and they will always set the ubit to show that the value lies within an ULP-wide range. Expressing π required 18 bits to the left of the utag.

On the other hand, expressing –1 never requires more than three bits to the left of the utag:

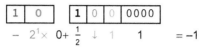

Notice that the hidden bit is **0** in this case since the exponent is zero. The number to add to the fraction is therefore 0 instead of 1, shown under the gap between the boxes. Just think about how often certain floats like –1 and 2 are used in computer programs; the ultra-concise way they can be expressed with unums starts to hint at how much storage and bandwidth they can save.

> **Exercise for the reader**: No matter what the utag is, there are *five exact numbers* that can always be expressed with just three bits to the left of the utag (and the utag set to all **0** bits). What are they? If the ubit is set to **1**, what are the *four open intervals* that can always be expressed with just three bits?

We need unums for inexact "positive zero" and inexact "negative zero" to fill in the gap between –*smallsubnormal* and *smallsubnormal*. The ubit gives us a reason to have "negative zero" representations. Here are two examples of open intervals next to zero in a {2, 3} environment. Using the maximum possible values for exponent size and fraction size with an inexact zero expresses the smallest possible ULP.

Here are the two bit strings that unum arithmetic uses instead of underflowing to zero:

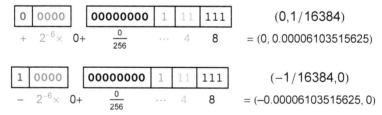

The largest possible ULP next to zero is when the exponent size and fraction size are as *small* as possible (one bit each):

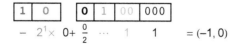

Unums provide a rich vocabulary for expressing the concept that a number is "small" even when it is not expressible as an exact number.

Before automatic computing, operators of desk calculators could track the significant digits in every calculation because numbers were entered by hand. With practice, operators could avoid the pitfalls of rounding errors by remembering which results had been rounded, and by how much. This disappeared with automatic computing, with intermediate scratch work *hidden* from the programmer.

5.1 The hidden scratchpad

Every computer has a hidden "scratchpad" for arithmetic. Perhaps the simplest example is the multiplier. Careful examination of the desk calculator on the previous page reveals eight digit positions that can be entered on the left half of the machine. Above where the keys are you can see two rows of displayable numbers, one with eight digits and one below with *sixteen* digits, enough to hold the exact product of two eight-digit fixed-point numbers. That is the scratchpad where extra digits are tracked temporarily. In those days, the scratchpad was certainly not hidden.

When computer designers build hardware for float arithmetic, they also allocate room to perform the multiplication of the fraction to twice as many bits as the fraction length. To do that quickly means being able to do all the partial products at once in the "parallelogram" for a multiply, and then collapse the sum with a double-length accumulator. For example, a simple 8-bit by 8-bit multiplier that produces an 8-bit rounded result can be envisioned like this, the binary version of the way people do pencil-and-paper multiplication with decimals:

```
          01111010
        × 10101101
        ───────────
          01111010
         00000000
        01111010
       01111010
      00000000
     01111010
    00000000
   01111010
   ───────────
  01010010 01110010
```

The bits shown in light blue are never shown to the user; they are part of the *hidden scratchpad*. (Some computers have instructions for retrieving the low-order half of the bottom-line product, however, for doing things like extended-precision arithmetic). There are similar structures for doing floating point addition and subtraction (where the bits must usually be shifted first because their exponents put their binary point in different places), divides, and square roots. If the rules for rounding are well defined, all computers will return the same result. There are no compatibility problems and no surprises.

> One layer of calculation is where all numbers are stored in some standard format. There is also an *internal layer*, a hidden scratchpad with extra bits, where the computer performs math perfectly, or at least accurately enough to guarantee correct representation in the standard format.

Computer history is littered with confusion having to do with hidden scratch pads that do *not* follow standard rules, or have nonstandard behavior that is poorly communicated to the programmer or user. Some historical examples point to a way to *avoid such confusion in a unum environment*.

5.1.1 The Cray-1 supercomputer

A Cray-1 at the Computer History Museum, Mountain View, California.

The Cray-1 (first shipped in 1976) was so advanced relative to the other computers of its era that it caused the coining of the term "supercomputer." The entire computer had only about 200 000 gates, yet it consumed 150 kilowatts of power. It was the first computer to have native hardware for 64-bit floating point. It predated the IEEE 754 Standard, and it was a stretch to achieve fast 64-bit arithmetic in that era, so the Cray floats took some shortcuts. The fraction was only 48 bits, and the multiplication scratchpad was incomplete, as shown by this diagram of the multiplication parallelogram from the original Cray-1 User's Manual:

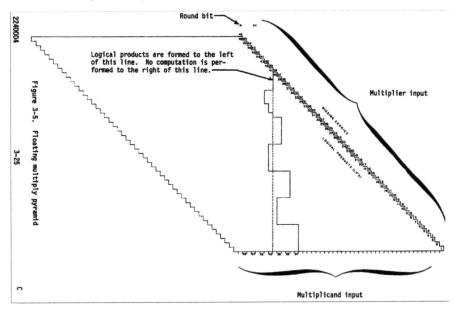

They literally *cut corners*. The corner of the multiplication parallelogram was cut off, leaving a ragged-looking "multiply pyramid" instead, to save gates and increase speed. No user can complain they were not warned, since it was explained in detail right there in the user's manual that the rounding would not always follow standard rules. This is a case of a scratchpad being *less* than what is expected, leading to surprise errors. The main unpleasant surprise users found was this one, a breaking of the *commutative law*:

$$a \times b \neq b \times a$$
(on the Cray-1)

Any course in numerical analysis will warn you that floats do not obey all rules of algebra, like the associative and distributive laws, but at least the *commutative* law is something floats can usually get right. Not on the Cray-1. Subtle changes to how a program was compiled could alter the answer, behavior which looks for all the world like a compiler bug. The bug was in the hidden scratchpad of the system. Cray eventually found a workaround: They *sorted* the inputs to the multiplier, so the smaller of *a* and *b* always went into the same input; the rounding was still wrong some of the time, but at least multiplication looked commutative again, and *a* times *b* comfortingly gave the same wrong answer as *b* times *a*.

5.1.2 The original definition of the C language

What gets done in the scratchpad is sometimes defined by software, not hardware. The first edition of *The C Programming Language* by Kernighan and Ritchie was published in 1978, and it contained some rules about how computations should be executed, rules that caused a great deal of grief.

Some of the rules are *still* causing grief, but one of them is scratchpad-related and has fortunately been removed from the ANSI standard definition of the language. To quote from page 41 of the first edition, where `float` here specifically means *single precision* floating point and not floats in general:

> "Notice that all `float`'s in an expression are converted to `double`; all floating point arithmetic in C is done in double precision."

This policy was probably envisioned as a wonderful protection feature, a way to automatically reduce (but not eliminate) cumulative rounding error within a single-precision routine. Early users of C could not *wait* to get rid of this rule. Suppliers of C language compilers quickly had to throw in options to disable the rule and leave each `float` (single precision) as a `float` unless explicitly converted to `double`. For one thing, the double precision policy was a big reason that C software involving floats ran much slower than an equivalent Fortran program. Double precision used more time, memory, and bandwidth, but the original C language defined it as mandatory, something the programmer could not turn off. Unlike the Cray example, this is a case of a scratchpad doing *too much* for the user.

Effectively, C was initially defined to "try to move floats into the scratchpad and keep them there as long as possible." Of course, as soon as a result gets stored in a `float`, the already rounded number leaves the hidden scratchpad and must be rounded to a lower precision. So if some common routine like summing a list of numbers is made into a separate function of type `float`, the program will inexplicably produce a different (less accurate) answer than if the summation is done *within* a routine, as a loop where everything stays in the hidden scratchpad. Having to convert between `float` and `double` takes extra time, too.

5.1.3 Coprocessors also try to be helpful

The Intel i8087 coprocessor has already been mentioned in the tall column of float types that can be represented by a {4, 7} utag, in Section 4.7. Motorola made a math coprocessor about the same time, also with 80-bit internal registers, to go with its 68000 series processors. The IEEE 754 Standard was being created at the same time as the i8087 architecture, and there was considerable interplay as the result of Intel participation in the IEEE committee.

The Motorola and Intel coprocessors both came out in the late 1980s and were very successful chips, commercially; the pent-up demand for a chip that could process floats at a higher speed surprised many people in Intel's marketing department.

Designers of both coprocessors fell prey to the temptation to invisibly "help out" the user. Once a calculation was inside coprocessor registers, it had a bigger scratchpad, with a larger exponent and a more precise fraction. There were really *two* levels of hidden scratchpad, since their 64-bit fractions had to be computed internally by the coprocessor chips to their mathematical accuracy of 128 bits to guarantee results correctly rounded, which then got rounded *again* in returning a conventional float to the program requesting the operation.

Programs run using coprocessors got different answers than ones run without coprocessors, which made users wonder: *Is that a bug?* Why is there a difference? Imagine trying to certify that a large application intended for a wide range of computer systems, like a circuit simulator or structural analysis code, is running correctly; what do you do with hardware configurations that have coprocessors and claim more accurate (different) results? Some of the applications are life-critical, like the safety of a bridge or a skyscraper; what do you tell customers who notice that the software gets different answers on different systems, despite the use of standard IEEE storage format?

Programmers and users were never given visibility or control of when a value was promoted to "double extended precision" (80-bit or higher) format, unless they wrote assembly language; it just happened automatically, opportunistically, and unpredictably. Confusion caused by different results outweighed the advantage of reduced rounding-overflow-underflow problems, and now coprocessors must dumb down their results to mimic systems that have no such extra scratchpad capability.

5.1.4 IBM almost gets it right: The fused multiply-add

Except in the Cray example, where the scratchpad did less than expected by the user, the recurring theme is designers trying to be helpful in the invisible layer by doing *more* than expected, creating havoc with incompatibility in the visible layer. IBM declared the "fused multiply-add" to be a good way to reduce rounding error: If a multiply is followed by an add, the full product from the scratchpad exact multiply is used as input to the adder, so the result is rounded only at the end of the pair of operations. This trick can often reduce rounding error: $y = ax + b$ becomes a fused operation that is accurate to within 0.5 ULP, just like multiplication and addition, by staying in the scratchpad for two operations instead of just one at a time.

The catch: Programming environments do *not* present the fused multiply-add (FMA) as *a distinct operation*. The approach has been more along the lines of "When we compile your program, we will look for places where you do a multiply followed by an add, and produce an instruction to use our fused multiply-add automatically." Computer systems that support FMA get different answers from those that do not.

If a standards committee had instead required *explicit* usage in various languages, like "`call fma[a,b,c]`" then at least a computer lacking built-in hardware for FMA would know to use software to mimic the result perfectly. If a programmer expresses the desire to compute $a \times b + c$, there is no communication mechanism that allows the computer to report "I calculated that with a fused multiply-add and then rounded" versus "I did the multiply and add separately, rounding after each operation." Invisible and inexplicable help invariably causes programmers to ask, "Is there some kind of compiler switch I can use to turn all this damned stuff off?"

At least IBM had the clarity of thinking to declare "fused multiply-add" to be *a distinct operation* from a multiply followed by an add. They were starting down the road of defining what goes on in the scratchpad layer and what goes on in the approximate side of things, the float layer. Kudos for that.

5.1.5 Kulisch gets it right: The exact dot product (EDP)

In the late 1970s, Ulrich Kulisch noticed that a large, but not impossibly large number of bits would suffice to accumulate the dot product of two float vectors with *no* rounding error. If a and b are lists of floats with n elements, the exact dot product (EDP) of a and b is $a_1 b_1 + a_2 b_2 + \ldots + a_n b_n$, with no rounding until the entire expression is finished. A binary float is a rational number, an integer times an integer power of two. Build a large enough integer accumulator, keep the entire multiplication parallelogram result (106 bits for double precision) instead of rounding it, and the dot product can be done *exactly* within the scratchpad layer, with perfect integer math. As Kulisch points out, there is historical precedent for computing the EDP as far back as G. W. Leibniz (1673). Kulisch has managed to get support for it as an option in the IEEE Standard. Just as with $+ - \times \div$, the EDP only gets rounded when the result moves from the hidden scratchpad layer to the float layer. Notice that having such a feature also allows *sums* of large lists of floats to be accumulated without rounding error until the end. The idea was patented and licensed to several computer companies, including IBM.

What Kulisch's invention finally got right was the declaration that the EDP was a *different operation*, one done exactly on the hidden scratchpad and returned as a rounded float. There was no intention to look for dot products and sums in existing programs and secretly try to help out with the accuracy loss by using the long accumulator; it had to be invoked *explicitly* and without ambiguity.

Also a good idea was that the definition allowed the EDP to be done in hardware or with extended-precision software, the only difference being speed. Some companies experimented with building exact dot products in hardware or software and discovered a surprisingly low penalty for doing the calculation exactly instead of the usual way. In 1994, Kulisch developed a coprocessor chip for x86 computers that computed the EDP exactly in *less* time than the x86 computed it with rounding error, even though the coprocessor technology was less advanced than that of the x86. How can that be, given that each double-precision product has to be done to 106 bits and shifted hundreds of bits to be added to the correct location in an accumulator?

The main answer is: *The memory wall*. As pointed out in the introduction, it takes at least sixty times as long to fetch an operand out of memory as it does to perform a multiply-add with it. If the two vector inputs to a dot product are coming from main memory, or even if only one input is, there are *dozens* of clock cycles in which to execute the instructions needed for the lossless accumulation.

Although it is an excellent idea, the EDP did not catch on in the 20 years after it was patented in 1980, possibly because it *was* patented, adding a barrier to the use of a mathematical idea. That possibility is a major reason that no ideas in this book have been protected by patent. It is difficult enough to persuade people to change to a better paradigm without also imposing a financial penalty for doing so!

Kulisch believes the bigger reason was that the introduction of IEEE floats and PC microprocessors in the 1980s made users complacent about the kinds of rounding error that his EDP is designed to battle.

> The lesson taught by all these attempts at clever hidden scratchpad work is that computer users want *consistent* answers, and want to be *permitted to decide the trade-off between speed and accuracy* depending on the situation. There is a subtle arrogance in any computer system that takes the position "You asked for a quick, cheap, approximate calculation, but I know what is good for you and will instead do a better job even if it takes more time and storage and energy and gives you a different answer. You're welcome."

5.2 The unum layer

5.2.1 The ubound

We still need to add something to the layer that works with the limited-precision representations (unums) before discussing the scratchpad layer. Unums give us the vocabulary needed for precise control of *sets* of real numbers, not just point values. In Section 3.4, we pointed out that adding a ubit still does not create a representation that is closed under the four basic arithmetic operations. The unum representing (0, 1) times the unum representing 2 should be a representation of (0, 2), but there is no such unum. We need to be able to define a range of real numbers where we can represent the endpoints carefully. The way to do this is the *ubound*.

> **Definition**: A *ubound* is a single unum or a pair of unums that represent a mathematical interval of the real line. Closed endpoints are represented by exact unums, and open endpoints are represented by inexact unums.

The set of ubounds is closed under addition, subtraction, multiplication, and division. Also square root, powers, logarithm, exponential, and many other elementary functions needed for technical computing. Ubounds are a superset of traditional "interval arithmetic" but are much more powerful, for reasons that will be shown in the next section. First, some examples of ubounds to give the reader a feel for how they work:

Suppose a unum *u* represents the exact value 3 and another unum *v* represents the open interval (4, 4.25). The pair {*u*, *v*} is a ubound, one that represents the mathematical interval [3, 4.25]. The left endpoint "(4" in "(4, 4.25)" is ignored. Any inexact unum that has a range that ends in "4.25)" will do, so we generally use the one that takes the fewest bits. The ubound is the *outermost endpoints*.

If we just wanted to represent (4, 4.25), we can write {*v*}, with braces to distinguish the ubound {*v*} from the unum *v*. The ubound {*u*} represents the exact number 3. Notice that the computer system will need an additional bit to indicate whether a ubound has a single unum or a pair of unums.

In the vocabulary of the prototype, the ubound {`neginfu`, `posinfu`} represents the set of all real numbers as well as the infinities, [−∞, ∞]. If we instead want to represent just real values, (−∞, ∞) is represented (in the prototype) by the ubound {`neginfu − ulpu`, `posinfu − ulpu`}. That is, each endpoint backs off just one ULP to get back to finite numbers.

As a preview of the type of math possible with ubounds, the cosine of the ubound {`maxrealu`, `posinfu`} is the unum pair that represents the closed interval [−1, 1]. Correct, concise, and very quick to evaluate. Details of operations on ubounds are provided in future chapters.

The one thing ubounds *cannot* do is represent *disjoint* sets of real values. (To do that requires a distinct data structure that is a *list* of ubounds, so such functions can still be created as ones that always produce *lists* of ubounds as output, instead of individual ubounds. This will be covered in Part 2.) Ubounds include *all* values between their endpoints. If the result of a computation produces two distinct regions of the real number line with a gap, the answer is the ubound for NaN, meaning "not a ubound," represented by {`qNaNu`}, the unum representation of quiet NaN.

With ubounds, we are ready to define the unum equivalent of the layer the user and programmer see directly when working with real values:

> **Definition**: The *u-layer* is the level of computer arithmetic where all the operands are unums (and data structures made from unums, like ubounds).

5.2.2 Processor design for the *u*-layer

Transistor density on microprocessor chips is now so high that using all the transistors at once on a chip will far exceed the electric power budget. It may look wasteful to have idle transistors on a chip, but think of them like the empty seats in a multi-passenger car, or the snowplows at a London airport shown above. Perhaps they are not in use for many hours in any given year, but when you need them, you *really* need them. When not in use, they do not add that much additional cost.

For that reason, it may be sensible to build a unum processor with fixed-size registers that accommodate unums up to some stated maximum environment setting. Chip designers certainly prefer fixed-size registers, and long registers need not consume full power unless every bit is in use.

Suppose we build a unum CPU that can do every existing IEEE float type, the {4, 7} environment shown in Section 4.7. The *maximum* unum size in that environment is 157 bits. The largest ubound would take twice that for a pair of maximum-sized unums, plus a bit indicating a pair instead of a single value: 315 bits total. Even a modest set of eight such registers would be capable of error-free arithmetic without a need to manage variable storage sizes. The smallest ubound would take only 16 bits, and since engineers are getting quite clever about not spending power except where there is active processing going on, the power savings possible with concise unum format may well benefit the *on-chip* power consumption, not just the reduced power needed for off-chip data transfer.

Instructions are needed that can pack and unpack unums without any wasted bits. A block of data like a cache line that contains packed unums is serially unpacked into however many bits of a fixed-size register are needed, leaving the other bits as "don't care" since the format is unaffected by them. Conversely, an instruction can concatenate the registers into a block of data for transport off-chip.

5.3 The math layer

5.3.1 The general bound, or "gbound"

To build a real computer system that uses unums, there must be a well-defined scratchpad layer. The scratchpad equivalent of a ubound is the *general bound*, or *gbound* (pronounced jee-bound).

Definition: A *gbound* is the data structure used for temporary calculations at higher precision than in the unum environment; that is, the scratchpad.

One way to build a gbound data structure is nine bits and four variable-length integer values. The first bit is whether the value is numerical (including $\pm\infty$) or NaN. If the NaN bit is set to **1**, no other entries matter since the result of any operation is simply NaN. The other values form a two-by-six table, as follows:

	$\lvert f_L \rvert$	$\lvert f_R \rvert$
	f_L negative?	f_R negative?
	$\lvert e_L \rvert$	$\lvert e_R \rvert$
NaN?	e_L negative?	e_R negative?
	open?	open?
	infinite?	infinite?

Single-bit values are those ending in "?" where **0** means False and **1** means True. The *f* and *e* values are subscripted with *L* or *R* to indicate whether they are left or right endpoint numbers. The top two row numbers store values as $f \times 2^e$, much like float or unum notation, but here the *f* values represent unsigned *integers* instead of signed fractional values between 1 and 2, and all values are stored with flexible precision. The top row values must be in numerical order; that is, $f_L \times 2^{e_L} \leq f_R \times 2^{e_R}$, with equality permitted only when both entries in the bottom row are closed. (An interval like "[23, 23)" makes no mathematical sense, since the set of *x* such that $23 \leq x < 23$ is the empty set.)

The "open?" bits are very much like ubits, indicating if the endpoint is open (`True`) or closed (`False`). The bottom row indicates infinite values, where a True value means the *f* and *e* values are ignored (but we still need to know if the infinity is an open or closed endpoint, and whether it is positive or negative). Most of the operations on gbounds begin by examining the NaN bit, and if that is False, then examining the infinity bit of each column, and only if neither bit is set is there any arithmetic to do using the *f* and *e* integers. The unsigned integers are stored as extended-precision numbers, using whatever scheme the designer prefers. A simple one is a fixed-sized integer that describes the length (in bits, bytes, or words) of the integer, followed by the integer as a bit string, located in a dedicated region of memory. A more sophisticated scheme could manage memory allocation and de-allocation for each operation by storing pointers for each integer.

Another approach is to use a stack architecture for the integers. The number of bits needed for f_L, e_L, f_R, and e_R depends on the environment settings and the operation being performed, but it is predictable and limited. Earlier we said that different computers will produce identical unum answers to identical problems, but that presumes that they have enough memory to *get* to an answer. There are many schemes for representing variable precision integers, and it does not matter which one is used in the gbound so long as it can make use of the available memory.

Even modest computing hardware can guarantee fast, *on-chip* calculation up to some unum environment setting; beyond that, it can "spill" to off-chip memory. Calculations have to be pretty outrageous for that to happen, however, using 2014-era chip technology. Some current microprocessors have on-chip caches with over a billion bits. If those bits were instead used to store extended precision scratchpad integers, a gbound could have endpoint values with *hundreds of millions of decimal digits without going off-chip.* To put that in perspective, the scratchpad space needed to support the EDP idea using the largest unum environment discussed so far (a {4, 7} environment with 16-bit exponent and 128-bit fraction) would take less than 200 000 temporary storage bits for a general bound accumulator. A calculation could support five thousand of those gigantic accumulators at once, entirely on-chip.

As an example of a gbound, the interval $(-0.046875, 4]$ can be written as $\left(-3 \times 2^{-6}, 1 \times 2^{2}\right]$, so the scratchpad would store it as follows, where the quantities in italics are single bits, **0** or **1**:

	$\|f_L\| = 3$	$\|f_R\| = 1$
	negative f_L	*positive f_R*
	$\|e_L\| = 6$	$\|e_R\| = 2$
not NaN	*negative e_L*	*positive e_R*
	open	*closed*
	not infinite	*not infinite*

It may look like gbounds are intended for exact integer arithmetic, but the arithmetic only has to be good enough to guarantee that the value that is returned to the *u*-layer is the same as if the arithmetic were exact. Since the scratchpad in a unum environment is based on gbounds, it is called the *g*-layer.

> **Definition**: The *g-layer* is the scratchpad where results are computed such that they are always correct to the smallest representable uncertainty when they are returned to the *u*-layer.

The *g*-layer knows all about the environment settings *esizesize* and *fsizesize*, so when converting it does the minimum work needed to get to the right *u*-layer answer.

When asked for the square root of the ubound for (9, 10], for example, the *g*-layer notes the open-closed qualities, then evaluates the square roots. A square root can be computed left-to-right just like long division, and that process stops either when the square root is exact, or the number of fraction bits is at the maximum, $2^{fsizesize}$. In the latter case, the "open" bit is set and for that endpoint, and the fraction is increased by one ULP to include all the unknown bits that follow the last one computed. Even though a closed endpoint for $\sqrt{10}$ was requested, the best the computer can do is produce an open bound that includes $\sqrt{10}$ as accurately as possible. The square root of 9 computes as an exact 3, but since 9 was specified as an open endpoint, the ubit in the unum representing 3 is set, to mark it as open.

In the prototype environment, a gbound is represented a bit differently from the above, since *Mathematica* can handle infinities and NaN as numbers. The prototype uses a two-by-two table:

$$\begin{array}{cc} \texttt{lo} & \texttt{hi} \\ \texttt{loQ} & \texttt{hiQ} \end{array}$$

where `lo` and `hi` are the left and right endpoints, and `loQ` and `hiQ` are `True` if the endpoint is open, `False` if it is closed. When the result of a prototype gbound is NaN, it is stored as $\begin{array}{cc} \textbf{NaN} & \textbf{NaN} \\ \textbf{open} & \textbf{open} \end{array}$. This color helps group the 2 by 2 table items visually.

5.3.2 Who uses NaN? Sun Microsystems gets a nasty surprise

The way NaN is stored in the *g*-layer makes it *very* fast and easy to deal with NaN values. There is nothing to do; every operation simply returns NaN. Some chip designers like to trap NaN exceptions and send them to a software routine instead of dealing with them in hardware. After all, who actually *uses* NaN, anyway?

Sun Microsystems chip designers in particular figured they could optimize away that part of the arithmetic hardware in one of their versions of the SPARC chip. The engineers did not consult with customers or application specialists; they simply did what they assumed was reasonable.

Surprise. Many customers running financial trading applications that used floats screamed in protest when they discovered their applications were running far *slower* on the latest chip, and wanted to know why. And they wanted it fixed. *What was special about financial applications?*

Imagine that you have a formula for stock trading that depends on knowing various numbers describing the stock, like its price and volatility and so on. If any numerical information is *missing*, the financial software **prudently uses NaN for that value** to prevent a spurious numerical calculation that could lead to a disastrous decision. A NaN is a great way to express "I don't know" to a computer, and NaN allows the computer to report "I don't know" back to the user. Sun had no way to fix the problem. They had engineered NaN handling out of their hardware, forcing it to be done very slowly, with software. And if there is one application where speed translates directly into money, it is automatic stock trading.

Computer users dislike being surprised by slightly inconsistent numerical answers, and they also dislike being surprised by highly inconsistent *performance* between different computers. Another thing that can produce inconsistent performance is when subnormal floats are handled with a software trap instead of calculated in hardware. Transistors are fast and cheap; software is slow and expensive. The only reason to shift the burden to software is to make life easier for a chip designer.

5.3.3 An explosion of storage demand in the scratchpad?

You may be thinking that the *g*-layer resembles the number-handling abilities of a symbolic algebra program like *Mathematica*, Maple, or dozens of other programs over the years that have offered nearly perfect numerical environments. That would mean slow operation, a very difficult hardware implementation, and probably not the hoped-for savings in energy and power promised by economical unum storage.

But the *g*-layer is *not* like any of those elaborate software environments; it is far simpler, and tiny compared to the feature set provided by symbolic algebra programs. The main difference is this: Calculations are **not allowed to dwell** in the *g*-layer, unless they are in a specific and limited set of "fused" operations permitted to chain in the scratchpad area before returning the final result to the *u*-layer. The *g*-layer does not need to do anything as elaborate as symbolic algebra.

For example: What if the *g*-layer needs to calculate

$$\sqrt{\tfrac{3}{10}} \quad \sqrt{4} \quad ?$$

open   closed

That kind of expression can never get *into* the *g*-layer, unless one of the permitted fused tasks is something we could call a "fused divide-square root," like IBM's fused multiply-add. A standardized set of tasks submitted to the *g*-layer must immediately return to the *u*-layer. That anchors the number of bits every operation requires to a manageable range for which a processor architect can plan. We will want some fused operations, including the multiply-add and the EDP, as well as several others that are too useful to pass up, but stop short when a temporary result cannot be expressed in the *g*-layer data structure. As arithmetic operations in the *u*-layer are defined in later chapters, their demands of the *g*-layer will also be discussed.

What this definition makes possible is *bitwise identical results* from computers from different vendors, including different math libraries for computing various elementary functions. There is exactly one legitimate value to return from a sequence of operations, just as $1 + 1 = 2$ is portable across computer systems. The real number line can finally be made as safe a place to compute in as the set of representable integers. (Maybe safer, since integers overflow and underflow and the point where they do so varies from system to system. Which is another thing the C language definition got wrong, since it allows integers to vary in size from system to system yet be called the same name, like `int` or `short`).

What is possible for the first time with unums is a mathematically correct data structure for describing operations on *subsets of the real number line.* That may sound a bit like what traditional interval arithmetic attempts to do, but the shortcomings of traditional interval representations lead us into a side discussion to explain the differences between ubounds and interval arithmetic.

5.3.4 Traditional interval arithmetic. Watch out.

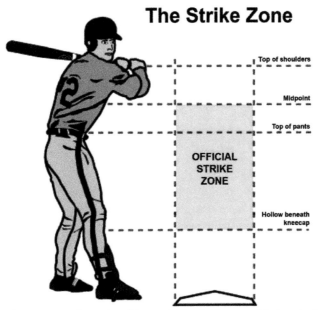

A familiar two-dimensional interval: The "strike zone" in baseball

What is called "interval arithmetic" dates back to the 1950s (Ramon Moore is usually credited with its introduction). It has been a partially successful attempt to deal with the rounding errors of floating point numbers. A traditional interval is always a *closed* interval, written [a, b], where a and b are exact floating point numbers, a less than or equal to b, and [a, b] means all real numbers x such that $a \leq x \leq b$. Traditional intervals always include their endpoints, which unfortunately makes them quite different from what mathematicians call an "interval." When a mathematician uses the term "interval," it means any of the following:

(a, b)	The set of numbers such that $a < x < b$
[a, b)	The set of numbers such that $a \leq x < b$
(a, b]	The set of numbers such that $a < x \leq b$
[a, b]	The set of numbers such that $a \leq x \leq b$

Traditional interval arithmetic can only express the last case. Unfortunately, this means we have to use the phrase "traditional interval arithmetic" because it is a corruption of the more general mathematical definition of "interval" shown above.

When the result of a traditional interval arithmetic calculation is inexact, the a and b are rounded in the direction that makes the interval larger, to ensure the interval *contains* the correct answer. But the rounded endpoints are incorrectly represented as *exact floats*, so what should be an *open* interval is represented as a looser bound, one that is closed.

Interval arithmetic falls short of what is needed for valid computer arithmetic in several important ways:

- **Need for expertise**. Traditional interval arithmetic requires considerable (and rare) expertise to use, even more rare than the numerical analysis expertise needed to get decent answers from floats.

- **Overly pessimistic bounds**. Traditional interval arithmetic often produces rapidly expanding, overly pessimistic bounds. It is easy to get the correct but utterly useless answer "somewhere in $[-\infty, \infty]$" to an interval calculation.

- **Inexpressible sets**. Traditional intervals cannot express sets like "All x less than 3." Should they return $[-\infty, 3]$? That says that 3 is less than itself, which is not acceptable math. Should the set of numbers less than 3 instead use the representable float that is one ULP less than 3, that is, $[-\infty, 3 - \text{ULP}]$? No, because that range leaves out all of the real number line in the open interval $(3 - \text{ULP}, 3)$, another math error. Traditional intervals cannot express the set of numbers that are less than 3. In other words, they *cannot express less-than and greater-than relationships correctly*.

- **Wasteful storage**.Traditional intervals take twice as many bits to store as a float of the same precision. If the bounds are tight, then most of their bits are identical, so the storage is redundant and wasteful; if the bounds are far apart, then you have lost accuracy and are wasting bits by storing inaccurate data to unnecessarily high precision.

- **Lost correlation**. Because it treats occurrences of a number as independent ranges instead of a single (unknown) point within a set, traditional interval arithmetic produces ranges when the correct answer is exact. For instance, $x - x$ should be identically zero, but interval math will instead return a range. If, say, x is the interval [23, 25], then traditional interval arithmetic says $x - x$ is [−2, 2] instead of zero. Similarly, the expression $\frac{x}{x}$ will return $\left[\frac{23}{25}, \frac{25}{23} \right]$ instead of the exact value 1.

Do traditional intervals at least provide *operation closure*, where arithmetic on them always produces another traditional interval? Not quite. Suppose x is the interval [−1, 2]. What is $\frac{1}{x}$? Some interval environments say the answer is [−∞, ∞], even though that interval range incorrectly includes a swath of real numbers that *cannot* be in the answer set. In other computing environments that support traditional interval arithmetic, including *Mathematica*, what you get is *two* closed intervals, [−∞, −1] and $\left[\frac{1}{2}, ∞ \right]$, which requires a different data structure. Division by an interval containing zero breaks the definition of "interval" as a pair of numbers that bound a connected region of the real line. Since it takes two floats to represent an interval, a divide operation can require *four* floats (two intervals) and thus represents a data structure unlike that used to represent single intervals. If you say, "Well, let's allow *pairs* of intervals," that soon breaks as well, with the result of some operations being *three* intervals. And four. There is a "Sorcerer's Apprentice" effect in play; two magical brooms might spring up for every one we try to deal with. There is no limit to the storage needed for the results. It is not a closed, limited-precision system.

There is another subtle dishonesty in using a *closed* interval. Suppose you ask for the square root of two, using IEEE half precision floats and interval arithmetic. Traditional interval arithmetic says this:

$$\sqrt{2} = [1.4140625, \ 1.4150390625]$$

That bound says "The square root of two could be equal to $\frac{1448}{1024}$, $\frac{1449}{1024}$, or any value in between." But we know $\sqrt{2}$ *cannot* be a rational number, just as we know that π cannot be rational. We can do better than such a sloppy bound. Computer algorithms that require real numbers are less treacherous to create if we have a number format that is inherently more careful, and rules out equality wherever no equality is possible. With the ubit, we can express the information that "The square root of two is somewhere between $\frac{1448}{1024}$ and $\frac{1449}{1024}$ but *cannot equal either endpoint*." Recall the example where taking the root of a number a few times produced a value that was mistakenly declared exactly equal to 1. Traditional intervals are just as prone to this mistake as floats. With ubounds, the error cannot occur.

Interval arithmetic has some good ideas about restoring rigor to computer arithmetic, and decades of research effort have been spent trying to avoid its numerous pitfalls. Some of those ideas apply to ubounds, and we will seek the ones that can be made automatic so computer users do not have to become experts in interval arithmetic.

5.3.5 Significance arithmetic

Another aside. Earlier we mentioned *significance arithmetic*, which attempts to track how many significant figures remain in a calculation. It is an estimate of the accuracy, not a guarantee like a ubound. For example, significance arithmetic says that "$2 + 3.481 = 5$," since the 2 has only one significant digit and thus the answer must be rounded to one significant digit. Significance arithmetic is a very lossy and unsatisfactory form of numerical error analysis. Otherwise, significance arithmetic would be in general use as a way to automatically check for rounding errors and help programmers choose the right precision for their floats. If "2" really means "2 ± 0.5" and "3.481" really means "3.481 ± 0.0005," then the sum "$2 + 3.481$" is actually "5.481 ± 0.5005," a tight bound on the accumulated uncertainty. *Significance arithmetic has no way to express such a range.* The only ranges it knows are digit-sized.

With ubounds, we can do *much* better than significance arithmetic, at a low cost. Ubound arithmetic is more powerful at tracking accuracy *because the endpoints can be different precisions and independently be closed or open.* Ubounds cling to every shred of information about how large or small a number set can be. Suppose we wanted to express the range $2.5 \leq x < 3.1$. The number 2.5 is exact, and the upper value 3.1 is expressed with a minimal-width ULP. In contrast, if we put that range into significance arithmetic, it would simplify it as "3, to one significant figure," since the range $2.5 \leq x < 3.1$ is contained in 3 ± 0.5.

Significance arithmetic throws away a *lot* of information with every operation. As a result, it usually announces complete loss of significance after a few successive operations, even though the calculation seems to be proceeding just fine.

5.3.6 Why endpoints must be marked open or closed

One reason endpoints must be marked open or closed is that we often need to know the *complement* of a set. If you request "the set of numbers larger than 8," traditional intervals return $[8, \infty]$, which incorrectly includes 8. If you now ask for the complement of that set, all the numbers that are *not* in $[8, \infty]$, you get $[-\infty, 8]$, which also incorrectly includes 8. Both errors are sloppy math that can be corrected with just a single bit of storage to indicate the open or closed quality of an endpoint.

Sometimes, we display mathematical intervals as a thick line, showing open endpoints as rounded and closed endpoints as squared-off, imitating the curve of a parenthesis ")" and the squareness of a bracket "]". For example, an interval $[-2, 3)$ on the real number line that is closed on the left but open on the right can be visualized like this:

Suppose x is $\left(-1, \frac{3}{2}\right)$; what is x^2? The plot of $y = x^2$ for $-1 < x < \frac{3}{2}$ is a section of a parabola, but if we plot it with unconventionally thick lines that can show open (rounded) or closed (square) endpoints, it helps visualize how the open or closed properties can change when doing arithmetic. The input range is shown in light magenta, and the output range is shown in light green:

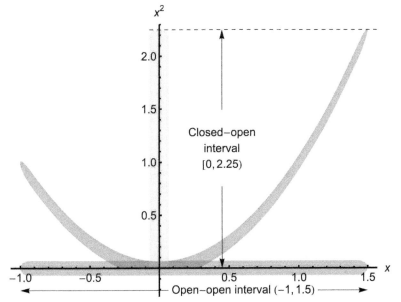

The upper bound of the range of the result is shown as rounded (open), but the lower bound is blunt to indicate that it **includes zero** (closed). The square of $(-1, 1.5)$ is $[0, 2.25)$. This illustrates why we need endpoint representations that are independently open or closed.

If all the fuss about open or closed endpoints seems like mathematical hair-splitting, consider a programming bug that has bitten just about every programmer who ever lived: The "off-by-one" bug. The bug stems from selecting one of the following loop constructs, expressed in an abstract computer language:

```
for i=1 to n
for i=0 to n
for i=1 to n-1
for i=0 to n-1
```

Do you stop when you *get* to **n**, or just before? Was that a "less than" or a "less than or equal" that determined when it was time to stop? Does the array index start at 0, or at 1? No matter what language you use, it is very easy to accidentally start or end at the wrong point, and the results of that mistake are usually not pretty. If you are a programmer who has ever experienced this error, **good**. Then you understand. Sometimes this kind of bug is referred to as an "edge condition" or "corner case" bug.

The equivalent of the off-by-one bug in programming with real numbers is whether you include the endpoint of intervals (closed) or include everything approaching the endpoint but not the endpoint itself (open). Floats and traditional intervals do not even allow this distinction. The ability of unums to track open-closed endpoint qualities eliminates many errors from calculations involving real numbers.

5.4 The human layer

5.4.1 Numbers as people perceive them

The chapter heading said there are three layers of calculation. Certainly there is a scratchpad layer and a layer where numbers are represented as bit strings with format rules. The third layer is the *human* layer, where numbers exist in forms that humans can send to a computer and experience the results.

> **Definition**: The *h-layer* is where numbers (and exception quantities) are represented in a form understandable to humans.

The term "input-output" is not a synonym for *h*-layer, because "input-output" often refers to moving data from one part of a computer system to another. The *h*-layer is input-output with a human on one end.

You may be thinking of digits 0 to 9, the decimal point, "+", "−", "···", "↓", and so on. Think more generally. Graphics, sound, haptic feedback, the furnace in your house coming on from a thermostat command, are common examples of calculations converted into physical events that humans perceive. When you turn the steering wheel of a car or play an electronic musical instrument, you send numerical information to a computer without ever typing in digits, so it works both directions.

Some have proposed that all floats should use base ten internal format (and there are IEEE decimal formats defined for floats). But that only helps for textual input and output of decimal character strings, which is a small fraction of the many ways computers receive and send float values. Decimal format floats run (at best) half the speed of binary floats, a harsh penalty to pay if the purpose of the float is to generate pixels or send music to a speaker system!

For now, consider the conversion of decimal character strings (the *h*-layer) to and from the *u*-layer. That conversion is an *operation*, just as prone to information loss as division or square root or any other operation; therefore, we need policies that prevent confusion about accuracy when numbers pass between layers. Even when there is no information loss and numbers pass perfectly between layers, we need a way to communicate a *range* of real numbers. In most computer languages, there is no standard way to convey whether an input or output value is exact or approximate, and if approximate, within what range. The vocabulary exists in conventional notation; we simply need to apply it.

5.4.2 Demands and concessions

Computers work best as tools when the user is able to communicate both what *is* and what *is not* required, or "demands and concessions" for short. If a computing system has no information about what does *not* matter, it makes conservative assumptions and often does more work and takes more time than needed.

One of the most spectacular examples of such unintended waste is the way standard math libraries compute trigonometric functions sine, cosine, tangent, cosecant, secant, and cotangent. For example, the cotangent repeats what it does between 0 to π, at every integer multiple of π:

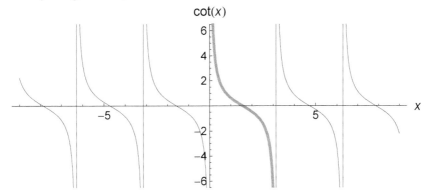

The software library routines that compute periodic functions for any real-valued input first use *argument reduction* to find an equivalent angle; then, tables or ratios of polynomials can accurately find the trig function of that reduced argument. Suppose you are writing a program in C or C++ and ask for the cotangent of an angle in radians represented by a single precision float (about seven decimals of accuracy). There is no way to communicate whether that angle is exact, since floats contain no accuracy information. The math library has to assume, conservatively, that *every* input value is exact. The library routine first has to find what that angle is modulo π. To produce a result that is within half an ULP of the correct answer, it has to first divide the seven-decimal "exact" float by π accurately enough that the remainder (the reduced argument) *also* has over seven decimals of precision.

Suppose the angle is one million radians. What is one million, modulo π?

The only way to "unroll" the value to sufficient accuracy is to store π accurate to *hundreds* of bits and perform extended-precision arithmetic requiring many opera-tions, and that is just what the standard math libraries do. However, in the very likely event that the float for "one million" is not an exact value, then "one million" is actually somewhere between 999 936 and 1 000 064. That range of angles goes all the way around the unit circle *dozens of times*. The cotangent is therefore anywhere from $-\infty$ to ∞, inclusive. Imagine the amount of useless work a computer does to produce a precise seven-decimal answer, because its *h*-layer has no way of expressing whether an input argument is exact or not.

Some vendors provide math libraries that use less precision and return computations faster, so the user who elects to use such a library is making a concession in exchange for a demand of higher speed. The selection of a lower-accuracy library is just about the only way programmers can presently communicate an accuracy concession to a computer.

Real numbers are usually programmed with decimal input values, which do not usually convert to exact unums any more than they convert to exact floats. Simply converting a decimal produces an inexact value. Without any concessions, "2.997" could mean 2.997 exactly, or it could mean 2.997 ± .0005. What is the best thing to do? Current practice is to find the closest binary approximation to the decimal number and silently substitute it for the requested value. With the *h*-layer for unums, we allow users to make concessions about the precision of inputs, express the demand for accuracy more explicitly, and let the computer concede when a calculated result is inexact.

If no concession is made, a decimal input like "2.997" is treated as the ratio of two exact numbers: $\frac{2997}{1000}$. Anywhere the value is used in a calculation, the ratio $\frac{2997}{1000}$ is calculated *according to the current environment settings*, but the value is never summarized as a single unum since that could discard information.

One rationalization you sometimes hear for the incorrect answers produced by floats is this: "The answer *is* the exact result, but for a problem slightly different from the one you posed." The rationalization even has a catch phrase: "backward error analysis." It is a variation on the Original Sin, and amusingly, it puts the blame back on the computer user:

> "I cannot give you the answer you requested, and it's all your fault, because you *should* have asked a slightly different question. I gave you a perfect answer to the question you should have asked."

Backward error analysis only makes sense if the user has some way of giving a computer *permission* to treat input values as approximate.

We can also give the user the symbols to optionally communicate exact and inexact decimal inputs. The aesthetic here is to reduce user "surprises." If the user simply specifies "6.6" with no accuracy information and gets the most compressed possible unum, it might be surprising (and annoying) to find the value turned into the most economical (and least precise) unum that contains the range 6.55 to 6.65. But if the user enters "~6.6" then we need the tightest possible ubound for [6.55, 6.65). The interval is closed on the left but open on the right because of round-to-nearest-even rounding rules.

In a {3, 4} environment, an input number written as "~6.6" becomes the ubound that encloses {6.55, 6.65} with the least accuracy loss, which is the ubound for

(6.54998779296875, 6.6500244140625).

The range is slightly larger than ±0.05, but expresses what the "~" means as closely as possible for this environment.

Here are some suggested "grammar rules" for the *h*-layer: Ways to express demands and concessions regarding inputs, and ways for a computer to express its knowledge of the accuracy of output results.

6.5⋯	There are more nonzero digits; the open interval $(6.5, 6.6)$, like **ubit** = 0 in binary.
6.5⁺	For output; there are more *known* digits, but insufficient room to show them.
~6.5	$6.4500⋯$ to $6.5499⋯$, the *open* interval that could have been rounded to 6.5.
6.5±.05	The *closed* interval centered at 6.5: $[6.45, 6.55]$
6.5↓	For input: Use the smallest unum containing $\frac{65}{10}$. For output: The result is exact.
6.5	The same as 6.5↓; a boldface last digit means there are no more nonzero digits.
6.5(2)	Concedes the last digit could be ±2. In this case, the interval $[6.3, 6.7]$.
6.5(0)	Another way of saying 6.5 exactly.

Using "..." instead of "⋯" is hard to read because the dots in the ellipsis "..." look just like decimal points and periods, so we prefer the slightly elevated "⋯" as a way to say "there are more digits after the last digit shown." That approach will be used throughout this book. The prototype does not support all the format annotations suggested above. They are presented here to illustrate that there are many ways people have expressed exactness and inexactness in human-readable formats. In any computer language that includes unums, the grammar should support ways to express input demands and concessions, and **accuracy information must *always* be included in the output**.

If a computed number is sent to mass storage for use later, then preserving the binary unum is obviously the best way to avoid losing information. If the number *must* be converted to characters representing the *h*-layer decimal form, then the above table provides some notations that help record the amount of accuracy. If we omit the accuracy information then we misrepresent the number as exact, which is what leads to so many problems with floating point arithmetic.

While there is usually some accuracy loss in expressing a decimal to the *u*-layer, it is always possible to express a unum or ubound precisely. The prototype includes a function **autoN**[*fraction*] that displays just the right number of decimals to express *fraction* exactly when *fraction* is an integer divided by a power of 2. For example:

```
autoN[19/2048]
```

```
0.00927734375
```

The **autoN** function can be combined with the open-closed information to display a unum or ubound exactly. While it is far more economical to store blocks of unums as binary, if it is important for a block to be human-readable, one could store, say, the following ubound mathematical meanings as character strings separated by spaces:

```
[0, 2.75) NaN 0 3 0.00927734375 [480,inf] (0,0.0000152587890625)
```

All of those character strings convert perfectly back into their original ubounds (with the same or larger environment settings). Just as unums have adjustable length that sometimes saves space, sometimes we are better off not using a fixed length format. Format specification like those in C and Fortran erroneously clip off decimals in some cases, and on the other hand makes us have to read "0.0000000000000000000" when "0" would serve just as well. Any "rounding" of output must be explicitly requested, and the output marked as a rounded result using a notation like those in the table above.

5.5 Moving between layers

5.5.1 Defining the conversion function

Section 4.11 presented the formula for converting a unum to its mathematical value. The other direction is more challenging: *Convert an arbitrary real number or range of real numbers into a unum or ubound.* It is much easier to turn a unum string into a real number than the other way around. The tricky part is that we seek the *smallest* bit string that does the job. If it happens that a real number can be expressed exactly by a unum string in a given environment, then we look for the smallest exponent and fraction that does the job so that the unum requires the minimum number of bits. If the real number lies *between* expressible exact values, then we use inexact unum that has the smallest ULP we can muster.

The devil is in the details, because there are exception cases for finite reals with magnitudes larger than *maxreal* or smaller than *smallsubnormal*. Also, infinity and NaN cases need to be dealt with.

The prototype function that converts an arbitrary real number x into a unum is called **x2u** ("x to unum"), and the code for it is in Appendix C.5, for anyone interested in the real nuts-and-bolts of unums. To make unum arithmetic easier to read, we simply write a hat "^" over a value to indicate its unum equivalent, like $\hat{511}$ for the unum form of the number 511. That is, **x2u[511]** is the same thing as $\hat{511}$. In the prototype, you can put a hat on top of any expressible real number, or positive or negative infinity, or NaN. Some **unumview** examples in a {2, 3} environment show how the fraction and exponent sizes adjust automatically to whatever is needed. First, set the environment to at most 2^2 exponent bits and at most 2^3 fraction bits:

---

setenv[{2, 3}]

With that environment, here is the unum for negative infinity, with every bit set to **1** except the ubit:

---

unumview[-$\hat{\infty}$]

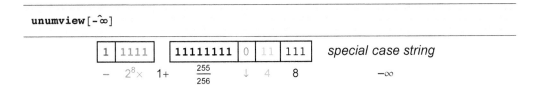

A unum environment will need to keep a few constants available, like π. The **x2u** conversion only needs enough knowledge about the value to assure correct conversion to the smallest possible ULP:

---

unumview[$\hat{\pi}$]

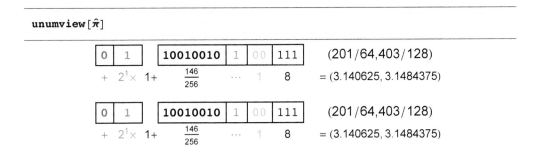

Notice that we can represent π correctly to three decimals using only ten bits to the left of the utag.

If computing with NaN, we assume it is the quiet NaN since otherwise we would have halted computing. That is why there is only one type of NaN in the *g*-layer.

unumview[NâN]

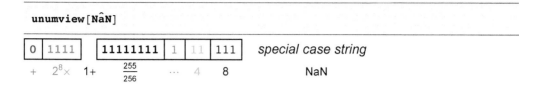

special case string

NaN

When the real number is a power of 2, the conversion routine favors putting bits into the exponent instead of the fraction, which minimizes the total number of bits needed:

unumview[1 /̂ 64]

| 0 | 0001 | | **0** | 0 | 11 | 000 | | $\frac{1}{64}$ |

$+ \quad 2^{-6} \times \quad 1 + \frac{0}{2} \quad \downarrow \quad 4 \quad 1 \qquad = 0.015625$

The above unum may look wasteful, with all those "leading zeros" in the exponent. There are quite a few ways to express $\frac{1}{64}$ in this environment, but every time you use fewer bits in the exponent, it creates the need for exponentially more bits in the fraction. The format shown above is as compact as it gets.

Numbers too small to express fall into the (0, *smallsubnormal*) unum:

unumview[0.00000000000000000̂00000000000000001]

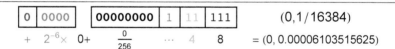

$(0, 1/16384)$

$+ \quad 2^{-6} \times \quad 0 + \frac{0}{256} \quad \cdots \quad 4 \quad 8 \qquad = (0, 0.00006103515625)$

Exercise for the reader: For a number too large to express exactly, like 100^{100}, how would it be expressed in the $\{2, 3\}$ environment used for the examples above? What inexact unum that contains 100^{100} has the *smallest number of bits*, and what open interval does it express?

5.5.2 Testing the conversion function

We demand mathematical perfection in the way we translate real numbers to unums and unums back to general intervals (prototype functions `x2u[x]` and `u2g[u]`, both detailed in Appendix C). For anyone developing their own unum environment, a good test of correctness is to draw graphs like the following, where we use a low-precision environment and sample the real number line to see if converting reals to and from a unum results in rectangular bounds that contain the line $y = x$:

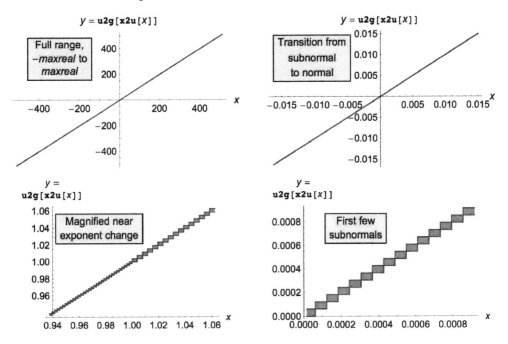

The full range test looks like the conversion functions `x2u` and `u2g` are ideal inverses within precision limits, but of course the plot could miss exceptions at that scale. The upper right graph checks the transition from subnormal to normal numbers, and still looks like a continuous line. If we magnify the plot to the point where there are just a few values, we can finally see the discrete unum rectangles; they contain the idealized $y = x$ line. The lower left plot has small rectangular regions on the left and larger rectangular regions on the right, because the exponent increases by one right in the middle of the plot. The lower right plot shows perfect conversion and inverse conversion for the very smallest subnormals.

How does this compare with what floats do? The graph for float conversion looks like a *stair step* instead of a series of rectangles. The stair step zigzags over the $y = x$ curve, missing everywhere but at one point per step. Whether the stair step is below, above, or centered on the correct value is determined by the "rounding mode." Mathematically, float conversion is incorrect "almost everywhere"; there are a lot more irrational numbers than rational ones, and float conversion errs unless converting a rational that just happens to be of a particular type.

5.6 Summary of conversions between layers in the prototype

In the prototype environment, **u2g**[*u*] converts a unum or a ubound *u* to its general interval equivalent. The **g2u**[*g*] function converts a general interval *g* to its most economical *u*-layer version. The code for both is in Appendix C.7.

To move a *u*-layer value to the *h*-layer means displaying the number in a form that humans can read. Usually, we just want the simplest mathematical notation, like "[1, 3.0625)." The prototype does that with **view**[*x*]. If you need to see all the under-pinnings of the unum format, the function is **uview**[*x*], which graphically displays the color-coded unum fields and their meanings, for unums, ubounds, and general intervals. The **gview**[*x*] function displays a unum, ubound, or general interval as a two-by-two table. The following diagram summarizes all the prototype functions for moving between layers.

The **view** functions are robust in that they look at the argument data type and automatically convert it as needed; that is why there are doubled arrows pointing left for the **view** functions, since values can come from either layer. Also note that it is possible to send a two-by-two general interval directly *to* the *g*-layer, which is often the most legible way to enter a general interval as opposed to an exact number.

Exercise for the reader: There are sixteen possible "fused" operations made from fusing an add, subtract, multiply, divide with another add, subtract, multiply, or divide before returning from the *g*-layer. Which fused pairs do *not* fit the internal scratchpad structure of a gbound shown in Section 5.3.1, if calculations must be mathematically exact? How might the gbound definition be modified to make all sixteen pairings storable exactly in the *g*-layer before returning to the *u*-layer?

5.7 Are floats "good enough for government work"?

Failed launch of the European Space Agency Ariane 5 rocket in 1996, caused by an attempt in the computer program to stuff a 64-bit float into a 16-bit guidance system. Loss: About $750 million in 2014 dollars. Computers, not programmers, should manage storage size. The mistake would have been avoided had unum arithmetic been used.
Credit: amskad/shutterstock

> **If you think rounding error, overflow, and underflow are only obscure academic worries, think again. The errors caused by floats can kill people, and they have.**

They also have caused billions of dollars worth of property damage. Computer users have developed a peculiar and fatalistic tolerance for the mistakes that floats make. Many only get uncomfortable when they discover that different computer systems get different "correct" results, since the IEEE standard allows (encourages?) hidden scratchpad work and tolerates varying amounts of such work on different systems.

It is just as important to define the three layers of computing and standardize what goes on in them as it is to let a number format track the range of real numbers it represents. With those ideas combined, it finally becomes possible to define a numerical standard that is mathematically truthful and assures *identical* results on any computer system. Even in the case where different algorithms are used to solve the same problem and results are not identical, the *u*-layer results will always contain the correct result and *intersect* unless there is a programming bug. The intersection is a refinement of the answer.

Instead of clamoring for computers that can perform ever more floating point operations per second, we should be asking for computers that can perform numerical operations… *correctly*.

6 | Information per bit

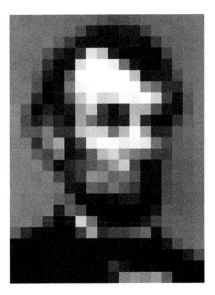

Abraham Lincoln, at low accuracy but high precision: 149 472 pixels. Most people would not even recognize who this drawing is supposed to depict.

Lincoln at high accuracy but low precision: Only 432 pixels, yet you immediately know who this is. Getting it *right* is more important than making it *precise*.

6.1 Information as the reciprocal of uncertainty

Remember the complaint about traditional interval arithmetic wasting bits: If you use two fixed-size floats to represent bounds, then either you have a tight bound (which means most of the bits in the two numbers are the same and therefore redundant), or the bound is loose (which means both floats have more precision than is justified). Unums and ubounds are not only more bit-efficient than traditional intervals, they also provide a way to manage the ratio of *information* to *storage*, either automatically or at the request of the programmer.

> *Information*, or answer *quality*, is inversely proportional to the width of a bound. If you have no bound at all on an answer, then your information is zero because the bound on the answer is $[-\infty, \infty]$.

We use computers to give us information about something, whether it is the chance of rain three days from now or the estimated value of an investment portfolio or where to find good coffee nearby.

The idea that information is measurable with a real number seems peculiar to computer people who only work with integer (or fixed point) problems, where you either get the answer or you don't. Too many computer science degrees have been earned by solving homework problems drawn only from the integer world, like "find all the ways to arrange eight chess queens on a chessboard such that none of them can attack each other."

Programs involving *real* numbers need to track and maximize information about the answer. They also need to optimize the ratio of information to compute time; that often means optimizing the ratio of information to the number of bits, since moving bits around takes most of the time. Readers familiar with Claude Shannon's definition of "information" may notice a concept similarity.

We need a couple more tools in our arsenal for allowing speed-accuracy trade-offs to be done, either by the computer automatically, or as a user choice.

6.2 "Unifying" a ubound to a single ULP

Suppose we are using the {4, 6} environment (capable of IEEE double precision) and the result of a unum calculation is the interval (12.03, 13.98), expressed as a ubound. Because the ubound is a pair of unums, it requires 122 bits of storage. That is more compact than traditional interval arithmetic, but it is still an awful lot of bits just to say a value is "between twelve and fourteen, roughly." The answer information is the reciprocal of the width of the bound, approximately

$$\frac{1}{14-12} = \frac{1}{2}.$$

Instead, we *could* opt to use a ubound with only one unum to represent the interval (12, 14), and accept the slight loss of information about the answer. The ubound that represents (12, 14) takes up only *17 bits*. From 122 bits to 17 bits with almost the same information about the answer means an approximately sevenfold increase in *information per bit*.

For intermediate calculations, it might not be a good idea to loosen the bound. However, when it comes time to declare something a final answer, and place it back into main memory or present it to the computer user, optimizing the ratio of information to bits might make sense. Unlike rounding of floats, the act of loosening a ubound should always be an *explicit*, *optional* command.

In an ideal unum computing environment, the unification function would be mostly or entirely baked into the hardware to make it fast. The prototype spells out the logic that would have to be turned into an equivalent circuit.

6.3 Unification in the prototype

6.3.1 The option of lossy compression

In the prototype, the `unify[ub]` function finds the smallest *single* unum that contains the general interval represented by a ubound *ub*, if such a unum exists. If no such unum exists, *ub* is returned without change. The result must be either exact, or one ULP wide. If inexact, the approach is to find the smallest ULP that describes the separation of the two bounds, express each bound as a multiple of that ULP value, and while they do not differ by exactly one ULP, double the size of the ULP. Eventually we find a one-ULP separation and can represent the range with a single inexact unum. We cannot unify any ranges that span 0, ±1, or ±2, or ±3 since the coarsest unums do not include those points. There are other special cases to deal with. The code for `unify` is in Appendix C.6.

When we construct unum arithmetic operations, we will typically compute the left bound and the right bound separately. The `unify` routine can be the final pass that notices if a *lossless* compression of the answer to a single unum is possible. For example, in computing the square root of the open interval (1, 4), the algorithm finds the square root of the left and right endpoints, marks both as open with the ubit, and initially stores it as a ubound pair $\{\hat{1} + \texttt{ubitmask}, \hat{2} - \texttt{ubitmask}\}$. However, that pair represents the same general interval as $\{\hat{1} + \texttt{ubitmask}\}$, because the ULP width is 1. Lossless compression requires no permission; unum arithmetic should do it automatically, whenever possible.

As an example from the prototype, we can unify a ubound that encloses the interval (23.1, 23.9). In the {3, 4} environment, it takes 57 bits to express the two-part ubound:

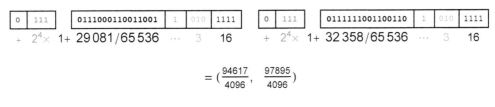

$$= (\frac{94617}{4096}, \frac{97895}{4096})$$

$$= (23.099853515625, \quad 23.900146484375)$$

The function `nbits[u]` counts the number of bits in a ubound *u*. It adds the number of bits in the one or two unums that make up the ubound, and adds 1 for the bit that says whether the ubound is a single unum or a pair of unums.

`nbits[{23.1,23.9}]`

57

Now if we compute `unify[{23̂.1, 23̂.9}]`, the result is a one-part ubound, with far fewer bits:

```
Module[{u={23̂.1,23̂.9},v},
v=unify[u];
Print[uview[v]]]
```

| 0 | 111 | | **0111** | 1 | 010 | 0011 |

$$+ \quad 2^4 \times \quad 1+ \quad \frac{7}{16} \quad \cdots \quad 3 \quad 4 \quad = (23, 24)$$

That looks like a 16-bit number, but again, we also count the bit that indicates whether the ubound contains one or two unums:

```
nbits[unify[{23̂.1,23̂.9}]]
```

17

Knowledge about the answer dropped because the bound is now wider. The bound expanded by the ratio $\frac{24-23}{23.9-22.1} = 1.25$, so the information decreased by the reciprocal of that: $\frac{23.9-22.1}{24-23} = 0.8$, a twenty percent drop. However, the bits needed to express the bound improved by $\frac{57}{17} = 3.35 \cdots$. So *information per bit* improves by a factor of $0.8 \times 3.35 \cdots = 2.68 \cdots$.

While the unum philosophy is to automate as much error control as possible, there is always the *option* of "switching to manual," since there may be a decision about the trade-off between accuracy and cost (storage, energy, time) that only a human can make, and it varies according to the application. There is a straightforward way to give the computer a guideline instead of the user having to decide each case.

6.3.2 Intelligent unification

A unum environment can have a policy regarding when to unify to replace precise bounds with looser ones in the interest of saving storage and bandwidth. For example, a policy might be that if unification improves information-per-bit by more than *r* (set either automatically or by the programmer), then it is worth doing. This is the kind of intelligence a human uses when doing calculations by hand. The routine that does intelligent unification in the prototype is `smartunify[u,r]`, where *u* is a ubound, and *r* is a required improvement ratio in information-per-bit.

Using a ratio of 1.0 says we have to get the same information-per-bit or better. An *r* ratio of 0.25 means we will accept a `unify` result even if it makes the information-per-bit as much as four times *less* than before, a valid goal if data compression is far more important than accuracy. A ratio of 1.5 would only unify if a 1.5 times improvement in the information-per-bit is possible. Many other more sophisticated policies are possible.

As an example of conditional unification, try **smartunify** on the interval used in the previous example, with a requested improvement factor of at least 3:

---

view[smartunify[{23̂.1,23̂.9},3]]

(23.099853515625, 23.900146484375)

It did not change the ubound, since it could not achieve the requested improvement factor. Try again, but this time ask for an improvement factor of at least 2.5:

---

view[smartunify[{23̂.1,23̂.9},2.5]]

(23, 24)

The prototype code for **smartunify** is shown in Appendix C.6, and the reader is encouraged to think about other policies that could automate the management of error and maximize information-per-bit. For example, information-per-bit can be improved while still keeping two separate unum endpoints, but using endpoints with fewer fraction bits. In the {3, 5} environment, a ubound representing (23.1, 23.9) demands the full 32-bit fraction size for each endpoint, yet the bounds are so far apart that it might make sense to relax the bound to, say, (23.0625, 23.9375) where the fractions only take 8 bits to express. Doing so reduces the storage from 91 to 43 bits, an improvement of about 2.1 times, while the bound only gets looser by about nine percent, so information-per-bit goes up by a factor of about 1.9.

One can imagine making the improvement ratio r part of the environment settings, just like *esizesize* and *fsizesize*. Just as in compressing music or video, sometimes you want the result to take less storage and sometimes you want higher fidelity. The programmer still has to exercise some initial judgment, but the computer does the hard part of the work managing the calculation after it is given that guidance.

Incidentally, asking for aggressive unification after every operation reduces a unum environment to something very much like significance arithmetic, with its much more limited ability to express the inexactness of left and right endpoints separately. Doing so might be useful in limited situations where input numbers are involved in only short sequences of calculations before output, but in general it is better to wait until the very end of a calculation to look for opportunities to increase information-per-bit.

6.3.3 Much ado about almost nothing and almost infinite

While there is only one way to express exact values $-\infty$ or $+\infty$, in any environment, there are generally many unums expressing *open* intervals that have $-\infty$ as the lower bound or $+\infty$ as the upper bound. For economy, we use the most compact unums that represent open intervals starting at $-\infty$ or ending in $+\infty$. In the prototype, they are **negopeninfu** and **posopeninfu**. Doing a better job than floats at handling "overflow" does not need to cost many bits.

There is a corresponding compact way to express the "almost zero" values, both positive and negative. If the *upper* bound is zero and open, then the smallest unum with that upper bound is $(-1, 0)$, so we will keep **negopenzerou** handy as well. These unum values are among those that are changed automatically whenever **setenv** is called. The positive equivalent of **negopenzero** is a number we have already seen: **ubitmask**. We show it again in the table below for the sake of completeness.

negopeninfu:	$(-\infty, -3)$
posopeninfu:	$(3, \infty)$
negopenzerou:	$(-1, 0)$
ubitmask:	$(0, 1)$

These values only require three bits to the left of the utag. One exception: If the environment is {0, 0}, the range of **negopeninfu** is $(-\infty, -2)$, and **posopeninfu** represents $(2, \infty)$.

> **Exercise for the reader**: What three bits left of the utag generate the intervals in the above table?

Suppose we need to express the range of values $(-0.3, \infty)$, in a medium-precision {3, 4} environment. The expensive way to express the open infinity on the right is to use the unum for *maxreal* but with **ubitmask** added, which represents

$$(680\,554\,349\,248\,159\,857\,271\,492\,153\,870\,877\,982\,720, \infty)$$

That unum consumes 33 bits in addition to the 8 bits in the utag, and we disregard the huge number on the left end anyway when using this as the right-hand endpoint of a ubound. If we instead use the unum for **posopeninfu**, it only requires *three* bits in addition to the utag.

So the compact way to express (0.3, ∞) as a ubound is with `{-0̂.3,posopeninfu}`:

`setenv[{3,4}]; view[{-0̂.3,posopeninfu}]`

(−0.3000030517578125, ∞)

That entire ubound takes only 39 bits to express.

6.4 Can ubounds save storage compared with floats?

Earlier we argued that unums can often save storage compared with floats. With ubounds, however, there is a good chance of needing two unums instead of one float. Is there still hope that, on average, ubound math will take fewer bits than fixed-size floats, even without doing frequent unification operations? First, let's take a closer look at the storage requirements of unums versus floats.

You might suspect that most numbers are inexact in actual applications, so that with unums you will almost always be using the maximum number of bits as well as the utag, and thus winding up using *more* storage than floats. However, this usually is not the case, for five reasons.

Reason 1: Exponent waste. The number of exponent bits in a float is almost always overkill in practical computations, so a unum will frequently use far fewer exponent bits. Even single-precision IEEE floats span *85* orders of magnitude; have you ever seen a program that legitimately needs that much dynamic range? If a single-precision or double-precision calculation stays mostly within the range of about −500 to 500, you will usually only need four bits of exponent. Staying in that range saves four bits versus single precision, seven bits versus double precision, and eleven bits versus quad precision.

Reason 2: No trailing zeros for exact values. If the fraction bits are random and the number is exact, then there is a 50% chance the last bit is 0 and can be compacted away, a 25% chance that the last two bits are 0 and can be compacted away, and so on. The average compression from this effect alone is $\frac{1}{2^1} + \frac{2}{2^2} + \frac{3}{2^3} + \frac{4}{2^4} \ldots = 2$ bits.

Just from these first two reasons, we get back most of the bits spent appending a utag to every bit string.

Reason 3: Shorter strings for common numbers. Many real-life calculations involve floating point versions of small integers or simple fractions. Browse through any physics or math textbook and you will see numbers like −1, 3, $\frac{1}{2}$, 4, etc., in the formulas. Like the "$\frac{1}{2}$" used when you compute kinetic energy as $\frac{1}{2} m v^2$, for example.

With conventional floating point, adding the real number 1 in double precision requires the 64-bit float representing 1.0, which has a *lot* of wasteful bits:

0 01 111 111 111 0 000 000 000 000 000 000 000 000 000 000 000 000 000 000 000 000 000

Whereas for unums, storing the value 1.0 always takes only *three bits* in addition to the size of the tag: 0, 0, and 1. In a {4, 6} environment capable of IEEE double precision, the unum for 1.0 only takes 15 bits to express:

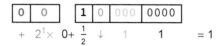

Unums are very concise when expressing small integers times a power of two, and such numbers arise more often than you might think in calculations.

Reason 4: Left-digit destruction saves space. Adding or subtracting numbers often destroys significant bits on the left or right side of the fraction field, and unum format is smart enough to exploit that loss by compressing away the lost bits. The savings from this effect depends on the calculation, which is why we need plenty of experiments on real problems to get a feel for the amount of savings that is possible.

Reason 5: No over-insurance. Computing with guaranteed bounds frees the user to ask only for a few decimals, instead of having to over-insure the calculation with far more decimals. Even half-precision floats start to seem excessive for many important applications, like the analysis of seismic data for oil exploration or the imaging of a medical scan.

In many cases, even a *pair* of unums needs fewer total bits to store than a single floating point number, so ubounds still wind up being a very economical way to express rigorous arithmetic. The comparison is not really fair, since floats usually produce an erroneous answer and unums produce a correct one, but it is intriguing that in many cases, it is not a trade-off; we can get *better* answers with *fewer* bits. And don't forget that when a result is finally ready to send to off-chip memory, it can usually be unified to a single unum. The need for two-unum ubounds mainly occurs in the on-chip calculations.

Some believe the solution to rounding error is to push for quad precision in hardware. That goes exactly the wrong direction, since it doubles the energy, power, bandwidth, and storage consumed by every operation, and it still provides no measure of information loss at the end of a calculation. What is interesting is to go the other direction, and ask:

*How **tiny** can a unum string be and still be useful for computing with real numbers?*

7 Fixed-size unum storage

Thomas Baines, *Thomas Baines with Aborigines near the mouth of the Victoria River, N.T., 1857.* Before the Warlpiri aborigines had much contact with other cultures, their counting system was "one, two, many."

7.1 The Warlpiri unums

There are aboriginal hunter-gatherer people (such as the Warlpiri in Australia's Northern Territory) whose language for numbers was limited to "one," "two," and "many" before they had contact with other cultures. Their original language is now almost lost, but we know they also had words for "none" (zero), "need" (negative), "junk" (NaN), and "all" (infinity), so in a sense their number vocabulary mapped very closely to the smallest possible float environment, augmented with a ubit. We can call this the "Warlpiri unum set." Each number takes only *four bits* to express.

Definition: The *Warlpiri unums* use the smallest format possible, a {0, 0} environment. They have a sign bit, an exponent bit, a fraction bit, and a ubit.

Later we will show that such primitive four-bit values can be superior to even very high-precision floats at avoiding wrong answers. Better to say "Two times two is 'many'" than to say "Two times two is overflow; let's call it infinity."

Hardware designers may be attracted to the fact that Warlpiri unums are fixed in size since they have no fields with which to express different *esize* and *fsize* values.

If you give up the economy of distinguishing between single and double ubounds and always use two unums per ubound, then every ubound can be represented with *one byte.* Imagine: A closed, mathematically unbreakable *real number* representation that requires only eight bits per value. Imagine how fast hardware could process such numbers.

There are sixteen four-bit strings, representing fifteen distinct values since we do not distinguish "negative zero" from zero. The precision may be ridiculously low, but given any real number or exception value, *exactly one* four-bit string will express it. Warlpiri unums can be useful in getting a very fast initial bound on an answer, or discovering that no answer exists. For example, in performing ray tracing for computer graphics, this low precision often suffices to prove that a ray cannot possibly hit the surface in question. The "maybe" cases that require more precision are surprisingly few. The three bits that form the "float" part of this tiny value obey IEEE rules for hidden bit, exponent bias, and so on, but remember that we deviate in selecting only two of the bit patterns to represent NaN: Signaling NaN and quiet NaN. Here is the Warlpiri number set, shown with four-bit strings in the top row, then conventional math notation for open intervals and exact numbers, then the real number line from $-\infty$ to ∞, and lastly, the English version of the way an aborigine might express the quantity:

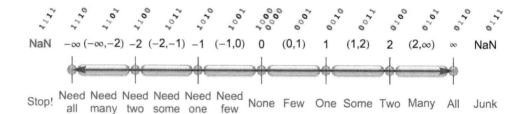

The addition and multiplication tables taught in elementary school usually are for numbers 0 to 10. Beyond that, you need the pencil-pushing algorithms for multi-digit arithmetic that make most students hate math starting at an early age. Imagine: With Warlpiri real numbers, we can define + − × ÷ tables that encompass *the entire real number line* and make statements that are mathematically unassailable. If you ignore the NaN and "negative zero," there are thirteen distinct unum input values, so each table has only 169 entries. Yet, they provide a closed mathematical system that covers all the real numbers and the exception cases like infinities and NaN. It is intriguing to think about a computer design that simply stores the tables and looks up results, which for some types of computer hardware can be faster than computing them with binary logic.

> **Exercise for the reader**: Which Warlpiri unums are unchanged by the square root operation?

As a preview of issues to come, consider: What is the open interval (0, 1) divided by (0, 1)? Conventional computing says "There is a zero divided by zero in the ranges, so the answer is NaN." Others might say "1 divided by 0 is infinity, so the answer is infinity." But the *open* interval (0, 1) is never 0 and never 1; it is only the numbers in between. That means $(0, 1) \div (0, 1) = (0, \infty)$. The result is somewhere between a number smaller than we can represent (but not zero), and a number larger than we can represent (but not infinity). The bound $(0, \infty)$ is the best we can do, in the sense that having more bits of exponent or fraction would not change the answer. Whenever that happens, we say the result is "tight." Notice that tight results are not necessarily *reversible*, because there is still possible loss of information. That is, addition and subtraction are not the inverse operations we would like them to be, where

$$(a + b) - b = a \text{ and}$$
$$(a - b) + b = a,$$

and neither are multiplication and division, where we wish we could always guarantee that

$$(a \times b) \div b = a \text{ and}$$
$$(a \div b) \times b = a.$$

Performing an operation on x followed by the inverse operation often gives a range that contains x but also contains other values. This can happen even when the operations are "tight," because zero and infinity can destroy information about real number ranges.

If we ask what is $1 \div 2$ in this ultra-low precision format, we must use the open interval (0, 1) since there is no string representing one-half. That is a "loose" result and information is *always* lost. It is still a true statement within the format, since one-half is *contained* in (0, 1), but if we had more bits, we could make the result tight. (Notice that if we use IEEE Standard float rules with no ubit, we get the false result that $1 \div 2 = 0$ with three-bit floats, since it underflows!)

7.2 The Warlpiri ubounds

The ubound set has the 15 that are single unums, the set shown above, and 78 paired unums, for a total of 93 Warlpiri ubounds. As mentioned earlier, it might be more economical to always use pairs, where the single-unum entries store two copies of the same value; that way, there is no need for a "pair bit" and the Warlpiri ubounds always fit in an 8-bit byte. The following page shows them in three forms: Math (as a general interval), English, and as a 4-bit string or pair of 4-bit strings.

The 93 Warlpiri Ubounds

Value	English	Binary	
NaN	Stop!	1 1 1 1	
−∞	Need all	1 1 1 0	
[−∞, −2)	Need all to need many	"	1 1 0 1
[−∞, −2]	Need all to need two	"	1 1 0 0
[−∞, −1)	Need all to need some	"	1 0 1 1
[−∞, −1]	Need all to need one	"	1 0 1 0
[−∞, 0)	Need all to need few	"	1 0 0 1
[−∞, 0]	Need all to none	"	0 0 0 0
[−∞, 1)	Need all to few	"	0 0 0 1
[−∞, 1]	Need all to one	"	0 0 1 0
[−∞, 2)	Need all to some	"	0 0 1 1
[−∞, 2]	Need all to two	"	0 1 0 0
[−∞, ∞)	Need all to many	"	0 1 0 1
[−∞, ∞]	Need all to all	"	0 1 1 0
(−∞, −2)	Need many	1 1 0 1	
(−∞, −2]	Need many to need two	"	1 1 0 0
(−∞, −1)	Need many to need some	"	1 0 1 1
(−∞, −1]	Need many to need one	"	1 0 1 0
(−∞, 0)	Need many to need few	"	1 0 0 1
(−∞, 0]	Need many to none	"	0 0 0 0
(−∞, 1)	Need many to few	"	0 0 0 1
(−∞, 1]	Need many to one	"	0 0 1 0
(−∞, 2)	Need many to some	"	0 0 1 1
(−∞, 2]	Need many to two	"	0 1 0 0
(−∞, ∞)	Need many to many	"	0 1 0 1
(−∞, ∞]	Need many to all	"	0 1 1 0
−2	Need two	1 1 0 0	
[−2, −1)	Need two to need some	"	1 0 1 1
[−2, −1]	Need two to need one	"	1 0 1 0
[−2, 0)	Need two to need few	"	1 0 0 1
[−2, 0]	Need two to none	"	0 0 0 0
[−2, 1)	Need two to few	"	0 0 0 1
[−2, 1]	Need two to one	"	0 0 1 0
[−2, 2)	Need two to some	"	0 0 1 1
[−2, 2]	Need two to two	"	0 1 0 0
[−2, ∞)	Need two to many	"	0 1 0 1
[−2, ∞]	Need two to all	"	0 1 1 0
(−2, −1)	Need some	1 0 1 1	
(−2, −1]	Need some to need one	"	1 0 1 0
(−2, 0)	Need some to need few	"	1 0 0 1
(−2, 0]	Need some to none	"	0 0 0 0
(−2, 1)	Need some to few	"	0 0 0 1
(−2, 1]	Need some to one	"	0 0 1 0
(−2, 2)	Need some to some	"	0 0 1 1
(−2, 2]	Need some to two	"	0 1 0 0
(−2, ∞)	Need some to many	"	0 1 0 1
(−2, ∞]	Need some to all	"	0 1 1 0

Value	English	Binary	
−1	Need one	1 0 1 0	
[−1, 0)	Need one to need few	"	1 0 0 1
[−1, 0]	Need one to none	"	0 0 0 0
[−1, 1)	Need one to few	"	0 0 0 1
[−1, 1]	Need one to one	"	0 0 1 0
[−1, 2)	Need one to some	"	0 0 1 1
[−1, 2]	Need one to two	"	0 1 0 0
[−1, ∞)	Need one to many	"	0 1 0 1
[−1, ∞]	Need one to all	"	0 1 1 0
(−1 0)	Need few	1 0 0 1	
(−1, 0]	Need few to none	"	0 0 0 0
(−1, 1)	Need few to few	"	0 0 0 1
(−1, 1]	Need few to one	"	0 0 1 0
(−1, 2)	Need few to some	"	0 0 1 1
(−1, 2]	Need few to two	"	0 1 0 0
(−1, ∞)	Need few to many	"	0 1 0 1
(−1, ∞]	Need few to all	"	0 1 1 0
0	None	1 0 0 0 or 0 0 0 0	
[0, 1)	None to few	"	0 0 0 1
[0, 1]	None to one	"	0 0 1 0
[0, 2)	None to some	"	0 0 1 1
[0, 2]	None to two	"	0 1 0 0
[0, ∞)	None to many	"	0 1 0 1
[0, ∞]	None to all	"	0 1 1 0
(0, 1)	Few	0 0 0 1	
(0, 1]	Few to one	"	0 0 1 0
(0, 2)	Few to some	"	0 0 1 1
(0, 2]	Few to two	"	0 1 0 0
(0, ∞)	Few to many	"	0 1 0 1
(0, ∞]	Few to all	"	0 1 1 0
1	One	0 0 1 0	
[1, 2)	One to some	"	0 0 1 1
[1, 2]	One to two	"	0 1 0 0
[1, ∞)	One to many	"	0 1 0 1
[1, ∞]	One to all	"	0 1 1 0
(1, 2)	Some	0 0 1 1	
(1, 2]	Some to two	"	0 1 0 0
(1, ∞)	Some to many	"	0 1 0 1
(1, ∞]	Some to all	"	0 1 1 0
2	Two	0 1 0 0	
[2, ∞)	Two to many	"	0 1 0 1
[2, ∞]	Two to all	"	0 1 1 0
(2, ∞)	Many	0 1 0 1	
(2, ∞]	Many to all	"	0 1 1 0
∞	All	"	0 1 1 0
NaN	Junk	0 1 1 1	

Here are a few examples of operations on Warlpiri ubounds:

$$[1, 2) - 2 = [-1, 0) \qquad\qquad -\infty + \infty = \text{NaN}$$

$$[-2, 2] \div \infty = 0 \qquad\qquad (-1, 2) \times (0, 1) = (-1, 2)$$

$$2 \times (-2) = (-\infty, -2) \qquad\qquad 1 \div 2 = (0, 1)$$

If we attempt closure plots, they will be very boring: A solid 93 by 93 grid of dots, indicating that *every operation on a Warlpiri ubound produces another Warlpiri ubound*. This proves that closure under the four arithmetic operations *is* possible using a bit string of limited length. Designing real number formats with very few bits forces us to face up to all that is wrong with the current way we store real numbers on computers, and shows us how to fix the shortcomings before demanding more precision. If something is not worth doing, then it is not worth doing well.

Another thing we can do with the Warlpiri ubounds is create flawless *elementary functions* for them. For example:

$$\sqrt{2} = (1, 2)$$

$$\text{cosine of } (-\infty, \infty) = [-1, 1]$$

$$\text{logarithm of } 0 = -\infty$$

Unlike math libraries for floats that strive to make almost all the results within half an ULP of the correct answer, there is no difficulty making every elementary function evaluation correct with Warlpiri ubounds. Ridiculously low precision, but correct. They can be used to prove the *existence* or *nonexistence* of an answer, which might save a lot of trouble. For example, if the mathematical question is "Does (*continuous function of x*) = 0 have any solutions for real numbers *x*?" then we can actually try *every* real number: Plug in $(-\infty, -2)$, -2, $(-2, -1)$, -1, $(-1, 0)$, 0, $(0, 1)$, 1, $(1, 2)$, 2, and $(2, \infty)$ for *x* and see if any result ranges contain zero. If they do not, there is no reason to do a much more tedious root-finding algorithm. (More about this in Part 2.)

Exercise for the reader: Suppose we wish to create an "exact dot product" (Section 5.1.5) for Warlpiri ubounds that works for vectors up to length one billion. How many bits are needed for the accumulator in the *g*-layer?

Notice that standard hardware storage can handle Warlpiri ubounds, **without the need to manage variable-length operands**. They can be treated as 8-bit values packed into whatever power-of-2 bit size the computer memory level uses, and managed as an indexed array of integers.

Chapters 7 to 11 develop basic operations on unums and ubounds. In trying to get a feel for how things work with unums, it is helpful to keep the Warlpiri unums and ubounds in mind as test cases, since there are so few of them and the arithmetic is trivial to do in your head.

7.3 Hardware for unums: Faster than float hardware?

7.3.1 General comments about the cost of handling exceptions

With IEEE floats, there is usually hardware support up to some precision, and beyond that the operations must be done with software. A few years ago, the dividing line for many computers was between single and double precision. Currently, the dividing line is usually between double and quad precision. Hardware for unums can follow the same pattern, guaranteeing the availability of dedicated hardware up to some maximum length for the bit strings, and a number larger than that length triggers a trap to a software routine. The first exception to handle is "This computation is too large to fit into the hardware; use software instead."

At some point, demanding too much precision or dynamic range will cause the system to run out of memory, as with all computer operations. The second exception to handle is "Even with software, there is not enough memory to perform this calculation."

Once it is determined that a value fits into the allocated hardware, the processing of representations that could be NaN or infinity or subnormal involves testing for those situations. For hardware, exception handling is usually a painful thing to build and always seems to be a serial bottleneck getting in the way of high performance. (Which is why the Sun engineers tossed out hardware handling of NaN.)

Hardware for integer operations is very straightforward because the exception handling is so simple. To add a pair of n-bit signed integers, the hardware jumps into the task immediately, since the only thing that can go wrong is that the answer requires more than n bits, indicating underflow or overflow.

For IEEE floats, detecting exception cases is complicated and expensive. Is it a NaN? Is it an infinity? Is it negative? You might think "Is it negative" is just a matter of testing the sign bit, but remember "negative zero" is not actually negative. To figure out if a number is negative means testing **all the bits** in a float, because of the exception for negative zero. Suppose a chip designer is building a square root instruction, and decides to trigger a NaN if the sign bit is set, since the instruction does not allow negative inputs. That would give a NaN for $\sqrt{-0}$, when the correct answer is 0. When the first version of a chip is tested, the errata sheets always seem to have quite a few cases similar to that one.

When the IEEE Standard for floats first came out, many chip designers grumbled that it was clearly not designed with circuit designers in mind since all the special values required tests. It is very much like decoding an instruction before a computer can begin to execute it. Float operations, unlike integer operations, involve crucial decisions that hinge on detecting special combinations of their bit patterns.

For example, if the logic for multiplication says "If x is infinity and y is strictly greater than zero, the product is positive infinity," what does the hardware have to do to detect infinity? To be even more specific, imagine the simplest possible case: Half-precision IEEE floats. There is exactly one bit string representing infinity (by IEEE rules): 0 111111 0000000000. That string needs to be detected by hardware, which means *every bit* has to be tested.

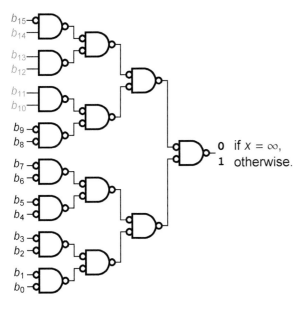

0 if $x = \infty$,
1 otherwise.

To do that quickly means building a tree of logic that collapses the state of all sixteen bits down to a single true-false bit. If the logic is composed of NAND gates (Not AND, meaning 0 if both inputs are 1, and 1 otherwise) and inverters (that turn 0 into 1 and 1 into 0), then the logic on the left (or something similar) is needed to test if x is the pattern representing the IEEE Standard representation of ∞. Even if you are not familiar with logic circuit symbols, this gives a flavor for the time and complexity involved.

If an inverter takes one gate delay and two transistors, and a NAND gate takes four transistors and one gate delay (CMOS technology), the tree of operations requires seven gate delays and 110 transistors. Imagine what it is like for a double-precision float; about four times as many transistors, and eleven gate delays. Of course, the way it is actually done is a little different. If the AND of all the bits in the exponent is 1, *then* the rest of the bits are checked to see if all of them are 0; otherwise, testing the other bits is not necessary. Is the hidden bit a 1? That test requires the OR of all the bits in the exponent, another tree structure like the one shown above.

There is a way that unums can make things *easier* for the hardware designer compared with IEEE floats. A simple principle can save chip area, execution time, energy, and design complexity all at once: *Summary bits*.

7.3.2 Unpacked unum format and the "summary bits" idea

If hardware dedicates 64-bit registers to unums, for example, environments up to {4, 5} will **always fit**, since *maxubits* is at most 59 for those combinations.

> **Exercise for the reader**: The environment {5, 4} fits, but might be a bad idea. About how many decimals are needed to express *maxreal* in that environment?

The {4, 5} maximum environment provides up to 9.9 decimals of precision and a dynamic range as large as 10^{-9873} to 10^{9864}.

Since the floats that unums seek to replace often are 64-bit, using the same 64-bit hardware registers for unums seems like a good place to start. Requests for an environment with exponent fields larger than 16 bits or fraction fields larger than 32 bits trigger the use of software.

Because *maxubits* is always 59 or less for a {4, 5} environment, five bits of the 64-bit register are never used by unum bits. That suggests a use for them that makes the hardware faster, smaller, lower power, and *easier to design*: Maintain a bit indicating if the unum is NaN, a bit indicating if it is a ±∞, a bit indicating the number is zero (exact or inexact), and a bit indicating if this is not the last unum in a ubound (so the second unum is in another register). This information is already contained in the unum, but making it a single bit simplifies the circuitry required for processing.

> **Definition**: A *summary bit* indicates a crucial feature of a unum that can otherwise be derived only by examining combinations of bits; it is used to save time and effort by "decoding" attributes of a unum.

The "hidden" bit can be unpacked into the "revealed bit," displayed at last just to the left of the fraction. That eliminates much of the exception handling for subnormal numbers by allowing the fraction (starting with the revealed bit) to be treated as an ordinary unsigned integer.

We also would rather not make bit retrieval a two-step operation, where we have to use the exponent size and fraction size to find out how far to the left to look for the sign bit, the exponent, and so on. Instead, the six sub-fields that make up a unum are given dedicated locations within the 64-bit register, right-justified bit strings for those with adjustable size. Here is one example of a fixed-size unum format:

An unpacked 64-bit unum register design
for high-speed, low-power arithmetic

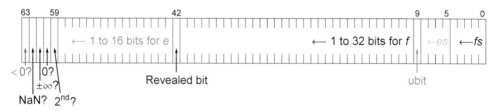

The creation of the summary bits occurs when a unum is unpacked from memory into a register by an instruction, or by the logic of an arithmetic operation. If someone mucks with individual bits in the above format, by using low-level instructions to set or reset them, it is perfectly easy to create a nonsensical, self-contradictory unpacked unum. The unpacked unum data type is not intended for programmer bit-twiddling, only for use by the computing system.

The "<0?" summary bit: Many operations need to check the sign of a value as one of the first steps. As a convenience, we can store it in the leftmost bit so that existing signed integer computer instructions that check "If $x < 0$..." work for the unpacked unum register format. It will save a lot of grief that "negative zero" is accounted for, and in the register; that way, testing "If $x < 0$" is answered correctly by checking the first bit and *not* also having to check for all of the fraction and exponent bits and the ubit being **0**. Of course, this assumes that the value represented is some kind of number, not a NaN; most of the arithmetic operations begin by checking for NaN, the second summary bit, but by putting the sign bit all the way to the left we get to use existing integer tests for "If $i < 0$" on the 64-bit unum string.

The "NaN?" summary bit: There really is only one type of NaN involved in any kind of operation: The quiet NaN. If it was a signaling NaN, there would not *be* any operation, because the task would have been stopped. There may be situations where a designer wants to use the unpacked format to indicate a signaling NaN, in which case the "< 0?" summary bit could, if set, indicate signaling NaN. We will here assume it is sufficient just to test bit 63, and if set, immediately return (quiet) NaN. Operations on NaN inputs should be the fastest execution of any possible inputs.

The rest of the bits of the register already store the unpacked form of NaN, providing a handy source for the bit string of the NaN answer. There is nothing to construct.

The "$\pm\infty$?" summary bit: This summary bit is set if all the bits in *esize*, *fsize*, the fraction, and the exponent are **1**. Notice that those four sub-fields have variable size and are right-justified within their dedicated location, so the environment settings still need to be taken into account when the unpacking is done.

b_{63} ⎯⎯⎯⎯⎯⎯ **0** if $x = \infty$,
b_{61} ⎯⎯⎯⎯⎯⎯ **1** otherwise.

Remember the 110-transistor logic tree to detect infinity for the half-precision IEEE float? Here is what the logic looks like for an unpacked unum.

That represents about a 95% reduction in chip area and a 71% reduction in the time necessary to detect the infinity value, compared to the detection logic for every IEEE operation or for unum unpacking. Perhaps chip designers will not be as opposed to unums as has been previously suggested, given that the unpacked format is of fixed size, there is no need for rounding logic, and exception handling can be far faster and simpler. Unpacking is easily done concurrently with a sequence of memory fetches, since each fetch takes longer than the unpacking operation.

Exception testing like "Is x a NaN?" "Is x infinite?" can be processing while the rest of the hardware proceeds under the assumption that the answers will be "No." That way, exception testing need not add time to the beginning of an operation for which actual arithmetic has to be done; it can overlap, and it can be *very* fast. When the answer turns out to be "Yes," the hardware can interrupt any arithmetic processing, but **energy is then wasted** because of unnecessarily having started the calculation. A power-conscious design would not overlap the arithmetic with the exception testing, but would instead *wait* to find out the result of the NaN and infinity tests before doing any speculative work. With summary bits, the wait will not be long at all.

The "0?" summary bit: Arithmetic operations on zero tend to produce trivial results, or exception value results: $0 + x = x$, $0 - x = -x$, $x \div 0 = \text{NaN}$, etc. There are many ways to express exact zero with a unum, and when zero is inexact, it represents different ULP sizes adjacent to zero. It is therefore extremely handy to be able to test a single bit since the decision is required so often by the arithmetic.

The "2^(nd)?" summary bit: This is what we have sometimes called the "pair bit" for describing a ubound: If it is **1** (`True`), it means there is a second half to the ubound in the next register up. (With *n* registers, the highest numbered register wraps around to the lowest numbered one). Otherwise, this is a single-unum ubound.

NaN?	$|f_L|$	$|f_R|$
	f_L negative?	f_R negative?
	$|e_L|$	$|e_R|$
	e_L negative?	e_R negative?
	open?	open?
	infinite?	infinite?

The summary bits help to load the true-false scratchpad data area in the *g-layer*. Recall what the *gbound* structure looks like, left. The summary bits make it trivial to populate the Boolean '?' entries with very few gate delays.

As an example, here is what the unpacked 64-bit register would look like representing the unum value for $(-smallsubnormal, 0)$ in a $\{2, 3\}$ environment, which is the open interval $\left(-\frac{1}{16\,384}, 0\right)$:

An unpacked $\{2, 3\}$ unum representing $(-1/16384, 0)$

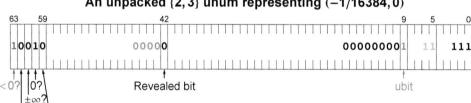

Revealed bit • ubit

<0? 0? ±∞? NaN? 2^(nd)?

Chip designers, note: The blank register regions are "don't care" bits that **need not draw power** while the environment stays at the same setting. Like snowplows at the airport during good weather; there is no need to keep their engines running.

Notice that the "<0?" summary bit is set, yet so is the "0?" summary bit. That is not a mistake; it is possible for inexact values of zero, of which this is an example, to be ULP-wide intervals on *either side* of exact zero.

Exercise for the reader: Create a similar unpacked format for a 32-bit register that works for environments up to $\{3, 3\}$. If you do not have a "revealed bit," create an unpacked format for environments up to $\{2, 4\}$. How much logic is needed to determine the hidden bit, since it is not revealed?

8 Comparison operations

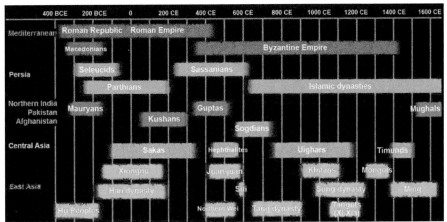

Comparing dynasty periods is like comparing unums. Sometimes they are distinctly before (less than) or after (greater than), and sometimes they overlap. If asked "Did the Roman Empire strictly precede the Byzantine Empire?" and the only answers allowed are "Yes" or "No," then we have to answer "No" because there was a period of overlap.

8.1 Less than, greater than

8.1.1 Conceptual definition of ordering for general intervals

The easiest operations to write for unums are comparisons. They are different from comparisons of floats, since unums and ubounds can represent ranges of numbers and thus overlap completely, partially, or not at all.

For most operations with unums and ubounds, the hardware approach is to move from the u-layer to the g-layer, perform the operation there, and convert the result back to the u-layer. The unpacked form of a ubound discussed in Chapter 7 provides a concrete example of a hardware approach for operations described here.

For a ubound u to be less than a ubound v, the maximum of u must be less than the minimum of v. This gets subtle, however, because endpoints can be closed or open. The maximum of u could **equal** the minimum of v if either endpoint is open, yet u would still be strictly less than v. The following type of construct comes up a lot when writing the logic for ubound operations:

If $x < y$ or ($x = y$ and (x is inexact or y is inexact)) then ...

In the prototype, the ltuQ[*u*,*v*] function (where the name comes from "less-than, u-layer Question") returns a Boolean True if *u* represents a numerical range strictly less than the numerical range represented by *v*. In all other cases, including ones where either ubound represents NaN, it returns False. Similarly, we have a greater-than comparison gtuQ[*u*,*v*]. The code for both functions is in Appendix C.8. Notice that these are "strictly less than" and "strictly greater than" functions; any overlap, even just an endpoint, makes the result False.

For example, the prototype command to find out if [1, 3] < [3, 100] is

---

$$\text{ltuQ}\left[\left\{\hat{1}, \hat{3}\right\}, \left\{\hat{3}, \hat{100}\right\}\right]$$

False

It returns False because the intervals both include the exact number 3. However, testing if [1, 3] < (3, 100] returns True, because one endpoint is open. Floats and traditional interval arithmetic cannot make this important distinction:

$$\text{ltuQ}\left[\left\{\hat{1}, \hat{3}\right\}, \left\{\hat{3} + \text{ubitmask}, \hat{100}\right\}\right]$$

True

The comparisons are performed by converting the *u*-layer arguments to general intervals (using u2g) and using the *g*-layer function ltgQ. This closely resembles the way hardware would do it. The "greater than" test is exactly like the "less than" test with the input endpoints interchanged. The prototype functions for the greater-than test are gtgQ in the *g*-layer and gtuQ in the *u*-layer.

With many operations on unums and ubounds, the prototype defines operators that resemble their mathematical equivalents, to make expressions look more like conventional math. The "less than" test for ubounds is "◁", a slightly curvy less-than symbol, so the test "−2 < −1" in the *u*-layer can be written in unum math as

---

$-\hat{2} ◁ -\hat{1}$

True

Similarly, the "▷" symbol can be used to make expressions easier to read than writing "gtuQ":

$\hat{\infty} ▷ \hat{\infty}$

False

8.1.2 Hardware design for "less than" and "greater than" tests

Recall what a gbound looks like, the scratchpad version of a ubound or unum. Suppose we have both u and v in the scratchpad and wish to test if u is less than v.

NaN?	$|f_L|$ f_L negative? $|e_L|$ e_L negative? open? infinite?	$|f_R|$ f_R negative? $|e_R|$ e_R negative? open? infinite?	? <	NaN?	$|f_L|$ f_L negative? $|e_L|$ e_L negative? open? infinite?	$|f_R|$ f_R negative? $|e_R|$ e_R negative? open? infinite?

We can ignore all the data shown in gray, since we just need to know if the right endpoint of u is less than the left endpoint of v. The "less than" comparison operation requires peeling away bits in the gbound in order of importance. The gbound parts are tested in the following order, breaking out with a "Done" as soon as a "less than" conclusion of True or False can be drawn:

- NaN? bits
- f negative? bits
- infinite? bits (with tie-breaking using the open? bits)
- Integer values of f after scaling by the difference in e values
- open? bits

If either value is NaN, a less-than or greater-than test returns False. Done. (A better answer might be "Unknown," but that is not in the vocabulary of a comparison test. There is some logic in answering "Is NaN less than 2?" with "No" since only numbers can be "less than" or "greater than" other numbers.)

Are the "f negative?" bits the same? If not, the decision is easy. If the left one is negative and the right one positive, the less-than test is True, otherwise False. Done. Otherwise, the comparison is between endpoints that have the same sign. Is just one of the "infinite?" bits set? That determines the ordering, done. Are *both* of the "infinite?" bits set? Then the "open?" bits need to be checked, since $(x, \infty) < \infty$ and $-\infty < (-\infty, x)$ give a True result, else False; done.

With exception testing dispensed with, the next step is more like arithmetic than bit-testing. The e values are integers; subtract them. Use the difference to shift one f value left and reduce its e value, so that both values have the same e value. In other words, **line up the binary points**, just as you would do for floating point addition and subtraction. Comparison of the scaled f values is then straightforward, but depends on how variable length is implemented. If there is a dedicated location that stores the length of f, then the different lengths quickly break the tie, done. Otherwise, the bit lengths match; go through bits from most significant to least significant until a tie-breaker bit is found. If one is found, done.

After all that, if the values are *still* tied (identical), then the open-closed flags are checked. If either one is open, then the less-than test is `True`; if both are closed, the less-than test returns `False` because *u* and *v* touch at that exact point. All it takes is one equal point to spoil a *strictly* less-than comparison.

A clever hardware designer will notice shortcuts that work directly on *u*-layer data for a comparison operation, instead of first converting *u* and *v* to gbounds. In particular, it is easy to compare two unums that have the same *esize* and *fsize* values just by treating the other bits (including the ubit) as signed integers. The trick is to be able to promote the exponent and fraction lengths of each value until they match, taking care not to change the value of inexact (open) endpoints used in the comparison. The prototype includes a function `promote[{u,v}]` that does this (Appendix D.1).

> **Exercise for the reader**: Find efficient logic for comparing two Warlpiri unums in the range `0 0 0 0` to `1 1 1 0`. (That is, every value other than signaling NaN). Use the bits directly instead of assuming the unum has been converted to a gbound. The value plots in Section 2.4 may provide insight.

8.2 Equal, nowhere equal, and "not nowhere equal"

8.2.1 Conceptual definition of "equal" for general intervals

Less than and greater than relationships are easy to understand even when the numbers are ranges instead of precise point values. A trickier proposition is to define "equal" when talking about ranges of real numbers. **What does it mean to say two general intervals are "equal"?**

Ubounds or unums are *equal* if they represent the same gbound. They must have the same endpoints **and** the same open-closed endpoint properties. The prototype command `sameuQ[u,v]` returns `True` only if those conditions are met. It also returns `True` if both *u*-layer inputs represent NaN, since we will have occasions where we need to test if a calculation has lost all validity. That is the *only* comparison involving NaN that can return a `True` result; `sameuQ` tests for "identical," numeric or not. The "≡" symbol is another way to write `sameuQ[u, v]` in the prototype: $u \equiv v$.

However, there is another level of equality that is typically the more useful one in calculations, and it may take some getting used to. It is the level we have to use in the *u*-layer as opposed to the scratchpad *g*-layer. Instead of thinking of "**ne**" as an abbreviation for "not equal," think of it as "*nowhere* equal": Two unums or ubounds are "nowhere equal" if one is strictly greater than or strictly less than another, by the rules in the previous section: **disjoint** ranges of numbers. The prototype comparison function `nequQ[u,v]` returns `True` if $u < v$ or $u > v$, and `False` for all other cases (including when either input is a NaN). There is actually an obscure math symbol for "less than or greater than": ⋚, but we will generally write out `nequQ` here, for "not equal u-layer Question," or use this infix symbol: "≠". Its version in the *g*-layer is `neqgQ`. However, `nequQ` is *not* the logical opposite of the equality test, `sameuQ`.

When the general intervals represented by the inputs **overlap**, we say they are *not nowhere equal*, which almost sounds like something a hillbilly would say. "Them ain't *nowhere* equal!" Here is a test of two ubounds that do not overlap:

$$\left\{\hat{1}, \hat{2}\right\} \neq \left\{\hat{3}, \hat{4}\right\}$$

True

But here is one where they *do* overlap, which means equality cannot be ruled out for all points in the two ranges. The "nowhere equal" test correctly returns `False`:

$$\left\{\hat{1}, \hat{3}\right\} \neq \left\{\hat{2}, \hat{4}\right\}$$

False

While "equal" means identical endpoints and identical open-closed qualities, "not nowhere equal" means the two intervals have some overlap and *could* represent the same result. If we use "not nowhere equal" where floats would use "equal" or "within some epsilon of each other," we can preserve mathematical rigor within the accuracy concessions of the *u*-layer. The prototype test for not nowhere equal in the *u*-layer is **nnequQ[**u**,**v**]**, and the test is so useful that we also define the operator "≈" to represent it. Do not confuse "≈" (as it is used here) with "approximately equal," which that symbol is sometimes used to mean.

A more explicit symbol is "$\cancel{\neq}$", meaning "not less than and not greater than," but "$\cancel{\neq}$" looks too different from an equals sign "=" or the common "==" equality test used in programming languages. Philosophically, "≈" means *equal within the expressiveness of the current u-layer computing environment.* If comparing unums or ubounds, $u \approx v$ means the same test as **nnequQ[**u**,**v**]**; True if u overlaps v, and False if they are disjoint ranges.

Here are a couple of examples of using "≈" to test if *u*-layer values have elements in common. An exact number is distinct from the ULP-wide range adjacent to it:

$$\hat{4} \approx \left\{\hat{4} + \textbf{ubitmask}\right\}$$

False

For the example where there was overlap between the ubounds, the "≈" test returns **True**.

$$\left\{\hat{-1}, \hat{3}\right\} \approx \left\{\hat{2}, \hat{4}\right\}$$

True

The code for all comparison tests mentioned so far is in Appendix C.8.

Algorithms that use floats often need to test when an answer has converged. Writing the stopping condition for an iterative method that uses floats is risky business. For procedures that look like "$x_{new} = f(x_{old})$ until (stop condition)," if you use "when the iteration is no longer changing" for the stop condition, that is, when $x_{new} = x_{old}$, you can still be quite far from the correct answer because the calculation is caught on the zigzag stair step of reals, rounded to floats. That is, the iteration cannot change the answer by more than an ULP, so it keeps rounding it back to where it was.

If the goal is to find solutions to $F(x) = y$ iteratively and you compute the *residual* as the absolute value of $F(x) - y$, then an obvious condition for stopping is "when the residual is zero." But the residual might never *get* to zero, again because of rounding error. That's when programmers set an "epsilon" of error tolerance, like "Stop when the residual is less than 0.001." But, why 0.001? What is the right number to use for epsilon? And should it be an *absolute* error, or a *relative* error?

With unums, expressing a stopping condition is far less prone to such errors and difficult choices. An iteration can stop when the ubound residual is "not nowhere equal" to zero, that is, *includes* zero. You also cannot get "stuck on a stair step" without seeing that the ubound is wide, indicating that more environment precision is needed to tighten the bound. Unum arithmetic automatically promotes precision up to the maximum set by *fsizesize*, so the algorithm can start with very storage-efficient approximate guesses, and the computer will escalate its efforts as needed. Later we will show ways the computer can adjust its own *fsizesize* automatically.

The not nowhere equal test says that two general intervals or ubounds intersect. Sometimes we want to know exactly what the intersection is, because that can be used to tighten a bound. If there are different ways to compute something, and we get different ubound answers, then **the correct answer lies in their intersection**. That can be used to improve the accuracy of the answer; the refined answer is that intersection. If the intersection is empty, that mathematically proves the two methods are *not* computing the same thing, which can be used to detect program bugs. Or flaws in the mathematical or physical theories. Or a reliability problem in the computing hardware, like a cosmic ray disrupting a memory bit.

This intersection refinement property is partly true for traditional interval arithmetic as well, but traditional intervals cannot detect, say, that 1.0 is *disjoint* from (1, 1.125). The example of pressing a calculator square root button repeatedly until the answer erroneously "equals" 1 can still happen with interval arithmetic, but not with unums. If you write a program to "Replace u with $\sqrt{u}$ until $u \approx \hat{1}$," it will run forever, which is the correct thing to do since the operation can never produce an exact 1. It gets to (1, 1 + ULP) using the smallest possible ULP, and stays there.

Exercise for the reader: Suppose u is the unum for exact 0, and v is the unum for the range $(-\infty, \infty)$. In terms of prototype notation for the comparison functions, what are the True-False values of $u < v$, $u \equiv v$, $u \approx v$, and $u \neq v$?

8.2.2 Hardware approach for equal and "not nowhere equal"

Different unum strings can represent the same value, but the gbound form is unique. To test if *u*-layer items represent identical values, the efficient approach is to see if both represent NaN, and if not, translate to the *g*-layer and compare bit-for-bit. The comparison can stop the instant a difference is found. If execution speed is favored over power efficiency, then all the bits can be checked in parallel and the results reduced with a fan-in tree to a single bit. If power efficiency is favored, then the best approach is to check, sequentially, the bits in the gbound in the same order as for the "less-than" and "greater-than" operations: Negative? Infinite? Different extended precision length? Different precision bits? Different open-closed bits? Only in the case where values are identical for every bit will the maximum time and energy be expended. Any difference that is detected along the way can save the effort of any remaining work. Those are the two extremes in the design, and many compromises are possible at the discretion of the hardware engineer, based on the application.

8.3 Intersection

When video games made the transition from 2D to 3D geometries, game developers struggled to get intersections right using floats. Notice the "leaks" in the house. The low graphics resolution of that era helped mask such artifacts. This could be viewed as a less-than or greater-than error, since the computer errs in figuring which of a set of polygons is closest to the viewer.

If we find that two ranges of reals are not nowhere equal, then they have an intersection that is not the empty set. We might want to know exactly what that intersection *is*; if the two ubounds come from algebraically equivalent operations, the intersection is a tighter bound on the answer.

The code for finding the intersection looks like an exercise in writing branch conditions! We have to consider every possible location of both endpoints of the first interval relative to the second, including the open-closed properties. While the code (Appendix C.8) is surprisingly tedious for such a simple operation, it illustrates how a unum environment makes life easier for the programmer by shifting the burden of carefulness to the provider of the computing system, where it belongs.

The prototype function for the intersection of two ubounds or unums *u* and *v* is `intersectu[`*u*`,`*v*`]`. The following shows the result of intersecting [−1, 3] with (2, ∞]:

$$\texttt{view}\Big[\texttt{intersectu}\Big[\big\{\hat{\texttt{-1}},\ \hat{\texttt{3}}\big\},\ \big\{\hat{\texttt{2}} + \texttt{ubitmask},\ \hat{\infty}\big\}\Big]\Big]$$

(2, 3]

Now for some actual unum *arithmetic*, starting with addition and subtraction.

9 Add/subtract and the unbiased rounding myth

In 1982, The Vancouver Stock Exchange established a stock index, scaled to initially equal 1000. The index steadily declined to 520 over the next two years even though the economy was healthy. The index was being updated by adding and subtracting *changes* in the stocks, using floats rounded to three decimals, so rounding error *accumulated*. Recalculating the stock index from the actual stock prices showed the index was actually 1098, not 520. The accumulation of approximations would have been obvious, had unum arithmetic been used.

9.1 Re-learning the addition table… for all real numbers

The prototype uses the "⊕" symbol to indicate addition of unums and ubounds. So the notation to compute "$-3 + 7.5$" with ubounds is "$\hat{-3} ⊕ 7\hat{.}5$", for example. It looks like bizarro arithmetic, with the notation for every number and every operation bent up a little, just to remind us of the difference between computer operations and perfect mathematical operations.

Now that we can start to do simple arithmetic with unums and ubounds, we can track the number of bits used and compare that number with the bit cost of IEEE floats. The bit count for unums varies, so the prototype tallies global quantities **ubitsmoved** and **numbersmoved** that can be initialized to zero just before starting a routine we want to measure. The easiest way to compare the cost is usually the ratio of **ubitsmoved** to **numbersmoved**, because that immediately tells the average *bits per number*. That allows easy comparison with IEEE float precisions of 16, 32, 64, or 128 bits per number. Operations like addition read two numbers from memory and store one result, so addition increases **numbersmoved** by 3. The **ubitsmoved** counter adds the size in bits of each ubound for the two inputs and one output, measured with the **nbits** function.

In a real unum hardware environment we would probably not care so much about measuring the communication cost, but here we need experimental data to answer the question:

> *When used to solve complete problems,*
> *do unums save storage, compared to floats?*

Adding traditional intervals is straightforward enough. The sum of [*a*, *b*] and [*c*, *d*] is simply [*a* + *c*, *b* + *d*], with *a* + *b* rounded toward −∞ and *b* + *d* rounded toward +∞. But with ubounds, there are the added complexities of open versus closed endpoints, dynamic exponent and fraction sizes, and correct handling of sums that exceed the largest magnitude number, *maxreal*. If you add, say, the open interval (−∞, 0) to ∞, the correct answer is ∞. The **open** −∞ endpoint means the left endpoint is *some finite number*, so adding ∞ will always result in ∞. In contrast, conventional interval arithmetic would treat the computation as containing −∞ + ∞, which results in a NaN.

What we need to add ubounds is two addition tables: One for the left endpoints and one for the right endpoints, where we write "[*x*" or "(*x*" to indicate a left endpoint and "*x*]" or "*x*)" to indicate a right endpoint. The tables tells us what to do in every possible situation. **Brace for some very odd-looking notation.** There are thought-provoking cases that anyone building a unum computer needs to handle carefully, such as:

- (−∞ ⊕ [∞ = [∞, and
  −∞] ⊕ ∞) = −∞]. Exact) ± infinity always "wins" over an open (inexact) ± infinity.

- "(∞" and "−∞)" aren't included in the table because they cannot occur. Inexact ∞ can only be a right endpoint, and inexact −∞ can only be a left endpoint. However, [∞ *can* be a left endpoint, which of course means ∞] is the right endpoint; similarly −∞] can be a right endpoint, which means [−∞ is also the left endpoint.

- If [*x* ⊕ [*y* is exceeds *maxreal*, the result is (*maxreal*, not [*maxreal*. If *x*] ⊕ *y*] is bigger than *maxreal*, the result is "∞)". Similarly for results that are less than −*maxreal*.

- If exact [closed] endpoints are added in the *g*-layer, they may still turn into inexact (open) endpoints when transferred to the *u*-layer because the closed endpoint lands inside of the smallest representable ULP in the *u*-layer. (The previous example with *maxreal* is just one example of this.)

The IEEE Standard cannot handle the subtle infinity cases, and simply produces NaN results. Since an overflow number becomes an infinity in the IEEE Standard, floats commit an infinitely large error in representing the result; unums correctly indicate that the number is too large to represent, but *finite*. The only time unum addition results in a NaN result is when adding exact −∞ to exact +∞. That represents the *ultimate* tug-of-war, and there is no way to resolve it.

Here are the tables for unum addition, covering the real number line and infinities, where **x** and **y** are exact floats expressible in the *u*-layer. If the table entry is simply the result of adding the values, it is shown in black, but exceptions are shown in magenta. Similarly, if the open-closed nature of the endpoint is simply the OR of the two inexact flags, the parenthesis or bracket is shown in black; exceptions are shown in magenta. First, the table for adding the **left** endpoints in the *g*-layer:

⊕ left	[−∞	(−∞	[y	(y	[∞
[−∞	[−∞	[−∞	[−∞	[−∞	(NaN
(−∞	[−∞	(−∞	(−∞	(−∞	[∞
[x	[−∞	(−∞	[x+y	(x+y	[∞
(x	[−∞	(−∞	(x+y	(x+y	[∞
[∞	(NaN	[∞	[∞	[∞	[∞

Next, the table for adding the **right** endpoints in the *g*-layer:

⊕ right	−∞]	y)	y]	∞)	∞]
−∞]	−∞]	−∞]	−∞]	−∞]	NaN)
x)	−∞]	x+y)	x+y)	∞)	∞]
x]	−∞]	x+y)	x+y]	∞)	∞]
∞)	−∞]	∞)	∞)	∞)	∞]
∞]	NaN)	∞]	∞]	∞]	∞]

The unpacked unum format makes it simple to look up the location in the above tables. For example, the test for "−∞]" is b_{63} (less than zero) AND b_{61} (±∞) AND NOT b_9 (inexact).

Returning to the *u*-layer is trivial except for the cases involving [**x+y** and **x+y**]. For those, if **x+y** is not an exact unum, the inexact unum that contains the sum is used instead, which changes the closed endpoint to an open one. This also takes care of the case where **x+y** is less than *−maxreal* or greater than *maxreal*, since the open intervals (−∞, −*maxreal*) and (*maxreal*, ∞) become the endpoint. The code for ubound addition (Appendix C.9) follows the above tables for left and right endpoints of a ubound separately, treating it as a gbound, and then converts the gbound to ubound form. It automatically checks if **unify** can make a lossless compression of the ubound, in which case the compressed form is given as the result.

The prototype function to negate a ubound is **negateu**. The ubound subtraction operation "⊖" is computed by simply adding the first argument to the negative of the second argument. That is, $x − y = x + (−y)$ is computed in the *g*-layer and converted back to the *u*-layer.

9.1.1 Examples and tests of data motion

Try out the tough case where endpoints look like ∞ but might be open or closed, which is crucial to getting the correct sum. For clarity, we express the inputs with general intervals, converted to ubounds. Closed infinities win over open infinities, so $(-\infty, 4) \oplus (maxreal, \infty]$ should evaluate to $(-\infty, \infty]$:

$$\mathbf{view}\left[\mathbf{g2u}\left[\begin{matrix} -\infty & 4 \\ \mathbf{open} & \mathbf{open} \end{matrix}\right] \oplus \mathbf{g2u}\left[\begin{matrix} \mathbf{maxreal} & \infty \\ \mathbf{open} & \mathbf{closed} \end{matrix}\right]\right]$$

$(-\infty, \infty]$

Adding opposite-signed *open* infinities produces the entire range of finite numbers.

$$\mathbf{view}\left[\mathbf{g2u}\left[\begin{matrix} -\infty & 4 \\ \mathbf{open} & \mathbf{open} \end{matrix}\right] \oplus \mathbf{g2u}\left[\begin{matrix} 0 & \infty \\ \mathbf{closed} & \mathbf{open} \end{matrix}\right]\right]$$

$(-\infty, \infty)$

However, $\infty - \infty$ is still NaN:

$$\mathbf{view}\left[\{\hat{\infty}\} \ominus \{\hat{\infty}\}\right]$$

NaN

Try adding $(2.25, 2.5)$ to 9 in a $\{3, 4\}$ environment, and measure the data motion. Adding **ubitmask** to the unum for 2.25 indicates an open interval of width 0.25. At the *u*-layer, each number looks like this:

0	1		**001**	1	000	0010	$(\frac{9}{4}, \frac{5}{2})$
$+$	$2^1 \times$	$1+$	$\frac{1}{8}$	$\cdots$	1	3	$= (2.25, 2.5)$

0	110		**001**	0	010	0010	
$+$	$2^3 \times$	$1+$	$\frac{1}{8}$	$\downarrow$	3	3	$= 9$

The unums take 13 and 15 bits, but we add the one-bit cost of recording whether they are one-part or two-part ubounds, so as ubounds they cost 14 and 16 bits. Reset the counters for bits moved and numbers moved, and view the sum as a *u*-layer bit string:

```
ubitsmoved = numbersmoved = 0;
uview[{2.2̂5 + ubitmask} ⊕ 9̂]
```

0	110		**01101**	1	010	0100	$(\frac{45}{4}, \frac{23}{2})$
$+$	$2^3 \times$	$1+$	$\frac{13}{32}$	$\cdots$	3	5	$= (11.25, 11.5)$

The result automatically expanded to 18 bits as a ubound. The prototype includes a function, `reportdatamotion[numbersmoved,ubitsmoved]`, that produces a table summarizing the data motion:

Numbers moved	3
Unum bits moved	48
Average bits per number	**16.0**

Adding numbers of the same sign tends to increase the number of fraction bits or exponent bits needed (though not always, since the sum can land on a fraction that reduces to lower terms). This is especially true if the numbers are of very different magnitude; if you were adding 2.0×10^{14} to 17 with pencil and paper and wanted the answer to be exact, you would have to write down all the digits: 200 000 000 000 017. Subtracting numbers, or adding numbers of opposite sign, can scrape digits off the left end of the fraction or reduce the exponent size needed, so sometimes an add or subtract operation creates a result that takes up *fewer* bits.

9.1.2 Three-dimensional visualization of ubound addition

Remember how a general interval could be visualized as a thick segment of the real number line with rounded ends for open endpoints and squared-off ends for closed endpoints? Now that we have functions of *two* general intervals, here is a similar way to visualize a two-dimensional general interval in *x* and *y* directions. The *x* interval is [−1, 5) and the *y* interval is (−2, 4], so that the four corners show all combinations of open and closed:

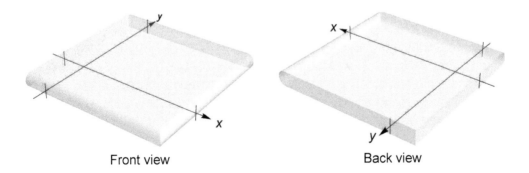

Front view   Back view

In the three-dimensional plots on the following page, the only square corner is the one at the bottom of the back view (where $x = -1$, $y = 4$); the others are rounded to remind us that the exact endpoint pair is not included. To get a square (closed) endpoint for a sum of two numbers, both input endpoints need to be square and the result needs to be representable exactly in the environment set for the *u*-layer.

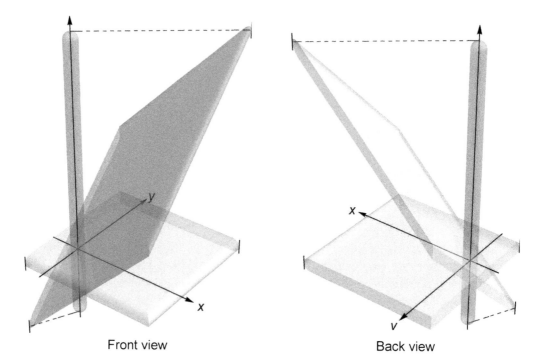

Front view   Back view

9.1.3 Hardware design for unum addition and subtraction

The addition and subtraction of unum endpoints is very much like that for conventional floats, about which much has been written. In both cases, once exceptions have been handled, the hidden bit is determined, the binary points are lined up, and the bits are added just as you add decimal numbers, from right to left. Since with inexact unums or ubounds there will be two endpoint pairs that need to be added, it makes sense to have logic for a *pair* of add/subtract units that can execute in parallel.

There are important differences, however, with respect to rounding. When a result has more bits than the fraction can hold, adding IEEE binary floats requires complicated hardware to select which incorrect float to present as the answer, especially since there are four "rounding modes" to support. Unums simply turn on the ubit to indicate "there are more bits after the last one" for answers that cannot be expressed exactly, a very simple thing to detect in hardware. This simplification, together with the use of pre-decoded exception bits and hidden bit in the unpacked format, suggest that each unum endpoint addition and subtraction could be faster and use less energy than required for floats. That helps compensate for the fact that two additions or subtractions must be done instead of one, for inexact ubounds.

Suppose a float has p fraction bits. Whenever the exponents differ by more than p, it signals that when the binary points are lined up there will be no overlap between the fractions, so the logic can take a shortcut to figure out how to round the number. In contrast the unum may well store fewer than p bits for the same value, yet have a maximum *fsize* larger than p that can capture the correct answer much of the time.

For example, here are the unums for $\frac{1}{256}$ and 30, in a {3, 4} environment:

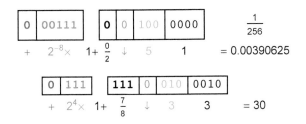

| 0 | 00111 | | 0 | 0 | 100 | 0000 | | $\frac{1}{256}$ |

$$+ \quad 2^{-8} \times \quad 1+\frac{0}{2} \quad \downarrow \quad 5 \quad 1 \quad = 0.00390625$$

| 0 | 111 | | **111** | 0 | 010 | 0010 | |

$$+ \quad 2^{4} \times \quad 1+\frac{7}{8} \quad \downarrow \quad 3 \quad 3 \quad = 30$$

The exponents differ by twelve $\left(2^{-8}\text{ versus }2^{4}\right)$. To add the numbers as IEEE 16-bit floats, hardware would detect that their exponents differ by twelve, which is more than the ten-bit fraction length. Float hardware ignores the smaller input value and says $30 + \frac{1}{256} = 30$. Unum hardware instead compares the exponent difference with $2^{fsizesize} = 16$, and increases precision so it can perform the addition exactly:

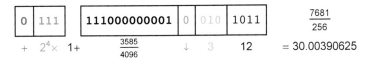

| 0 | 111 | | **111000000001** | 0 | 010 | 1011 | | $\frac{7681}{256}$ |

$$+ \quad 2^{4} \times \quad 1+\frac{3585}{4096} \quad \downarrow \quad 3 \quad 12 \quad = 30.00390625$$

Had we added two exact numbers that had exponents differing by more that $2^{fsizesize}$, unum hardware can also take a shortcut: It simply sets the ubit to indicate the presence of bits beyond the last one shown. For example, $1\hat{0}00 \oplus \frac{\hat{1}}{256}$ gives this correct (inexact) result:

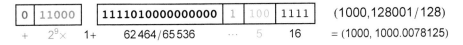

| 0 | 11000 | | **1111010000000000** | 1 | 100 | 1111 | | (1000,128001/128) |

$$+ \quad 2^{9} \times \quad 1+ \quad 62464/65536 \quad \cdots \quad 5 \quad 16 \quad = (1000, 1000.0078125)$$

9.2 "Creeping crud" and the myth of unbiased rounding

There are four "rounding modes" in IEEE binary floats: Round up, round down, round toward zero, and round-to-nearest-even. Synonyms for these are, respectively: Round toward infinity, round toward negative infinity, truncated rounding, and unbiased rounding.

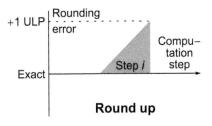

Round up

We can visualize "rounding up" as shown on the left. The left end of the triangle is an exact, correct value; after a computation step, the true answer is replaced by an exact number that is the high end (closer to infinity) of the 1-ULP interval of uncertainty.

In decimal math rounded to integers $13 \div 6 = 2.1666 \cdots$ would round up to 3. Had we divided 13 by –6 instead, the result $-2.1666\cdots$ would have been rounded up to –2. If the rounding error is random, then error will increase by an average of 0.5 ULP per computation step.

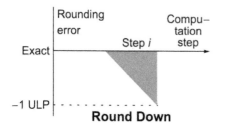

Rounding down, or toward $-\infty$, is what was used by the Vancouver Stock Exchange mentioned at the chapter beginning. They used three digits to the right of the decimal, but always rounded down after updating the value of their stock index. The graphic, left, is just the mirror image of rounding up.

In some applications, like signal processing or simple kinds of computer graphics, these rounding modes are not as terrible as they might seem because *only a few arithmetic operations are performed on each data item* and the rounding errors therefore do not mount up. For the Vancouver exchange, however, here is what the sequence of **cumulative** rounding errors looks like, where each rounding is statistically independent:

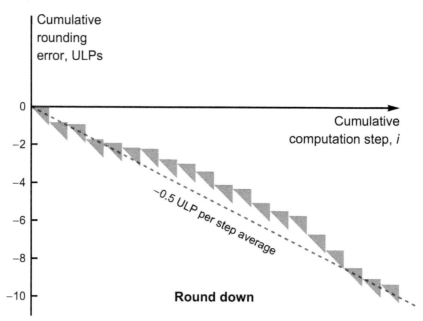

"Round toward zero" can be tolerated better than round-up or round-down *if the calculations have an even chance of being positive or negative numbers.* For example, the dot product of vectors $\{a_1, a_2, \dots a_n\}$ and $\{b_1, b_2, \dots b_n\}$ is $a_1 b_1 + a_2 b_2 + \dots + a_n b_n$. If the a and b elements are equally likely to be positive or negative, then rounding toward zero will round up about half the time and round down about half the time.

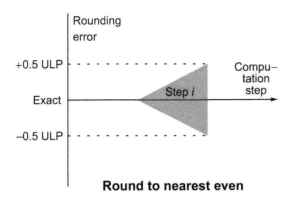

Round to nearest even

The most widely-used rounding mode is "unbiased rounding," which replaces the correct answer with a float that is within half an ULP. In the case of a tie, the nearest even number is selected, so the tie cases go up half the time and down the other half. Unbiased rounding takes extra circuitry and time compared to the other rounding modes but produces this more comforting-looking result.

Unbiased rounding reduces the *average* error from 0.5 ULP to 0.25 ULP, but it can be in either direction. This brings us to what statisticians call a "random walk" problem. Suppose you flip a fair coin repeatedly; if heads, you step forward, and if tails, you step backward. After n coin flips, about how many steps will you be from where you started? We expect the number of heads to be $n/2$, on the average, but with some deviations that approximate a bell curve. For $n = 100$ coin flips, here is the distribution of the number of times a fair coin will come up heads:

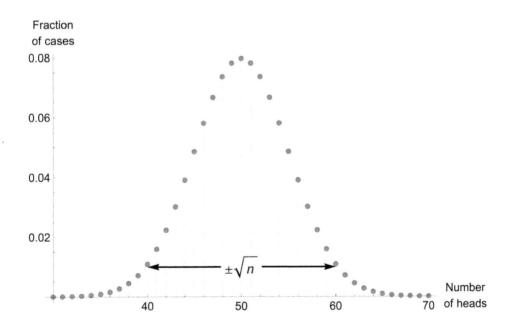

The variability about the average grows as the *square root* of the number of independent coin flips.

So the reasoning goes that the cumulative rounding error with unbiased rounding should be perfectly tolerable even if accumulated over a great many calculations:

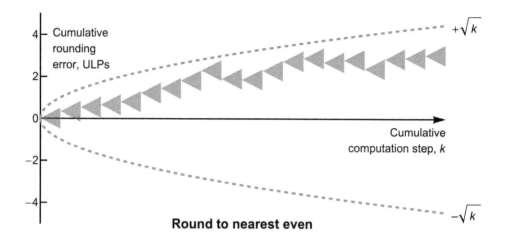

Round to nearest even

Even if we do, say, a billion (10^9) adds to accumulate a sum, this reasoning says that the number of ULPs we will be off should usually be within $\pm\sqrt{10^9}$ quarter-ULPs of the correct sum, which is about 7906 ULPs. If the numbers being accumulated are in single precision, an ULP is $\frac{1}{8\,388\,608}$ of the number, so we should be within $\frac{7906}{8\,388\,608}$ or about 0.001 of the right answer: One billion, plus or minus a million.

Do you buy it? It certainly sounds well-reasoned. Single precision can represent numbers up to 10^{38} or so, and single precision certainly has enough fraction bits to represent 10^9 as an exact integer.

> The random walk protects us from cumulative rounding errors. Right?

Not even close. Here is a simple C program that tries to count from 1 to a billion:

```
#include < stdio.h >
  float sumtester () {
        float sum; int i;

        sum = 0.0;
        for (i = 0; i < 1000000000; i++) {sum = sum + 1.0;}

        printf ("%f\n", sum);
  }
```

The result you get (with IEEE single precision for the `float` type) is

```
16777216
```

which is too small by a factor of about *sixty*! The error is not because single precision cannot accurately represent numbers this large; it can. So what went wrong?

What went wrong was the assumption that rounding errors are independently up or down. Rounding errors are often *not* independent. In the above sum, they all go the same direction: *Down*. After the sum reaches $2^{24} = 16\,777\,216$, an ULP width is more than 1, so adding 1 to the sum changes nothing. Round-to-nearest-even repeatedly rounds the correct answer, $16\,777\,217$, down to $16\,777\,216$.

One nickname for the accumulation of rounding errors in long-running calculations is "creeping crud." As the example shows, it does not always creep; sometimes it *races*. The Myth of Unbiased Rounding assumes that rounding errors are independent and randomly up or down. **There is no logical basis for that assumption**, and the above example shows how unfounded it can be in even the simplest problems.

A common technique in computing is the Monte Carlo method, which typically requires accumulating sums of functions of random numbers. If you add a billion random numbers that are evenly distributed between zero and one, statistics say the result should be about half a billion.

Try it in single precision, and again the result will be $16\,777\,216$, not some number close to $500\,000\,000$. One application of Monte Carlo methods is in modeling the financial value of stock portfolios, and deciding which stocks are overvalued or undervalued. Imagine the consequences of being off by a factor of *eight* when advising a client about a major investment, as the result of invisible rounding error!

9.3 Automatic accuracy control and a simple unum math test

Now that we can add unums, we can experiment with how well unums handle the challenge of accumulated error. If we use a {3, 5} environment that can imitate IEEE single precision on the preceding example, it is not a fair fight. The unums simply compute the exact answer, using a minimal exponent size and adjusting the fraction as needed to represent every integer add without rounding. The real issue is what happens when unums *have* to mark a result as inexact. Is that better than rounding?

To study the issue, use a very low precision {3, 3} environment to exaggerate the problem. In a {3, 3} environment, integers up to 512 can be expressed, but notice what happens if we add the unum for 1 to the unum for 512:

```
setenv[{3, 3}];
view[5̂12 ⊕ 1̂]
```

(512, 514)

This is better than floating point would have done; floats of the same precision would have returned the number 512, exactly. Unum math at least *warns* us that we have a "creeping crud" problem. The range (512, 1952) is alarmingly large, suggesting that unums may suffer the problems of interval arithmetic. However, because unum environments record accuracy in the numbers themselves, they offer the possibility of *detecting, analyzing, and correcting loss of accuracy automatically*, using the following rules:

> If a result contains *maxreal* or *–smallsubnormal* as its lower bound, or *–maxreal* or *smallsubnormal* as its upper bound, the calculation environment needs a larger *esizesize*. If the relative width is larger than a set tolerance, the calculation needs a larger *fsizesize*.

Since there is no way to represent 513 exactly, we get an open interval that contains 513, instead. This looks like traditional interval arithmetic, but the ubound uses only 22 bits instead of 128, and it is careful to track the fact that the sum is *not* equal to 512 nor is it equal to 514, but is a number strictly between those bounds. Try adding one again, and the ULP expands to a width of 4; it is still efficiently expressed with a 20-bit unum, but the accuracy has dropped to about two decimals:

$$ +\quad 2^9 \times \quad 1+ \quad \frac{0}{128} \quad \cdots \quad 5 \quad 7 \quad = (512, 516) $$

0	11000		0000000	1	100	110

With that understanding of how accuracy is being lost, here is a "scale model" of the 10^9 summation problem, this time just to 1000 since that suffices to force the use of inexact unums:

```
sumu = {0̂};
For[i = 0, i < 1000, i++, sumu = sumu ⊕ 1̂];
view[sumu]
```

(512, 1952)

Both shortcomings can be tracked and corrected automatically by a computing system. Automatic testing for the first case is easy; the **needmoreexpQ** function returns True if a result is too small or too large for the dynamic range, as described. For example, adding 1 to *maxreal*/2 will not exceed the range of the environment:

needmoreexpQ$\left[\text{maxr}\hat{\text{e}}\text{al} / 2 \oplus \hat{1}\right]$

False

But adding 1 to *maxreal* will, since the only way to represent the sum is with the open interval (*maxreal*, ∞).

needmoreexpQ$\left[\text{maxr}\hat{\text{e}}\text{al} \oplus \hat{1}\right]$

True

A relative width of 0.05, say, indicates one decimal of accuracy. The interval (10, 11) has one decimal of accuracy because $\frac{11-10}{10+11} = 0.047 \cdots$ which is less than 0.05. However, the interval (9, 10) does not quite make the cut, since $\frac{10-9}{9+10} = 0.052 \cdots$, which is greater than 0.05.

The prototype function that computes relative width for a ubound *u* is **relwidth[u]**. Ideally, a computer system would optionally calculate the relative width with built-in hardware, and thus monitor each calculation without slowing it down. Then, guidelines for optimum environment settings can be generated automatically and a programmer can place them explicitly in the source code.

As for the bound on a result being "too large," we can measure the relative width in a bound in various ways. The following approach does a good job of handling the exception cases. If *g* is a general interval, where g_{lo} is the lower limit and g_{hi} is the upper limit (and we ignore whether endpoints are open or closed), then a possible definition of relative width is

$$\text{Relative width of } g = \begin{cases} \infty & \text{if } g \text{ is NaN,} \\ 1 & \text{if } |g_{lo}| = \infty \text{ or } |g_{hi}| = \infty, \\ 0 & \text{if } |g_{lo}| = 0 \text{ and } |g_{hi}| = 0, \\ \frac{|g_{hi}-g_{lo}|}{|g_{lo}|+|g_{hi}|} & \text{otherwise.} \end{cases}$$

If *g* represents a NaN, the relative width is considered infinite. If *g* has any infinite magnitude endpoints, then the relative width is 1. The ratio of real numbers at the bottom will also be 1 if the endpoints are of opposite sign. (If we cannot figure out whether a number is positive or negative, we really do not know much about it at all.)

Recall the "Demands and Concessions" discussion (Section 5.4.2) regarding accuracy. A computer user should be allowed to demand, say, "I need four decimals of accuracy in my result"; this is also a concession, because it means the computer can save energy and time by not calculating *more* than four decimals of accuracy. What computer users should *not* specify is that a calculation be done with a certain precision throughout, because that confuses means with ends. Just as the computer user could set the policy for compressing numbers with `smartunify`, the user could set an environmental variable `relwidthtolerance` to specify the accuracy needed in an answer. The `needmorefracQ` function then uses that tolerance to detect if the calculation needs a larger *fsizesize*. The relative width tolerance also sets a lower bound for *fsizesize*, since there have to be at least enough fraction bits to express the required accuracy. Setting the `relwidthtolerance` to 0.005 says we want two decimals of accuracy, which is possible with a {3, 3} environment:

```
relwidthtolerance = 0.005;
```

The interval [10, 11] has only one decimal of accuracy, indicating the need for more fraction bits:

$$\texttt{needmorefracQ}\left[\left\{\hat{1}\hat{0}, \hat{1}\hat{1}\right\}\right]$$

True

The interval (500, 504) is a one-ULP wide number in the {3, 3} environment, and because its relative width is about 0.004, the `needmorefracQ` test returns False. The number has sufficient accuracy.

$$\texttt{view}\left[\hat{5}\hat{0}\hat{0} + \texttt{ubitmask}\right]$$

$$\texttt{needmorefracQ}\left[\left\{\hat{5}\hat{0}\hat{0} + \texttt{ubitmask}\right\}\right]$$

(500, 504)

False

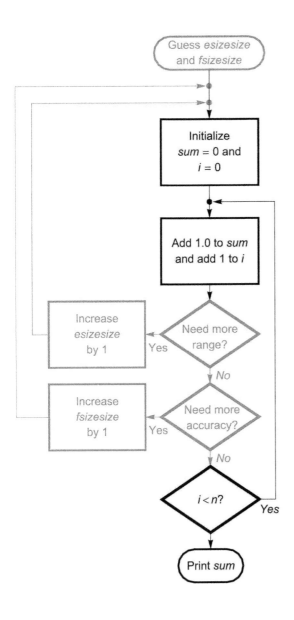

The flowchart for this approach looks like this, where the parts shown in blue can be requested by the programmer *or performed automatically*.

The strategy is very similar to the checkpoint-restart method used to protect long-running supercomputing tasks from hardware faults, which can also be done at the application level or automatically by the operating system.

With these tools, we will try the sum again but let the computer manage the accuracy and dynamic range automatically. We could wait for the final sum and then let the computer check to see if it needs more exponent or fraction, or we could monitor the calculation throughout, and restart it as soon as the need for an upgrade is detected. The continuous monitoring eliminates time wasted continuing a calculation that has clearly "gone off the rails."

Here is a software version of that type of monitoring, where we try to get away with a {2, 2} environment at first and let the computer decide if it needs more bits. Some old-fashioned "go to" statements start the computation over whenever the computer detects that the environment needs an upgrade. The statements are color-coded to the corresponding flowchart coloring, with automatic monitoring shown in blue:

```
Module[{n = 1000}, setenv[{2, 2}];
  Label[begin];
  sum = 0̂;
  For[i = 0, i < n, i++, {
      sum = sum ⊕ 1̂;
      If[needmoreexpQ[sum],     (* then *)
        setenv[{esizesize + 1, fsizesize   }]; Goto[begin]];
      If[needmorefracQ[sum],    (* then *)
        setenv[{esizesize,     fsizesize + 1}]; Goto[begin]];
  }];
  view[sum]]
```

1000

The answer is exact. Where did the environment wind up?

```
{esizesize, fsizesize}
```

{3, 4}

Since we only asked for two decimals of accuracy with **relwidthtolerance** of 0.005, the relative width test would have passed for any ubound contained in [991, 1010]. When it is reasonable to expect an exact result, as it is here, the programmer can set **relwidthtolerance** to zero to demand perfection.

What actually happened was this: Setting the environment to {2, 2} only allowed for four exponent bits and four fraction bits. The sum was exact up to 32, but adding 1 to 32 led to the inexact unum for (32, 34). The relative width was $\frac{1}{33}$ which is certainly larger than $0.005 = \frac{1}{200}$, so the environment was upgraded to {2, 3} and the summation started over after just a few wasted operations.

On the second try, however, the summation exceeded *maxreal*, which is only 510 in the {2, 3} environment. So adding 1 to *maxreal* resulted in the unum for (510, ∞). Need more exponent! The computer detected the use of the limits of the dynamic range, raised the environment to {3, 3}, and restarted the summation. On the third try, the summation got to 512 with exact results, but then the iterations became inexact, and ULP widths started to accumulate:

$$512, \quad (512, 514), \quad (512, 516), \quad (512, 518)$$

The relative width of (512, 518) is about 0.0058, which set off the **needmorefracQ** test. The environment was improved to {3, 4} and the computation remained exact all the way to completion. This example shows at least *some* numerical analysis can be automated, if accuracy loss is made visible.

10 Multiplication and division

In the first Gulf War, the U.S. prepared Patriot missiles for immediate launch if needed, and kept them ready for several days before they were used. Their ability to hit targets depends on accurate calculations using elapsed time in tenths of a second. After about 100 hours had passed (3.6 million ticks), a Patriot was launched to defend against an attacking Scud missile. The guidance system multiplied the ticks since boot by a float value of 0.1, or *about* 360 000 seconds. That big number to the left of the decimal point only left room for *one decimal* to the right of the decimal point; the important bits were crowded out, magnifying rounding error to about 0.34 second. A Scud missile can travel about 570 meters in a 0.34 second error, so the Patriot missile missed the Scud, which struck a U.S. Army barracks and killed 28 soldiers. Patriots have since been modified to use higher-precision floats. The incident stands as the worst disaster ever caused by floating point rounding errors, in this case from multiplication. A unum calculation would have made the inaccuracy visible, long before anyone pushed the red button.

10.1 Multiplication requires examining each quadrant

Multiplication of general intervals is also more complicated than for traditional intervals. With closed intervals, the product is simply the minimum and maximum of every possible endpoint product:

$$[a, b] \times [c, d] = [\text{Minimum}(a \times c, a \times d, b \times c, b \times d),$$
$$\text{Maximum}(a \times c, a \times d, b \times c, b \times d)]$$

But for general intervals, consider the following weird case: The product of $[-1, 2)$ and $(-0.5, 1]$. The four combinations of endpoint products are -0.5, -1, -1, and 2. So the left endpoint is -1 and the right endpoint is 2, but the two -1 cases are *not* the same. Here is a graphical view of this, using rounded-square endpoints as before. The 3D plot of $x \times y$ shows the product makes a saddle-shaped surface:

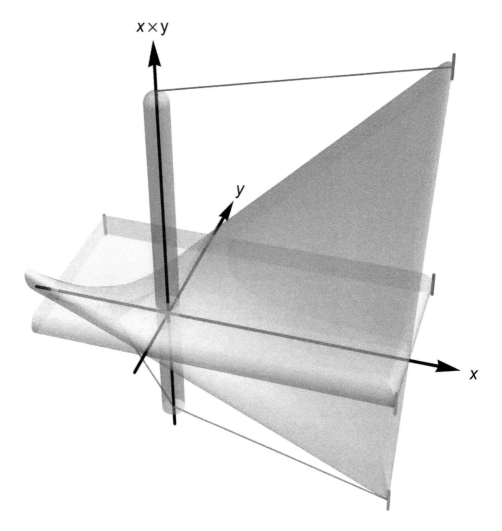

That view, from above the *x-y* plane, shows the product maximum (blue line) as an open (rounded) endpoint at a height $2 \times 1 = 2$. To get a better look at what is happening at the minimum (orange line), here is the view of the saddle-shaped surface from below:

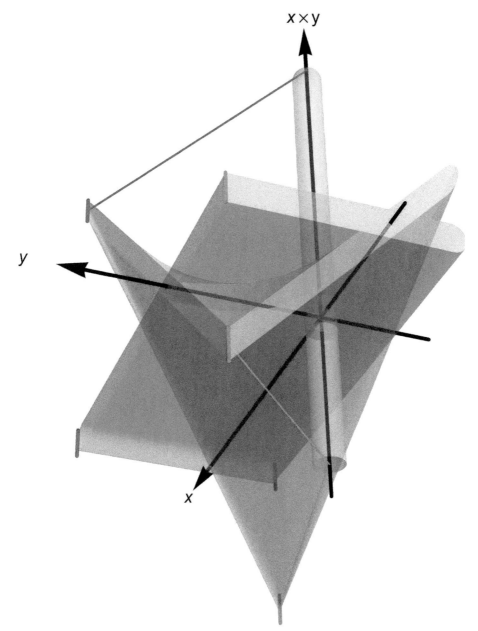

The closest corner at height −1 is squared-off, because it is the result of a product of a closed endpoint and a closed endpoint, $-1 \times 1 = -1$. The farther minimum ties the value at $2 \times -0.5 = -1$, but one of the input endpoints is open, so it does not actually *include* −1; it approaches it as an open extreme. If both intervals span zero, then the products in opposite quadrants contend for being the extrema.

We also have to watch out for $0 \times \infty$ forms, which for conventional intervals are NaN. Because we have more information about endpoints, we have the following combinations in the case where both arguments are non-negative:

- If you multiply an infinity times a range with nonzero endpoints of opposite sign, the result is $[-\infty, \infty]$, not NaN.
- An inexact zero times any finite number still approaches but does not contain zero.
- A zero times inexact infinity (that is, "almost infinite") is zero. Inexact zero (that is, "almost nothing") times infinity is infinity.
- A positive inexact zero times inexact infinity is the entire positive real number line.

Here are the *g*-layer multiplication tables for the left and right endpoints, **for non-negative inputs**. As with the addition tables, we use magenta for cases that take a little thought and need to be handled carefully in the multiplication logic.

⊗ left	[0	(0	[y	(y	[∞
[0	[0	[0	[0	[0	(NaN
(0	[0	(0	(0	(0	[∞
[x	[0	(0	[x·y	(x·y	[∞
(x	[0	(0	(x·y	(x·y	[∞
[∞	(NaN	[∞	[∞	[∞	[∞

⊗ right	0]	y)	y]	∞)	∞]
0]	0]	0]	0]	0]	NaN)
x)	0]	x·y)	x·y)	∞)	∞]
x]	0]	x·y)	x·y]	∞)	∞]
∞)	0]	∞)	∞)	∞)	∞]
∞]	NaN)	∞]	∞]	∞]	∞]

If *x* and *y* are both negative, the preceding tables work if we apply them to $-x$ and $-y$. If only one of *x* or *y* is negative, then **we reverse the sign of the negative argument and also reverse the roles of left and right.** See the 3D plots to see why this is so; in the second and fourth quadrants of the *x-y* plane, the saddle-shaped surface goes more negative as *x* and *y* get farther from zero, so we have to flip the ordering of the extrema. Nothing like this happened when simply adding two numbers because the 3D plot of $x + y$ is simply a flat, tipped plane. More challenging to picture is what happens when we multiply a right endpoint by a left endpoint, like *x*] times (*y*. It again depends on the quadrant.

If both x and y are positive or both negative, then the product cannot be an extremum of the bounding interval. The x] notation means there are smaller products on the left as you multiply by y. Then when you get to exactly x and consider x times $(y,$ the product continues to get larger. So we can save work in doing trial products of intervals that go from xlo to xhi and ylo to yhi. In the following list, "×" means the value from the "⊗ **left**" table above for lower bounds, and from the "⊗ **right**" table for upper bounds. All we have to do is flip the sign of the input.

xlo × ylo	is a possible lower bound if xlo, ylo are both non-negative;
(−xlo × −ylo)	is a possible upper bound if xlo, ylo are both negative;
−(−xlo × yhi)	is a possible lower bound if xlo is negative, yhi is non-negative;
−(xlo × −yhi)	is a possible upper bound if xlo is non-negative, yhi is negative;
−(xhi × −ylo)	is a possible lower bound if xhi is non-negative, ylo is negative;
−(−xhi × ylo)	is a possible upper bound if xhi is negative, ylo is positive;
(−xhi × −yhi)	is a possible lower bound if xhi, yhi are both negative.
xhi × yhi	is a possible upper bound if xhi, xhi are both non-negative.

The prototype code for the product of two general intervals, `timesg`, is in Appendix C.10. The u-layer version is `timesu`, which is assigned the symbol, "⊗"; it tracks how many bits and numbers were moved to perform the operation.

Try ubound multiplication on the curious example mentioned at the beginning of this section, the product of $[−1, 2)$ and $(−0.5, 1]$, using a {2, 4} environment:

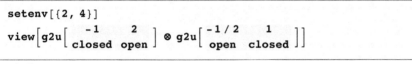

$[−1, 2)$

As with traditional floats, zero times infinity is Not-a-Number:

```
view[0̂ ⊗ ∞̂]
```

NaN

However, with unums we can rescue the following cases from falling into the land of NaN, from which there is no return:

```
view[smallsubnormalu ⊗ ∞̂ ]
view[0̂ ⊗ posopeninfu]
```

∞

0

As with addition/subtraction, when values are moved from the *g*-layer to the *u*-layer, they may change from exact to inexact and cause closed endpoints to become open ones. The new twist with multiplication is that it is possible for a calculation to land in the range (0, *smallsubnormal*) or (−*smallsubnormal*, 0).

10.2 Hardware for unum multiplication

Every good tool collection has a range of sizes of screwdrivers, pliers, hammers, and so on. Even though wrenches can be adjustable like the ones above, it still makes sense to have more than one size. So it is with unum multipliers.

10.2.1 Multiplies that fit the standard hardware approach

As with addition and subtraction, the quick access to exception bits in the unpacked unum form make for quick, minimal hardware logic to dispense with all the exception cases shown in Section 10.1. The main hardware discussion here, therefore, is how to perform **x·y** in the scratchpad, the *g*-layer, when none of the NaN-zero-infinity cases determines the result.

The processing in the scratchpad need not be exact, of course, if a result is obviously not going to be exact when returned to the *u*-layer. For example, the multiplication involves adding the exponent field bit strings of each operand as integers; if the sum is less than 0 or greater than $2^{esizesize}$, it indicates the result falls in (0, *smallsubnormal*) or (*maxreal*, ∞) (or the negative versions of those), and there is no need to compute the actual product of the fraction bit strings.

A trick that is easier for unums than for floats is that unums always track how many bits are in each fraction for exact normal numbers; they start with a hidden 1 bit and, because of lossless compression, they also end with a 1 bit.

The product of two bit strings

1…1 (*m* bits) and
1…1 (*n* bits) will be
1…1 (*m* + *n* bits).

If $m + n$ is greater than the maximum number of fraction bits (including the hidden bit), the result will have its ubit set to indicate there are nonzero bits that cannot be expressed in the *u*-layer. That is simpler than trying to figure out how to round the number to a nearby float.

The 64-bit unpacked unum format has a maximum exponent size of 16 bits and a maximum fraction size of 33 bits, including the revealed bit. For environments up to {4, 5}, the *g*-layer needs logic for adding/subtracting 16-bit integer exponents with a carry/borrow bit to indicate integer overflow and underflow (magnitude greater than *maxreal* or smaller than *smallsubnormal*). Ideally, it would have a 66-bit multiplication parallelogram so that extended-precision multiplication methods are not needed.

The parallelogram is why high-precision multipliers tend to consume chip area roughly proportional to the square of the precision, up to some point when more sophisticated techniques are used. For any unum pair, only a fraction of the parallelogram is needed, and the rest can stay powered down to avoid energy waste. The red dots show hardware in use for the product of a 24-bit fraction and a 17-bit fraction; the cyan dots show unneeded hardware, powered down.

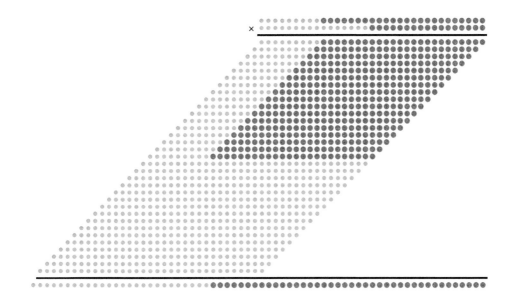

Historically, designers have occasionally made use of the trick of splitting the multiply parallelogram for lower-precision products. With unums, it could be split anywhere, not necessarily in half:

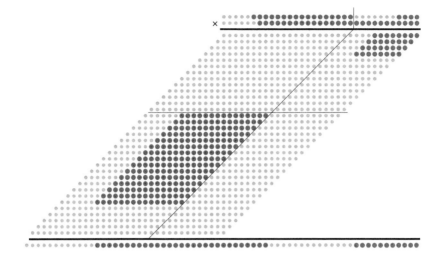

This helps provide parallel computing to multiply both ends of an inexact ubound when both fraction bit strings are less than the maximum length. Though frankly, there is no shortage of parallel multipliers on chips lately. A current-generation Intel Xeon has *hundreds* of multipliers, for example, and they take up just a few percent of the chip area. Minor modifications to recognize the left-right multiplication tables of the previous section would provide high-speed unum multiplication up to some environment size, after which extended precision multiplies would be needed.

If there are no exception cases and **x·y** must be computed, the ubits of the unums representing **x** and **y** are ORed together. In other words, the product can only be exact if both inputs are exact. As the parallelogram of partial products is being done, the result ubit is also set with an OR if there are any bits beyond the maximum expressible fraction. As the summation is performed of all the partial products from right to left, doing an OR of each sum bit with the ubit suffices to take care of both normal and subnormal cases.

If the product is 66 bits or fewer, unum multiplication as described here should be as fast as or faster than float multiplication and will consume less power. The fact that the fraction length of the unpacked unum is a *maximum* of 33 bits (almost 10 decimals) means less multiplication time than for IEEE double precision fractions of 53 bits (almost 16 decimals). Two unum multiply parallelograms (one for each endpoint) have 2178 entries for partial products, of which only a fraction are usually used each time. In contrast, an IEEE double precision multiply parallelogram has 2809 entries, requiring 29 percent more chip area and must supply electrical power to every entry for every multiplication.

10.2.2 Extended precision and the "complete accumulator"

Extended-precision integer multiplication requires partitioning the bit string repre-senting each input integer x and y into the largest pieces that fit the hardware multi-plier, and accumulating the product of all possible combinations of pieces of x times pieces of y, shifted by the appropriate number of bits, as exact integer adds. The "shift" actually looks more like a conventional memory address operation; the amount of the shift, divided by the bits per word, determines where the integer will land.

That is what Kulisch's Exact Dot Product does. Kulisch refers to the accumulator for the EDP as the "complete accumulator." An extended precision multiply is the dot product of all pieces of x times one piece of y, accumulated with all pieces of x times the next piece of y, and so on until all pieces of y have been multiplied and added to make the extended precision product. Chapter 11 talks about which fused unum operations should be in a standard computing environment, but EDP is a shoo-in for a number of reasons. One reason is that it provides the next "wrench size" up from products that fit into the 66-bit parallelogram for unpacked unum data.

10.2.3 *Really* big multiplies

At the risk of frightening the reader, there *may* be some reasons to want the ability to compute scratchpad products of bit strings that are many thousands of bits long. If the work increases as the square of the bit length, the practicality of such rigor dwindles rapidly. For one thing, on-chip power consumption also goes up as the square of the precision, even if parallel execution can make the multiplication fast.

Fortunately, many mathematicians have labored for decades to find ways to get the multiplication time of n-bit integers down to something nearly proportional to n, not n^2. Depending on what method is used, and there are many (see Knuth's Section 4.3.3, "How Fast Can We Multiply?" in Vol. 2 of *The Art of Computer Programming*), at some trade-off point the more complicated method will take less time to do extended precision than simply using fused multiply-add units. Bailey estimates the trade-off point is when the input values have a thousand bits or more. This means additions and multiplications will take comparable time if the number of bits gets temporarily huge in the scratchpad layer.

> **Exercise for the reader**: Using logic operations AND, OR, and NOT on the four bits in Warlpiri unums, find expressions for the left and right multiplication table entries shown in Section 10.1.

10.3 Division introduces asymmetry in the arguments

A favorite from the despair.com web site.

10.3.1 The "bad boy" of arithmetic

Division has long been the "bad boy" of the four + − × ÷ arithmetic operations. More things can go wrong with division than with the other three operations; the division tables are more populated with NaN cases than the previous tables. Long division is the hardest hand calculation to learn in elementary school. Division is the hardest one of the four operations for computers as well; it is not uncommon for computer hardware to take almost ten times longer to perform a divide than a multiply! And it is the operation least likely to produce an exact result; the closure plots in Chapters 2 and 3 bear this out.

The subtraction operation is asymmetrical in its two arguments, of course. But it was trivial to negate the second argument and then create the operation from the logic for addition; negation is a "perfect" operation in that it never introduces information loss (widening of a bound). It might be tempting to create a divide operation by finding the reciprocal of the second argument and multiplying, but **doing so can lead to loss of information** about the result because the reciprocal is returned to the u-layer before multiplying, thus creating two opportunities for the bound to widen instead of just one.

For example, if we did that with Warlpiri arithmetic to divide 2 by 2, the reciprocal of 2 becomes the open interval (0, 1) since that environment cannot express the number $\frac{1}{2}$. Multiplying that reciprocal by 2 then gives the true but recklessly approximate result that 2 divided by 2 is somewhere in the open interval (0, 2). Information was lost. Building the algorithm for division will require more care than for multiplication.

Some cases to watch out for:

- $0/0 = \text{NaN}$

- $\pm\infty/\pm\infty = \text{NaN}$

- $\infty \; / \, (y, \infty) = \infty$ where $y > 0$

- $x / (0, y) = \begin{cases} (x/y, \infty) & \text{if } x > 0 \\ 0 & \text{if } x = 0 \\ (-\infty, x/y) & \text{if } x < 0 \end{cases}$ where $y > 0$.

Some number systems attempt to assign meaning to division by zero, like assigning $\frac{1}{0}$ to "$\pm\infty$" or to the entire real line, $[-\infty, \infty]$. With unums, dividing by zero produces a NaN, period. So does dividing by a ubound that straddles zero. It is not tolerated since dividing by zero is simply a nonsensical act within the realm of real numbers, and the best thing for a computer to do is inform the programmer or the user that something has gone completely wrong when division by zero occurs in a calculation. There are traditional interval computing systems that try to accommodate division by zero, perhaps succumbing to some mathematician's visceral desire to make operations work on *every* possible input. Even the interval support in *Mathematica* makes this mistake. For example:

```
Interval[{1, 2}] / Interval[{0, 0}]
```

```
Interval[{-∞, ∞}]
```

That answer breaks the rules for what an interval is supposed to mean mathematically: It returned *all* the real values between the endpoints. The interval data type has no way to communicate to the user "What I meant was, both $-\infty$ and $+\infty$ are the result of dividing by zero, but *not*, um, the real numbers in between. In this case." With unum arithmetic, we take the hard line that division by exact zero is an indication that the computation has lost all meaning, and should produce a NaN. Other than NaN, *a ubound result is always a connected set.*

The prototype uses the ⊙ symbol here to mean unum division. (For some reason, *Mathematica* does not have a symbol that looks like a "/" inside a circle.) Again we create tables for left and right cases, but notice the peculiarity that the endpoints of the two input values face *opposite directions*. Just as negation reverses inequality, so if $x < y$ then $-y > -x$, reciprocation also changes order: If $x < y$, then $\frac{1}{x} > \frac{1}{y}$ for positive x and y. Here are the "⊙ **left**" and "⊙ **right**" tables for computing the left and right bounds of a quotient in the *g*-layer.

⊙ left	0]	y)	y]	∞)	∞]
[0	(NaN	[0	[0	[0	[0
(0	(NaN	(0	(0	(0	[0
[x	(NaN	(x/y	[x/y	(0	[0
(x	(NaN	(x/y	(x/y	(0	[0
[∞	(NaN	[∞	[∞	[∞	(NaN

⊙ right	[0	(0	[y	(y	[∞
0]	NaN)	0]	0]	0]	0]
x)	NaN)	∞)	x/y)	x/y)	0]
x]	NaN)	∞)	x/y]	x/y)	0]
∞)	NaN)	∞)	∞)	∞)	0]
∞]	NaN)	∞]	∞]	∞]	(NaN

As usual, thought-provoking cases are indicated in magenta. The open or closed endpoint types are colored [ ] ( ) whenever they are not simply the "OR" of the open state of the two inputs, just as they have been in previous tables. And just as with the previous tables, if the result of the arithmetic operation on an exact **x** and an exact **y** cannot be expressed as an exact unum, the conversion of the *g*-layer result to the *u*-layer has its ubit set to indicate an inexact range.

Here is a divide operation, $[3, 4) \div [2, 3)$. The largest value is $4 \div 2$, the smallest value is $3 \div 3$, and both extremes involve at least one open endpoint so both result endpoints will also be open.

$$\texttt{view}\left[\texttt{g2u}\left[\begin{matrix} 3 & 4 \\ \textbf{closed} & \textbf{open} \end{matrix}\right] \odot \texttt{g2u}\left[\begin{matrix} 2 & 3 \\ \textbf{closed} & \textbf{open} \end{matrix}\right]\right]$$

$(1, 2)$

The prototype code for *g*-layer and *u*-layer division (`divideg` and `divideu`) is in Appendix C.11.

> **Exercise for the reader**: In a {2, 0} environment, there are 63 distinct exact unum values. If unums u and v are exact, in **approximately** what fraction of the cases will $u \div v$ also be exact? (*Hint*: The answer is similar for a {3, 0} and a {4, 0} environment, and the closure diagram for subtraction in Section 2.2 is part of the answer.)

10.3.2 Hardware for unum division

The elementary school approach to long division can eke out only one digit at a time, working from left to right. Computers using the binary form of long division (repeated subtracts and shifts until a remainder is determined) will always be much slower at divides than multiplies, since much of the work in a multiply can easily be done *in parallel*; each partial product in the parallelogram can be created at the same time, and then summed using circuit tricks like "carry look-ahead."

A variety of tricks exist to make division faster, and the reader is encouraged to check out the literature on those tricks if interested. There is very little difference between how IEEE floats compute a divide and how unum hardware does it, since both are required to know whether the result is exact or not. (The IEEE Standard says to set the "inexact" processor flag instead of incorporating the information into the number the way unums do.) As with multiplication, additional bits are needed in the scratchpad, but the number is predictable based on the maximum number of bits in the fractions of the two inputs.

Pre-decoded exception bits in the unpacked unum, and the lack of a need to round (if the remainder is zero, result is exact; otherwise, set the ubit) help reduce the cost of a unum divide compared to a float divide of equivalent precision. The precision of the unum will usually be lower than that of the float, since it dynamically adjusts up and down as needed. Only actual design experiments will determine if the *on-chip* effect is to reduce the amount of power needed for divide operations compared to floats, since most ubound operations will require a pair of divides to the nearest ULP, one for each endpoint, instead of just one divide.

It is tempting to think about a more capable scratchpad format that stores not powers of 2 for scaling, but unsigned integer numerators p and denominators q, both stored using extended precision:

	p_L negative?	p_R negative?
	p_L	p_R
NaN?	q_L	q_R
	open?	open?
	infinite?	infinite?

This is a more expensive internal scratchpad, but superior at dealing with division operations since it supports exact rational arithmetic. Since the above scratchpad looks like a big step backward for energy efficiency, we will stick with signed integers scaled by powers of 2 in the g-layer structure. It would be interesting to experiment with both approaches using a field-programmable gate array (FPGA); there may be cases where finishing a division in less time would save enough energy to make it worthwhile.

This completes the definition of addition, subtraction, multiplication, and division as individual operations. We need more than just the operations of a four-function calculator, however. The next two chapters will show unum versions of most of the functions you expect to find on a scientific calculator.

11 Powers

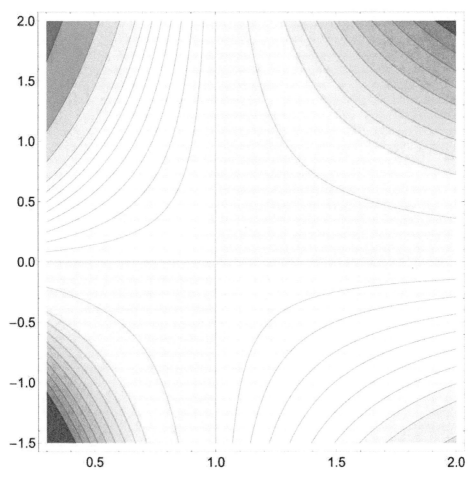

Section 10.1 showed a three-dimensional visualization of the saddle-shaped surface for multiplication, $z = x \times y$. A similar saddle-shape occurs in the plot for $z = x^y$, shown in a different way above, with a contour plot. The red corners are increasing values of the general power function x^y and the blue corners are decreasing values. Computing the function means navigating a minefield of exception cases, as well as facing a previously unsolved math conundrum.

11.1 Square

Compared to the four basic arithmetic operations, finding x^y for general values of x and y is a big jump up in effort. We start this chapter with a *really* soft toss, the creation of a routine for x^2, just because squaring is such a common operation. Even that simple task provides a taste of a challenge that will be discussed in more detail in Part 2. Why not just evaluate the square of x by multiplying x times itself?

Suppose *u* is the unum representing $[-1, 2)$. Here is why we should **not** simply use $u \otimes u$ to find the square of *u*. The square *should* be $[0, 4]$, but a mindless multiplication lacks the information that the two values *are the same variable*, and produces the looser (albeit correct) bound $(-2, 4]$:

$$\text{Module}\left[\left\{u = g2u\left[\begin{array}{cc} -1 & 2 \\ \text{closed} & \text{open} \end{array}\right]\right\}, \text{view}[u \otimes u]\right]$$

$(-2, 4)$

Squaring a number should never produce any negative numbers, but that is what the multiply routine will produce since it incorrectly assumes its input values are *independent*. So the code for squaring needs to watch for ranges that straddle zero. In a well-made unum environment, writing "$x \otimes x$" should trigger an error and a question to the user, "Did you mean to square *x*?"

It is also still possible to get weird ties, as with the multiplication of general intervals: The square of $(-2, 2]$ is $[0, 4]$, not $[0, 4)$. Closed endpoints "win" over open ones.

In the prototype, the *g*-layer function to square a general interval *g* is `squareg[g]`, and the *u*-layer version is `squareu[u]`. The code for both is shown in Appendix C.14. Here is a quick test that shows `squareg` does the right thing:

$$\text{view}\left[\text{squareg}\left[\begin{array}{cc} -4 & 4 \\ \text{open} & \text{open} \end{array}\right]\right]$$

$[0, 16)$

The *u*-layer version is `squareu[u]` where *u* is a unum or ubound. It tallies the numbers moved and bits moved; for unary operations (operations with one input), the count of numbers moved increases by two (one in, one out) and the cost of bits moved is the sum of the bits in the input and the output.

$$\text{view}\left[\text{squareu}\left[g2u\left[\begin{array}{cc} -2 & 3 \\ \text{open} & \text{closed} \end{array}\right]\right]\right]$$

$[0, 9]$

> **Exercise for the reader**: What is the algorithm for finding the *cube* of a general interval?

The square function works on all real numbers and infinities. The hardware demands are no greater than that for multiplication. If we do higher integer powers of *x* like x^3, x^4 and so on, we will often need the extended-precision multiplier, or even the *really* big multiplier. Before we discuss that, consider another very common power function: $x^{1/2}$, the square root of *x*.

11.2 Square root

Square roots that are not exact are irrational numbers; they cannot be expressed as a ratio of two integers $\frac{p}{q}$. They are not, however, *transcendental* numbers, which may sound like a distinction that only mathematicians would ever care about. Square roots are *algebraic* numbers, defined as numbers that are roots of algebraic equations (polynomials with rational coefficients). Algebraic numbers are easy to evaluate to within one ULP, because there are plenty of fast ways to find the root of an algebraic equation. For the square root, there are techniques that double the number of correct digits with each iteration. It is really no harder to perform a square root in hardware than to compute the reciprocal of a number. That is why non-mathematicians might care about the difference between algebraic numbers and transcendental numbers: **Speed**.

The square root function will return NaN if used on a negative number, since we are restricting computations to real numbers. However, the square root allows an exact "negative zero" to work, instead of declaring the result imaginary, since the sign bit is ignored if the value is exactly zero. An *inexact* negative zero would always produce an imaginary result, though, because it includes numbers strictly less than zero. Hence, we have to return NaN for those cases. The unpacked unum format makes it simple and quick to check for values that trigger exceptions.

Incidentally, the IEEE Standard specifies that the square root of negative zero should be... *negative zero*. What in the world were they thinking?? It was Berkeley, and it was the 1980s, so perhaps some controlled substances were involved. Unum math ignores the sign of zero when it is an exact float.

In the prototype, the square root function in the *g*-layer is **sqrtg[g]** where *g* is a general interval. The square root in the *u*-layer is **sqrtu[u]**, where *u* is a ubound. Try it on the square root of $(1, \frac{25}{16}]$. The open end of the interval remains open, and because the closed end has an exact square root, it remains closed. As usual, the endpoint value of 1 is made open by setting its ubit (by adding **ubitmask** to the unum bit string.) The square root of $\left(1, \frac{25}{16}\right]$ should be $\left(1, \frac{5}{4}\right]$:

$$\texttt{view}\left[\texttt{sqrtu}\left[\left\{\hat{1} + \texttt{ubitmask}, \frac{\widehat{25}}{16}\right\}\right]\right]$$

(1, 1.25]

Square roots on general intervals are easy because the function is monotone increasing. That means simply evaluating the square root of each endpoint like the example above; it preserves the less-than, greater-than ordering. That property did not hold for the square function, which caused complications.

Most square roots are numbers not expressible as exact floats, so just as with every other operation, we create an open interval that *contains* the true value. If an endpoint is closed and the square root is exact, the endpoint remains closed. Otherwise, ubound arithmetic finds the smallest possible ULP that contains the square root, and uses the unum for that to contain the correct value at that endpoint.

In the following example, we display the underlying unum machinery to show how the left endpoint is exact and compact (three bits in addition to the utag), whereas the $\sqrt{3}$ endpoint demands all the fraction bits available and is marked inexact.

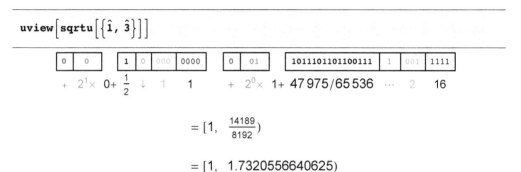

$$= [1, \ \frac{14189}{8192})$$

$$= [1, \ 1.7320556640625)$$

The square root of three is $1.7320508\cdots$ so the upper bound is accurate to over five decimal places. If we square this result, we will get a range slightly larger than the $[1, 3]$ range that we started with.

$$\texttt{view}\Big[\texttt{squareu}\big[\texttt{sqrtu}\big[\{\hat{1}, \hat{3}\}\big]\big]\Big]$$

$[1, 3.000030517578125)$

Notice that the upper bound of the result is open. The correct answer $[1, 3]$ is contained within the ubound result.

> **Exercise for the reader**: The non-negative exact values of a $\{2, 2\}$ environment are listed in Section 4.10; how many of them have exact square roots?

Except for exception values, the endpoints represented by ubounds are like floats, a number of the form $f\,2^e$ where f is a fraction in the interval $[1, 2)$. Notice that we can always make e an even number by allowing f to be in the interval $[1, 4)$ instead. So if the hardware sees that the exponent is an even integer, it evaluates $\sqrt{f}\,2^{e/2}$; if the exponent is odd, it instead evaluates $\sqrt{2f}\,2^{(e-1)/2}$, so the exponent is very easy to compute and is always representable. A unum square root operation differs from that of a float only in that unums automatically minimize the number of bits used to represent the exponent .

11.3 Nested square roots and "ULP straddling"

This may seem like a strange direction to take the discussion of x^y, but there is a reason why we might want to find powers where $y = \frac{1}{4}, \frac{1}{8}, \frac{1}{16} \ldots$ as the next step to making the x^y function general.

Since square root is such a well-behaved operation, can we use it to find $x^{1/4} = \sqrt{\sqrt{x}}$, either as an exact number or the smallest ULP that contains it? The answer is yes, *if* you watch out for "ULP straddling" and correct it when it occurs. The first square root can produce an inexact unum, of course. The second square root then treats each endpoint separately, and the square roots of *those* might be different inexact unums, because the interval answer *straddles* an exact value instead of lying in the open interval between adjacent unums. Information loss could accumulate with each square root.

Fortunately, the problem is easily dispensed with, since *squaring* a number can be done exactly in the scratchpad. Here is a case where that happens in a {3, 3} environment when we try to find $266^{1/4}$ by taking its square root twice:

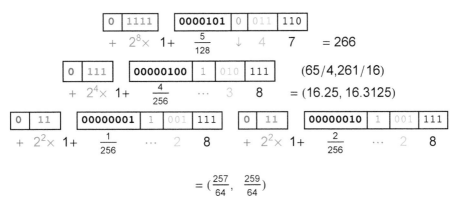

$$= (4.015625,\ 4.046875)$$

The result is two ULPs wide because the result of the second square root straddles the exact value $\frac{258}{64}$. But **only one of them can be correct**. So raise the endpoints and the straddled point to the 4th power by squaring them twice to check (actually, we only need to do this for the midpoint, but we show the 4th power of all three values here for purposes of illustration):

$$\left(\frac{257}{64}\right)^{2^2} = 260.023 \cdots \qquad \left(\frac{258}{64}\right)^{2^2} = 264.094 \cdots \qquad \left(\frac{259}{64}\right)^{2^2} = 268.212 \cdots$$

Aha. Discard the lower ULP range, the one that goes from $\frac{257}{64}$ to $\frac{258}{64}$, since its fourth power does not contain 266. The answer is the unum that represents $\left(\frac{258}{64}, \frac{259}{64}\right)$.

This technique is easily extended to nested square roots:

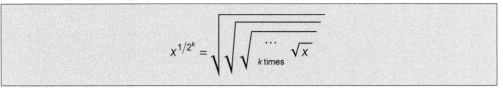

$$x^{1/2^k} = \underbrace{\sqrt{\sqrt{\sqrt{\cdots \sqrt{x}}}}}_{k \text{ times}}$$

Whenever a square root straddles an exact value and thus is two ULPs wide, just square the two ULPs repeatedly to see which one is the correct one, and continue. So our repertoire for finding x^y is now "whenever y is an integer power of 2." For positive powers of 2, do nested *squaring*. For negative powers of 2, do nested *square roots* with pruning of "bad ULPs" along the way. If the exponent is negative, simply compute with its absolute value and take the reciprocal at the end.

11.4 Taxing the scratchpad: Integers to integer powers

There is a reason it is called the *power* function. Just because an expression is easy to jot down does not mean it should be easy to compute. The factorial function $n! = 1 \times 2 \times \ldots \times n$ is a good example of this; even for a number as low as $n = 70$, $n!$ is bigger than a googol. A googol is also easy to jot down with only five characters: 10^{100}. The power function is in a whole different category from $+ - \times \div$; it easily generates a gigantic amount of work to compute to the maximum precision. How big does the scratchpad have to be?

It is a large number of bits, but not as bad as it first looks because the separate storing of scale factors (exponents) means we only deal with integers expressed by the fraction bits. To get this down to numbers small enough to grasp, suppose the unum environment is {2, 2}, so there are at most $2^2 = 4$ bits to hold the fraction. With the hidden bit, the fraction represents integers 0 to 31, scaled by powers of 2. The scratchpad would need to hold integers as large as 31^{31} if we need the integer to be expressed exactly. That may look scary, but it only requires 155 bits at most. Expressing 31 takes five bits, so 31^2 takes ten bits, 31^3 takes 15 bits, and so on, therefore 31^{31} takes at most $5 \times 31 = 155$ bits.

If we built a unum environment with direct hardware support up to an environment of {4, 5}, the maximum fraction length plus the hidden bit is 33 bits long. The integer represented by the fraction bits ranges from 0 to $2^{33} - 1$, or 0 to $8\,589\,934\,591$. The largest integer power would take (at most) this many bits to store:

$$33 \times 8\,589\,934\,591 = 283\,467\,841\,503 \text{ bits.}$$

In more common terms, about 33 gigabytes. In 2014, that much storage lies between the main memory size of a portable computer and that of a server.

So it *is* a big number, but such hardware exists. A current-generation microprocessor takes less than one second to read or write that much external data, quick enough by human standards but a billion times slower than an add or multiply. Power functions x^y for general x and y are not operations we should demand casually if we want the same absolute standard of accuracy as the four arithmetic operations.

Now that we know what we are up against, try computing x^y for a small environment.

11.5 A practice calculation of x^y at low precision

We can express numbers $x = 17$ and $y = \frac{3}{8}$ exactly in a $\{2, 2\}$ environment. Try computing $x^y = 17^{3/8}$, by using $17^{3/8} = \left(17^3\right)^{1/8}$. The cube of 17 takes a couple of scratchpad multiplications, and then raising that to the $\frac{1}{8}$ power requires nested square roots as described in Section 12.4.

The number 17 occupies five bits, so $17^3 = 4913$ can require no more than fifteen bits in the scratchpad. This is a small enough number of bits that we can actually look at the binary string for 4913:

```
setenv[{2, 2}]
IntegerString[17³, 2]
```

1001100110001

With square roots, we needed the exponent to be an even number. To take the 8th root of a scale factor 2^e, we need e to be an integer multiple of 8. So slide the bit string for 4913 eight places to the right:

$$1\,001\,100\,110\,001 = 10\,011.00110001 \times 2^8$$

Of course, the eighth root of 2^8 is simply 2^1. So to find $17^{3/8}$, we seek an ULP-accurate value for $(10\,011.00110001)^{1/8} \times 2^1$. As a unum, the binary value $10011.\cdots$ is

0	111		0011	1	10	11

$$+ \quad 2^4 \times \quad 1+ \quad \frac{3}{16} \quad \cdots \quad 3 \quad 4 \quad = (19, 20)$$

Now to perform three successive square root operations, checking each one for a straddle and pruning if necessary. (The notation "**2^^**" followed by zeros and ones allows direct entry of a bit string. They are color-coded with the usual system to make it easier to see the unum bit fields.)

```
uview[sqrtu[2^^0111001111011]]
```

0	11		0001	1	01	11

$$+ \quad 2^2 \times \quad 1+ \quad \frac{1}{16} \quad \cdots \quad 2 \quad 4 \quad$$

$$\left(\frac{17}{4}, \frac{9}{2}\right)$$

$$= (4.25, 4.5)$$

That one landed inside an ULP, so no pruning is needed. The square root operation shrinks the *relative width* of a value by about half, so about half the time the new range does not straddle two ULPs. Perform the second square root:

`uview[sqrtu[2^^011000110111]]`

0	1		0000	1	00	11		$(2, \frac{17}{8})$
$+$	$2^1 \times$	$1+$	$\frac{0}{16}$	$\cdots$	1	4		$= (2, 2.125)$

Notice that the exponent field keeps shrinking. Again, it landed inside a smallest-possible ULP; for the final square root we will not be so lucky.

`uview[sqrtu[2^^010000100011]]`

0	01		011	1	01	10		$(\frac{11}{8}, \frac{3}{2})$
$+$	$2^0 \times$	$1+$	$\frac{3}{8}$	$\cdots$	2	3		$= (1.375, 1.5)$

That straddles an exact value, since the fraction has less than four bits. There are two quick ways to know that a straddle has occurred: The fraction is not at its maximum length, or the ubound for the answer is a unum pair instead of a single unum. This is a case of the former, since the fraction field is only using three of its four bits maximum.

The computed range (1.375, 1.5) straddles the following exact unum:

0	01		0111	0	01	11		$\frac{23}{16}$
$+$	$2^0 \times$	$1+$	$\frac{7}{16}$	$\downarrow$	2	4		$= 1.4375$

Find the 8th power of the straddled exact value by repeated squaring. If it is greater than the original number `10011.00110001` then we prune away the larger of the two ULPs; otherwise we prune the smaller one. Again for illustration, here are all three values raised to the 8th power:

$$\left\{ 1.375^{2^{2^2}}, \ 1.4375^{2^{2^2}}, \ 1.5^{2^{2^2}} \right\}$$

{12.7768, 18.2332, 25.6289}

The range {12.77$\cdots$, 18.23$\cdots$} does not contain `10011.00110001` $= 19.19140625$, so it is pruned away.

Exercise for the reader: If a straddle occurs *every time* when taking the 2^k root of a number using k nested square roots, what is the total number of multiplications required for the pruning of the spurious ULP-wide ranges?

Finally, scale the final ULP-wide unum by the exponent, 2^1, to get $17^{3/8}$ to the maximum available precision (all four bits of the fraction field in use):

0	1		**0111**	1	00	11	$(\frac{23}{8}, 3)$

$$+ \ 2^1 \times \ 1+ \ \frac{7}{16} \quad \cdots \quad 1 \quad 4 \qquad = (2.875, 3)$$

In studying the problem of computing the power function, the author came across this extraordinary and oft-quoted statement in the literature on the subject:

> "Nobody knows how much it would cost to compute y^w correctly rounded for every two floating-point arguments at which it does not over/underflow... No general way exists to predict how many extra digits will have to be carried to compute a transcendental expression and round it correctly to some preassigned number of digits. Even the fact (if true) that a finite number of extra digits will ultimately suffice may be a deep theorem."
> —*William Kahan*

Really??

The first part of that quotation is certainly no longer true, for anyone who has read this far. We just showed the maximum cost and the maximum time needed to compute a power function for two floats. The high cost of doing it that way makes us anxious to find a better way, but it *is* a predictable finite cost and time, guaranteed. It is just as deterministic a method as the ones for multiplying and adding bit strings.

The quotation is misleading, probably through omission. The "..." part of this quotation probably left out an explanation that y^w **is not a transcendental function if y and w are floats**, because then the problem is *algebraic,* and Kahan knows that. The second part of the above quote remains true, and practical approaches for computing the transcendental function e^x will be discussed in Section 11.7.

11.6 Practical considerations and the actual working routine

11.6.1 Why the power function can be fast

At this point, some readers may be thinking: *Is this guy insane??* Potentially computing hundreds of millions of digits exactly, just to guarantee the correctness of x^y? That is not what is proposed here. The argument about the maximum number of digits and the maximum number of computational steps was to prove that there *is* a finite upper bound to both that is known in advance. That is not a practical way to compute x^y, but something more like an "existence proof."

In practice, we are willing to tolerate two things that can make the job of finding x^y *much* easier:

- We do *not* require that the routine always takes the same amount of execution time for any two input values.

- We do not require provably minimum ULP width for every result.

Almost every operation with unums involves variable amounts of execution time; it would be ridiculous to find the maximum time, say, for a multiplication of any precision, and insist that the processor always allow that many clock cycles to pass even when multiplying a finite number by zero. Every operation should be allowed to go as fast as possible.

As for requiring "provably minimum ULP width," there are efficient methods for computing x^y that seem to do just that, but *proving* that they always are so accurate is tough. Unums offer the perfect solution to this, however, because they *contain the accuracy information*. If there is a very rare case where the calculation of x^y is two ULPs wide instead of one, *the ubound will say so.* There is no possibility of being misled by the result, since the unum format allows communication of the width of a rigorous bound for the answer.

11.6.2 The prototype power function

The prototype approaches the power function similarly to the approach for multiplication. The contour plot that began the chapter shows a very similar need to do "left facing" and "right facing" endpoint calculations for a saddle-shaped surface. In this case, the flat point at the center of the saddle is not at the origin, but at $x = 1$, $y = 0$. The two tables below define what to do for general interval endpoints in the *x-y* plane where $x \geq 1$, $y \geq 0$.

powg[x,y] left	[0	(0	[y	(y	[∞
[1	[1	[1	[1	[1	(NaN
(1	[1	(1	(1	(1	[∞
[x	[1	(1	[x^y	(x^y	[∞
(x	[1	(1	(x^y	(x^y	[∞
[∞	(NaN	[∞	[∞	[∞	[∞

powg[x,y] right	0]	y)	y]	∞)	∞]
1]	1]	1]	1]	1]	NaN)
x)	1]	x^y)	x^y)	∞)	∞]
x]	1]	x^y)	x^y]	∞)	∞]
∞)	1]	∞)	∞)	∞)	∞]
∞]	NaN)	∞]	∞]	∞]	∞]

If $0 \le x < 1$, we compute $1/(1/x)^y$ instead, reversing the roles of left and right endpoints. If x or y is negative, the result is often a NaN (a complex number instead of a real) unless y is an integer. Here are some of the weird cases that can happen when computing the power function:

- $0^{-2} = \infty$. It looks like it should be a NaN since we divide by zero, but for *even* negative integer powers of zero, the result is $(\pm\infty)^2 = \infty$.

- $1^{(maxreal, \infty)} = 1$ but $1^\infty = \text{NaN}$.

- $(-\infty)^\infty = \text{NaN}$ but $(-\infty)^{-\infty} = 0$.

> **Exercise for the reader**: In the *g*-layer, what is the correct value for $\left(\frac{1}{2}, 2\right]^{[-1,1)}$?
>
> (The contour plot at the beginning of the chapter may help answer this.)

Here is an example of the function **powg** evaluating $\left[\frac{81}{256}, \frac{625}{256}\right)^{0.75}$ in the *g*-layer:

$$
\texttt{view}\left[\texttt{powg}\left[\begin{array}{cc} \frac{81}{256} & \frac{625}{256} \\ \textbf{closed} & \textbf{open} \end{array}, \texttt{u2g[0.}\hat{7}\texttt{5]}\right]\right]
$$

[0.421875, 1.953125)

The *u*-layer power function for u^v is **powu[u, v]**, and it tracks bits and numbers moved. The code for **powg** and **powu** is in Appendix C.15, and it probably has the record for complexity of all the operations in the prototype unum environment, because of all the special cases.

11.6.3 A challenge to the defenders of floats

Here is a challenge problem for people who think they have a slick way to find x^y within 0.5 ULP using floats: Compute

$$5.9604644775390625^{0.875}$$

Based on the way most math library routines work, the result with IEEE double precision will probably be something close to

$$4.76837158203125$$

and the library routine will round the result to that 15-decimal number. It will also set the "rounded" bit in the processor as required by IEEE rules, but guess what? It is incorrect to do so! **The result is *exact*,** something conventional methods of computing x^y have no way to detect.

We only need a {3, 5} environment to get this result with the prototype:

```
setenv[{3, 5}]
uview[powu[5.9604644̂775390625, 0.8̂75]]
```

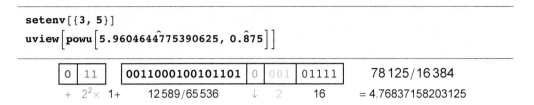

0	11		0011000100101101	0	001	01111	78 125/16 384
+	$2^2\times$	1+	12 589/65 536	↓	2	16	= 4.76837158203125

Interval methods are often criticized for their tendency to produce overly conservative statements about a result. In this case, it is floats that deserve that criticism. There is a joke from the last century about someone asking an engineer "What is two times three?" He pulls out his slide rule and fiddles with it for a few seconds, then announces, "It's about... six point zero." If values can be posed exactly in the *u*-layer and an elementary function of them is also exactly expressible, it should be considered an *error* if the exact result is mislabeled as "approximate."

Exercise for the reader: Construct another example like the one above where an exact exponent *y* has at least three nonzero digits to the right of the decimal and the base *x* has at least ten decimals when expressed exactly, and x^y is also exact.

11.7 Exp(x) and a solution to "the Table-Maker's Dilemma"

11.7.1 Another dilemma caused by the dishonesty of rounding

The exponential function is e^x, also written exp(x), where e is the base of the natural logarithm, 2.718281828 ⋯. Unlike x^y where both *x* and *y* are floats, the exponential function is transcendental and cannot be computed as the root of a polynomial equation with rational coefficients. Fortunately, there are fast ways to evaluate it at any precision; the algorithms produce a tightening bound on the digit expansion until it finally falls within a minimum-sized ULP. However, the number of iterations needed could be anything, since it could just happen to be an irrational number that is *very* close to an exact value, forcing the evaluation to "go into overtime" to break the tie.

Here is an example of a transcendental calculation that shows this type of behavior in decimal form: $e^{\pi\sqrt{163}}$ looks an *awful* lot like an integer. If evaluated to thirty decimals accuracy, it is

$$e^{\pi\sqrt{163}} = 262\,537\,412\,640\,768\,743.999999999999 \cdots$$

If the uncertainty in the last place is even as small as 10^{-12}, it is unknown at that point whether the correct ULP to land the result in is *just above* the integer or *just below* it. Early makers of tables of logarithms first noticed that rounding for a few entries was extraordinarily hard to determine.

Professor Kahan coined the phrase "The Table-Maker's Dilemma" for this phenomenon. It stems from the desire to assure the user of a table, "I promise you that none of the incorrect entries in this table are off by more than half an ULP." The Table-Maker's Dilemma is a direct consequence of the "original sin" of computer arithmetic that float users are asked to commit (Section 3.1).

> **Unums completely solve The Table-Maker's Dilemma.** Unums do not worry about "correctly rounding" a result because they *do not round*. If a value is not known to the smallest possible ULP size, it is simply returned with a larger ULP size (or a ubound) guaranteed to hold the value; unums *label the size of the bound* instead of returning an erroneous value, admitting what they do not know.

That said, there is certainly an easy way to make every single value of a function like e^x land in the smallest possible ULP size. Either tolerate variable-length calculations with the (very rare) occasional long evaluation time, as Thomas Lynch's high-radix online arithmetic does, or simply identify in advance which floats are prone to the problem and handle those with a table.

Suppose the scratchpad computes a transcendental function with twice as many bits as there are in the fraction; the number of cases that happen to straddle two ULPs in the *u*-layer will be, statistically, about *one* at that precision. It is like keeping track of the "bad spots" on a hard disk drive. Find out where they are and work around them, making a small table part of the library function that has the handful of difficult tie cases precomputed so they do not have to be decided at run time.

11.7.2 The prototype exponential function

The code listings for the *g*-layer version **expg** and *u*-layer version **expu** are in Appendix C.15. Because it is a monotone increasing function, all the ubound logic has to do is evaluate the function at each boundary endpoint and track the open-closed properties. Here is an example of evaluating $e^{(-\infty,0]}$:

$$\texttt{view}\left[\texttt{expu}\left[\texttt{g2u}\left[\begin{matrix} -\infty & 0 \\ \texttt{open} & \texttt{closed} \end{matrix}\right]\right]\right]$$

$(0, 1]$

There are times that people get into the habit of using e^x when in some cases they could have used an exactly representable number for the base instead, like 10^x or 2^x. Those are available in the prototype by using $\textbf{powu}\left[\hat{10}, u\right]$ or $\textbf{powu}\left[\hat{2}, u\right]$, for example. Doing so has the advantage that they have far more cases where the answer is *exact*. With the transcendental number e as the base, the only cases for which e^x is exact are when x is $-\infty$, 0, or ∞.

Exercise for the reader: If we put a limit of at most five bits to the left of the utag in a {2, 2} environment, there are 66 nonnegative possible unum values:

$$0, \frac{1}{128}, \frac{1}{64}, \frac{3}{128}, \frac{1}{32}, \frac{3}{64}, \frac{1}{16}, \frac{3}{32}, \frac{1}{8}, \frac{3}{16}, \frac{1}{4}, \frac{5}{16}, \frac{3}{8}, \frac{7}{16}, \frac{1}{2}, \frac{5}{8}, \frac{3}{4}, \frac{7}{8}, 1, \frac{9}{8}, \frac{5}{4}, \frac{11}{8}, \frac{3}{2},$$

$$\frac{13}{8}, \frac{7}{4}, \frac{15}{8}, 2, \frac{17}{8}, \frac{9}{4}, \frac{19}{8}, \frac{5}{2}, \frac{21}{8}, \frac{11}{4}, \frac{23}{8}, 3, \frac{25}{8}, \frac{13}{4}, \frac{27}{8}, \frac{7}{2}, \frac{29}{8}, \frac{15}{4}, \frac{31}{8}, 4, \frac{9}{2}, 5, \frac{11}{2},$$

$$6, \frac{13}{2}, 7, \frac{15}{2}, 8, 10, 12, 14, 16, 20, 24, 28, 32, 48, 64, 96, 128, 192, 256, 384$$

If we include the negative versions of those, that makes 131 distinct exact unums. For which of those will 2^x also be exact and in the above set? (*Hint:* It happens for more than ten percent of the values!)

12 Other important unary operations

12.1 Scope of the prototype

To perform tests of the merits of unums, we will need the sort of functions you find on the buttons of scientific calculators. One category of functions *not* in the prototype is inverse trigonometry functions like arctangent. Those are left as an exercise for the reader to develop; they are not needed for any of the experiments in this book.

"Unary" operations are operations that take just one input value and produce one output value. Unlike the basic $+ - \times \div$ operations and the x^y function, unary operations increment the **ubitsmoved** tally by just the bit counts in the one input and one output ubound, and increment **numbersmoved** by 2 instead of 3.

The section on trigonometric functions contains a somewhat radical proposal, one that neatly solves the problem of argument reduction mentioned in Section 5.4.2 but may take some getting used to.

12.2 Absolute value

Usually, the absolute value function is the most trivial task imaginable: Set the sign bit to zero.

With unums and ubounds, absolute value has a few things to watch out for. We have to remember to reverse the order of the interval endpoints if both are less than zero. If zero is inside the interval, then a closed zero is the lower bound, and the maximum of the endpoint absolute values is the right-hand endpoint. The devil is again in the details, and the details are in Appendix C.16.

Notice that when negating reverses the left-right positions of the values, they carry their open or closed state with them. So, the absolute value of the ubound that represents [−3, −2) should be (2, 3] :

```
view[absu[{-̂3, -̂2 + ubitmask}]]
```

(2, 3]

Even though the operation seems trivial, **absu** updates the **ubitsmoved** and **numbersmoved** tallies. When the endpoints switch places, it *is* necessary to move numbers and bits, which takes energy. It also may be unnecessary to have a *g*-layer version of something that can clearly be done in the *u*-layer without any scratchpad facilities, but it is included in the prototype for completeness.

12.3 Natural logarithm, and a mention of log base 2

The inverse operation for e^x, the natural logarithm, has three special exact values. The logarithms of 0, 1, and ∞ are exactly −∞, 0, and ∞, respectively. Of course we return NaN for any negative real argument, since the logarithm is complex-valued there. The code for **logg** and **logu** is in Appendix C.16. Here is an example of finding the log of the range [0, 1], a result with peculiarly exact bounds:

$$\text{view}\left[\text{logu}\left[\left\{\hat{0},\ \hat{1}\right\}\right]\right]$$

[−∞, 0]

As with the exponential function, we can get many more exact results from logarithm functions if we use 2 instead of e as the base. There is a recurring theme here: Why use functions that are inexact almost everywhere when there are similar functions that have the promise of exact answers for some exact inputs? If that philosophy is applied to trigonometric functions, there are a number of computational advantages.

12.4 Trig functions: Ending the madness by degrees

When students first learn trigonometry, the subject is usually introduced with angles measured in degrees. Everyone knows angles like "180 degrees" and "45 degrees" even if they have never taken trigonometry. Later, courses at the calculus level and above teach that *radians* are a better way to measure angle because formulas for derivatives and differential equations look simpler that way. Because of the clout of higher math, all the standard language libraries use radians as the arguments for sine, cosine, tangent, and so on. If you want to compute the sine of 90°, you have to get as close to the radian equivalent as you possibly can, $\frac{\pi}{2}$, within the vocabulary of floats. Imagine asking for the *tangent* of 90°; using the float that is a little below $\frac{\pi}{2}$ produces an enormous positive number, but being the one a little above $\frac{\pi}{2}$ produces an enormously *negative* number.

The use of radians discards useful opportunities to express exact answers to exact questions. **So why not define the trigonometry functions with degrees for the arguments instead of radians?** If you ask for the sine of 30 degrees, you get *exactly* $\frac{1}{2}$. The tangent of 135 degrees is *exactly* −1. The unum function for tangent knows to return NaN as the tangent of an exact 90° angle.

Once you commit to using radians, you discard any hope of landing on such a special angle value again using numbers you can represent as floats or unums. Computers are like humans in this respect: They have an easier time working with degrees than with radians!

The benefit of using degrees is bigger than this, however: It solves the problem of *argument reduction*. Recall the example in Section 5.4.2 where evaluating a trig function of a large number required "unrolling" its value with a huge number of bits to store π to sufficient accuracy. What if we simply had to know the remainder of dividing a big float by the integer 180? That does not take much work at all, and certainly it does not require storage of hundreds of bits of the value of the transcendental number π. If anyone makes the goofy request that a computer evaluate the cosine of 10^{100} treated as an exact number, in *degrees* the argument 10^{100} is the same as ten degrees, exactly.

The prototype includes several trig functions, though not a complete set:

Function	*g*-layer version	*u*-layer version
cosine	**cosg**[*g*]	**cosu**[*u*]
sine	**sing**[*g*]	**sinu**[*u*]
tangent	**tang**[*g*]	**tanu**[*u*]
cotangent	**cotg**[*g*]	**cotu**[*u*]

The approach for interval bounds is to reduce the lower bound argument to a manageable range near zero, then reduce the upper bound argument to something no more than two periods greater than the lower bound. This catches cases where the function has a minimum or maximum somewhere within the input interval.

For example, imagine going 2/3 of the way around the unit circle, from 60° to 300°, with the angular interval open at both ends:

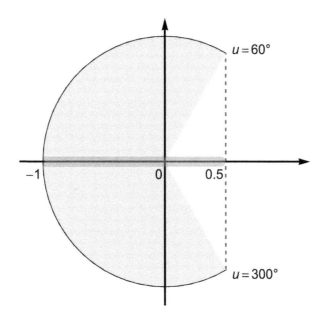

The cosine is the horizontal projection of that angle interval, which ranges from −1 to $\cos(60°) = \frac{1}{2}$ exactly, but the function value is open on the right (the blue bar is rounded on the right, to show this).

$$\texttt{view}\Big[\texttt{cosu}\big[\big\{\hat{60} + \texttt{ubitmask},\ \hat{300} - \texttt{ubitmask}\big\}\big]\Big]$$

[−1, 0.5)

The sine is the vertical projection of that angle interval, and while sin(60°) and sin(300°) are not expressible as exact unum values, that does not matter because the overall bound is the entire range of the unit circle, a closed interval:

$$\texttt{view}\Big[\texttt{sinu}\big[\big\{\hat{60} + \texttt{ubitmask},\ \hat{300} - \texttt{ubitmask}\big\}\big]\Big]$$

[−1, 1]

With the basic operations defined, we are ready to start experimenting with test problems to compare unums and floats. However, there is one bit of unfinished business: Definition of operations that may be *fused*, that is, computed in the *g*-layer such that the value returned to the *u*-layer is at the highest accuracy possible for the *u*-layer environment. This is what assures portability (bit-identical results across computer systems) while raising the accuracy. It is the opposite of the approach used for Java, where bit-identical results were assured by dumbing down the operations to match that of the least-accurate system. One consequence of having well-defined fused operations is that it can greatly diminish the need for large fraction and exponent fields in the numerical representation.

13 Fused operations (single-use expressions)

Standard library routines for performing complex arithmetic are crippled by a fear of "falling off the edge of the world," since straightforward arithmetic can overflow or underflow in the scratch calculations. These routines typically disassemble the float inputs into exponent and fraction parts to avoid this problem, which destroys most of the speed advantage of having hardware for float operations. This chapter presents a better way to avoid "falling off the edge" of the set of representable real numbers, by declaring *fused operations* and making their use explicit.

13.1 Standardizing a set of fused operations

This chapter is about the possibility of establishing standard fused expressions where the input values are only used *once*, or "Single-Use Expressions." Part 2 will explore the perfect calculation of fused expressions like polynomials or $x + \frac{1}{x}$ where an input value is used more than once, because such calculations introduce a new challenge that has plagued interval arithmetic ever since its introduction.

There have to be strict rules about how *long* a chain of calculations is permitted to remain in the scratchpad layer, and what *type* of calculations, to avoid impractical scratchpad storage demands. Otherwise a programmer would be tempted to ask that an *entire program* run in the scratchpad, approximating the answer only when completely finished!

Even simple calculations can generate billions of bits very quickly, so we proceed with caution in considering which operations should be available in a fused form.

13.2 Fused multiply-add and fused multiply-subtract

As discussed in Chapter 5, fused instructions should always be expressed as *different operations* so that programmers know which one is being used. That avoids the problem of inexplicably different answers from different computing systems. If a clever compiler wants to substitute a fused multiply-add for a multiply followed by an add, that substitution must be made **explicit in the source code and presented as an option to the programmer prior to compilation**.

For adds, the accumulator needs to be scarcely larger than the number of fraction bits. For *fused* multiply-add, the accumulator needs twice as many bits, since it will be holding the complete product (66 bits maximum, for the {4, 5} environment) before doing the addition. That is a small price to pay in hardware.

Computer makers have tended not to support a separate fused multiply-subtract instruction, since it is identical to negating the operand and doing a fused multiply-add. Since negation is a perfect operation with no information loss, that seems like a reasonable decision. Throughout this chapter, we will ignore subtraction as a distinct operation since it can always be accomplished perfectly with addition of a negative.

Notice that fused multiply-add is much like Kulisch's EDP idea, where a product is accumulated in extended precision and then returned to the working precision level. Instead of a dot product, however, there already is a number in the accumulator, and the vectors in the dot product have length one.

In the prototype, the *g*-layer function `fmag[`*ag, bg, cg*`]` computes $(ag \times bg) + cg$ exactly for general intervals *ag*, *bg*, and *cg*. The *u*-layer function `fmau[`*au, bu, cu*`]` is its equivalent, returning an answer that is exact if possible, or within the smallest available ULP width if not. As usual, the *u*-layer routine also tallies numbers and bits moved. With three inputs and one output, the **numbersmoved** tally is increased by 4. The **ubitsmoved** tally is increased by the number of bits in the ubound output, and the number of bits in the three ubound inputs; notice that the motion of the bits in the scratchpad layer is not tallied.

That means that `fmau[`*au, bu, cu*`]` uses *less bandwidth* than doing the operations separately as $(au \otimes bu) \otimes cu$, which moves $(au \otimes bu)$ twice (from *g*-layer to *u*-layer as a result, then back to the *g*-layer as an input). The cost of moving bits in a computer increases the farther you go from the processor, and we use the approximation here that scratchpad bit motion is cost-free whereas *u*-layer bit motion costs something, whether to off-chip storage or on-chip caches. The code for `fmag` and `fmau` is shown in Appendix C.13.

To illustrate the advantage of a fused multiply-add, even the Warlpiri unums suffice. If we compute $(2 \times 2) - 2$ as separate operations, the 2×2 becomes "many" when returned to the *u*-layer, and although a correct bound is returned, information is lost:

```
setenv[{0, 0}]
view[2̂ ⊗ 2̂ ⊖ 2̂]
```

$(0, \infty)$

As a fused operation, the answer remains exact and expressible in the *u*-layer:

```
view[fmau[2̂, 2̂, -̂2]]
```

2

13.3 Solving the paradox of slow math for complex numbers

Even early versions of Fortran had support for COMPLEX (a pair of single-precision floats) and DOUBLE COMPLEX (a pair of double-precision floats) to represent numbers $a + bi$ where $i = \sqrt{-1}$. More recent languages that support complex numbers (the C language standardized the complex data type in 1999) share a curious shortcoming with those early Fortran versions:

> Performance is much better if you do *not* use the COMPLEX type, but instead spell out how to get the real and imaginary parts of complex operations using REAL and DOUBLE variable types!

Application programmers have scratched their heads for a long time about this; how can it possibly be *slower* to use optimized libraries supplied by computer vendors?

The product of complex numbers $z = a + bi$ and $w = c + di$ is

$$z \times w = (a \times c - b \times d) + (a \times d + b \times c)i$$

This operation should take four float multiplications, one float subtraction, and one float addition to determine the real and imaginary parts. Why would writing "z*w" in a program cause it to, bafflingly, run considerably *slower* than if the operations are spelled out as complex(a*c-b*d,a*d+b*c)?

The reason is a lack of communication between programmer and computer system about demands and concessions, resulting in unwanted help from the hidden scratchpad. The optimized library routine is designed with fear of overflow or underflow, which is much like the fear of sailing off the edge of the earth.

Suppose a, b, c, and d are very large or very small magnitude numbers that a float can represent, whatever the precision. What if $a \times c$ overflows or underflows? What if $a \times c - b \times d$ is a representable number, but the *temporary* calculations $a \times c$ and $b \times d$ fall off the edge of the set of representable values? Most programmers do not insist that their complex calculations work even if the numbers have very large or very small magnitudes, but they have no way to express that concession to the computing environment. Hence, library designers assume the user demands that complex operations work for the full dynamic range of the floating point type. That means the complex operations must carefully *dissect the floats to re-scale the exponents to prevent overflow or underflow, do the arithmetic, and then replace the exponents*.

Little wonder that such machinations run so slow: They bring fast hardware down to the speed of software, because of fear of falling off the edge of what can be represented. The user has no way to tell the computer system, "I have no plan to use numbers outside the dynamic range of 10^{-150} to 10^{150}, so please simply do the complex math directly."

A fused operation completely solves this problem. Both $a \times c - b \times d$ and $a \times d + b \times c$ are dot products, of vectors of length 2. That means any system with a fused dot product instruction is already equipped to do the complex multiplies within the scratchpad, thereby protecting against spuriously landing outside the dynamic range for intermediate calculations. The Kulisch idea of an EDP operation is looking better all the time; the next section looks at what it takes to build it with unum hardware, specifically using the unpacked register form for unums.

Exercise for the reader: Create a fused operation for complex divide $\frac{a+bi}{c+di}$, assuming a scratchpad that can hold $c^2 + d^2$ and other temporary values. How many exact accumulators are needed to avoid having to compute anything twice?

13.4 Unum hardware for the complete accumulator

13.4.1 Cost within the unpacked unum environment

To accumulate products without accuracy loss, imagine the accumulator to be a very large fixed-point number, with enough bits to the right and the left of the binary point that the product of *any* two floats lands somewhere within the range. It takes about 600 bits to do this in IEEE single precision, and 4000 bits to do it for IEEE double precision. The main thing that determines the size of the exact accumulator is the dynamic range. If you think of it as the number of ULPs of size *smallsubnormal*, the bit string has to be large enough to express the integer $\frac{maxreal}{smallsubnormal}$. The extended-precision arithmetic will not have to operate on all those bits every time. The result of any product of exact unums is an integer with at most $2^{fsizesize+1} - 1$ bits.

The "shift" of bits into the right location is much like writing to a memory address, and if you think of 4000 bits as a block of memory, addressing 4K of memory does not seem difficult. The register has to be even larger than $\frac{maxreal}{smallsubnormal}$ to allow for the possibility of adding the largest number to the accumulator repeatedly; it only adds a few dozen bits to the length of the accumulator to protect against overflow, even for an astronomically large number of accumulations.

If we have direct hardware support for unums up to the {4, 5} environment, the exponent range is even larger than for quadruple precision IEEE floats. Use of the maximum dynamic range (about 10^{-9874} to 10^{9864}) means that the exact accumulator needs about 66 000 bits. The technology used for that much storage would most likely be the kind used for primary (Level 1) cache, not the kind used for registers.

Ideally, there are two such registers so the left and right interval endpoints of an EDP can be accumulated in parallel. The products that are accumulated are always treated as exact, even when they are from unums marked with a ubit set to indicate an inexact range. This is another example of how different unum arithmetic is from significance arithmetic. If you were doing calculations by hand and discovered that an accumulation with 200 digits was approximate past digit 17, say, it would be tempting to throw out every digit to the right of digit 17 from that point on as junk that is not worth tracking. But it is *not* junk. Every product produces a very precisely defined range of real numbers, open or closed at the endpoints. Only at the conclusion of all the accumulations is it possible to know the lower and upper bound for the total, and *then* return it to the *u*-layer where it may or may not be representable exactly but will lose at most the smallest ULP amount of width for the bound.

Recall that for the {4, 5} environment, the multiplier hardware can be built to take the product of unsigned integers as large as 33 bits and produce a result as large as 66 bits. The exponents are summed to determine the "shift" of the bits, but that really looks more like a memory indexing operation to find where in the accumulator to start adding the bits from right to left. It may look like there is potential for the need to propagate a carry bit all the way from the lowest to the highest bit of the exact accumulator. There is, but it does not need to take linear time; hardware designers figured out long ago how to "look ahead" to propagate carry bits much faster, in time more like the logarithm of the total number of bits. The accumulation can happen concurrently while the next product is being calculated.

The same hardware suffices to compute the exact sum (in the *g*-layer) of a list of unums of arbitrary length. Mathematically, the sum is the same as a dot product with a vector of all 1 values, but the sum operation deserves a separate instruction to spare the computer from having to invoke the multiplier hardware to multiply by 1. As mentioned previously, the long accumulator is exactly what is needed for extended-precision multiplication as well.

13.4.2 Fused dot product and fused sums in the prototype

In the prototype, the *g*-layer fused dot product is `fdotg[ag,bg]`, where *ag* and *bg* are equal-length lists of general intervals. The *u*-layer version is `fdotu[au,bu]`, where *au* and *bu* are equal-length lists of unums or ubounds. Note, however, that the arithmetic in `fdotu` is performed entirely in the scratchpad as an exact calculation, and returned as an exact or inexact ubound only at the end. Similarly, the *g*-layer fused sum of a list of general intervals *ag* is found using `fsumg[ag]`, and its *u*-layer equivalent is `fsumu[au]` where *au* is a list of unums or ubounds. The `fsumu` function tallies bits moved and numbers moved.

Having a fused sum is liberating. The *associative law* of addition says that for numbers *a*, *b*, and *c*,

$$(a + b) + c = a + (b + c).$$

This identity does *not* hold for floats. Floats flunk elementary algebra. For example, $\left(-10^8 + 10^8\right) + 1 = 1$ in single precision, but $-10^8 + \left(10^8 + 1\right)$ incorrectly evaluates as 0. The float operation $10^8 + 1$ commits a rounding error, returning $10^8 = 100\,000\,000$ exactly instead of $100\,000\,001$, since the latter is not a number that a single precision float can represent. Unums obey the associative law at two levels: First, what is sometimes called the "sub-associative law" when it arises with interval arithmetic, since the equality sign is replaced with the "not nowhere equal" sign:

$$\left(\hat{a} \oplus \hat{b}\right) \oplus \hat{c} \approx \hat{a} \oplus \left(\hat{b} \oplus \hat{c}\right).$$

Second, the *g*-layer sums produce identical results for any ordering of their operands, which is the liberation provided by obeying the associative law. Ingenious numerical analysts have gone to great lengths to design algorithms for adding up floats so that the rounding error is a minimum. Which is all well and good, but the simple act of adding numbers should not require looking up and encoding a sophisticated method! Having a fused sum that is immune to ordering and supported in hardware means *not having to think*. It just works.

As an example of a summation that might not go well if done in the straightforward way, suppose we write a loop to find the sum of the reciprocal squares from 1 to 100, that is, $1 + \frac{1}{2^2} + \frac{1}{3^2} + \ldots + \frac{1}{100^2}$, using only $2^3 = 8$ bits for the fraction. (The sum from 1 to ∞ is equal to $\frac{\pi^2}{6} = 1.64493 \cdots$, by the way):

```
setenv[{3, 3}]
Module[{su = 0̂},
  For[i = 1, i ≤ 100, i++,
    su = su ⊕ (1̂ ⊙ squareu[î])];
  view[su]]
```

(1.5703125, 1.94140625)

The bound on the sum is much looser than it needs to be; it is 95 ULPs wide! Some numerical analysis experts say the way to improve accuracy is to find ways to group sums so that the numbers being added are close to the same magnitude, and you can imagine how much programming *that* takes. Instead, just use a fused sum to add the list exactly in the *g*-layer, converting to an inexact value one ULP wide only at the end:

```
Module[{au = Table[1̂ ⊙ squareu[î], {i, 1, 100}]},
  view[fsumu[au]]]
```

(1.6328125, 1.63671875)

That code is easier to write, easier to read, executes *faster* (since it does not have to go back and forth between *g*-layer and *u*-layer with each addition), and is as accurate as possible in this environment (one ULP wide at maximum precision). It wins on every count. Best of all, no numerical expertise is required.

13.4.3 Consistent results for parallel computing

The Tianhe-2 supercomputer is representative of a 2014-era computer cluster. It has over three million processor cores, capable of communicating pairwise at about 50 gigabits per second after a startup latency of 9 microseconds.

> *The lack of an associative law for floats deters the use of parallel computing.*

When converting a sequential program to run in parallel, we often need to sum a list of floats. The sequential program would add four elements this way:

$$s = x_1 + x_2 + x_3 + x_4$$

which really means adding from left to right:

$$s = ((x_1 + x_2) + x_3) + x_4.$$

If you do the sum in parallel, one processor can sum $s = x_1 + x_2$ and the other can sum $t = x_3 + x_4$ at the same time, which you can then combine by adding s and t. But that means the sum evaluates as

$$s = (x_1 + x_2) + (x_3 + x_4)$$

and the float result is a *different answer.* The usual reaction to this type of discrepancy is, "Parallel processing doesn't work; it gives wrong answers. I'm going back to the sequential version." Or, if the programmer is determined enough, he or she will write a sequential version that mimics parallel operation ordering, which creates peculiar-looking, complicated code. We already know unums eliminate the problem *within* a processor, since they can do the sum exactly within the g-layer. Do unums offer any help for the problem of getting different answers when using multiple processors to find the sum?

If the partial sums are combined, they will generally produce a different bound from the one obtained from serially adding terms from left to right. However, the bounds will always *intersect* unless there is a real bug in the code. That means the programmer can immediately spot the difference between a bug and loss of precise information about a result. Unums have a built-in "quality-of-service" measure, namely the reciprocal of the width of the bounds in the answer; that means jobs submitted to abstract machines on a network can be run on a wide range of hardware configurations. The more powerful the machine, the tighter the bound, but a unum program free of logic bugs will always return the right answer to some accuracy. However, *we can do even better than this.*

There is no rule that says values in the g-layer cannot be **exchanged between processors**. If the partial sums are exchanged in their extended precision form (possibly as large as two sets of 66 000 bits for a {4, 5} environment), there will be *no* difference in the sum, and the system is free to use any number of processors without creating discrepancies in the result.

If the idea of sending that many bits off-chip sounds like a real energy-waster and something that will take too much time, consider the numbers for a current vintage cluster computer like the one shown at the start of this section, using message-passing to coordinate parallelism: It takes about **nine microseconds** of latency to exchange the smallest possible message, because of the time to set up the link. Connections between core processors can sustain about 50 gigabits per second once the link is set up. At that rate, even the worst case of having to send both *g*-layer bounds at maximum internal   length would only take **less than two microseconds** after the nine microseconds of setup! (We very rarely need the entire long accumulator, so it will usually be less than this.) This seems a small price to pay for perfect, reproducible results.

Just as unums and the careful distinction between *g*-layer and *u*-layer eliminate the phenomenon of different computers producing mysteriously different answers to real-valued computations, **they can also eliminate inexplicable differences between serial and parallel computing**.

13.5 Other fused operations

13.5.1 Not every pairing of operations should be fused

What other pairs of arithmetic operations make sense to fuse? The main criterion is that a fusing can significantly *reduce the information loss* compared to doing the individual operations, yet uses scratchpad resources that we *already have*. Fused divide-add and add-divide, for example, add a lot of complexity to the hardware but usually provide little benefit. A secondary consideration is that some fused operations allow the use of smaller maximum size exponents, since only the scratchpad needs to handle very large or very small numbers.

The fused operations that follow are arguably worth the trouble.

13.5.2 Fused product

Multiplication is not as prone to catastrophic loss of information as is addition or subtraction, *unless the dynamic range is exceeded*. We could define a "fused multiply-multiply," but it makes more sense to go all the way: A fused product of *n* numbers in a list, the multiplication analog of a fused sum.

It may be that $a \times b \times c$ is within the dynamic range even if the intermediate calculation of $a \times b$ or $b \times c$ would have fallen into the regions $\pm(0, smallsubnormal)$ or $\pm(maxreal, \infty)$ had it been returned to the *u*-layer before the second multiplication. The fused product solves that issue.

Another reason for wanting a fused product is that it obeys the associative law:

$$(a \times b) \times c = a \times (b \times c).$$

The first thing the hardware should do in computing a fused product is check *all* input values for exceptions, since they might allow a quick shortcut to the *u*-layer result. The second thing the hardware should do is simply add all the exponents; if they fall outside the dynamic range of the environment, then one of the regions $\pm(0, smallsubnormal)$ or $\pm(maxreal, \infty)$ is the answer, and again no arithmetic needs to be done. In such cases, the fused product can actually *save* considerable time and energy compared to doing the multiplications separately.

In other cases where all the arithmetic has to be done, it is easy to imagine running out of memory when the list of numbers is long. The hardware demands of fused multiply-multiply need not be more than we already have, but in the set of "adjustable wrenches" for multiplication, it potentially forces use of the next size up, where some extended precision is needed. As with all unum multiplication, hardware first notes which tool to use: The dedicated parallelogram of partial products, an extended precision multiply using the long accumulator, or a sophisticated routine efficient for very large numbers. A user should expect a fused product to be slower and more energy-consuming than separate multiplications, just as with selecting higher precision operations. Using a fused product in just a few cases where the dynamic range might be exceeded could spare the programmer from having to change the environment settings to a larger *esizesize*.

In the prototype, the *g*-layer fused product of general intervals in a list *ag* is **fprodg[**ag**]**. The *u*-layer function is **fprodu[**au**]**.

To see why a fused product might be useful, try a small environment {2, 3}, for which *maxreal* is only 510 and the fraction only provides about two decimals of precision:

```
setenv[{2, 3}]
maxreal
```

510

The product $400 \times 30 \times 0.02 = 240$ is evaluated reasonably well in the *g*-layer, given the fact that 0.02 converts into an inexact unum:

$$\texttt{view}\left[\texttt{fprodu}\left[\left\{4\hat{0}0,\ 3\hat{0},\ 0.\hat{0}2\right\}\right]\right]$$

(239.5, 240.5)

Contrast that with an evaluation of the product using separate multiplications, grouped two ways:

$$\texttt{view}\left[4\hat{0}0 \otimes \left(3\hat{0} \otimes 0.\hat{0}2\right)\right]$$
$$\texttt{view}\left[\left(4\hat{0}0 \otimes 3\hat{0}\right) \otimes 0.\hat{0}2\right]$$

(239, 241)

(10.15625, ∞)

Information loss can be the usual ULP-wide increases for calculations inside the dynamic range, or severe if an intermediate product falls into the (*maxreal*, ∞) range as in the second case above.

13.5.3 Fused add-multiply

At first glance, this looks like a loser. Computing $(a + b) \times c$ exactly in the scratchpad means doing an exact add of a and b, which would require the same huge accumulator as the EDP, and then multiplying *that*. But the distributive law of arithmetic says

$$(a + b) \times c = (a \times c) + (b \times c)$$

and the expression on the right side is simply the dot product of {a, b} with {c, c}. Which means fused add-multiply is simply a special case of the EDP, and far easier to compute than most exact dot products since the add can be skipped if the two products are of such different magnitude that the scaled bit strings do not overlap. (That situation simply sets the ubit to indicate there are more 1 bits after the last one in the fraction of the larger number.) The same trick is used with any add; the only difference is that fraction bit string products are up to twice as many bits long in the g-layer as the ones in the u-layer.

The prototype functions are **famg**[ag, bg, cg] for general intervals and **famu**[au, bu, cu] in the u-layer. As usual, the u-layer version tallies the bits moved and numbers moved for purposes of assessing the cost of unum math relative to float math.

Incidentally, if the add-multiply is done as *separate* unum operations, unum arithmetic obeys the "sub-distributive law":

$$(au \oplus bu) \otimes cu \approx (au \otimes cu) \oplus (bu \otimes cu)$$

which does not hold for float arithmetic. The ubounds produced by either side of the equation will always overlap, and if they are not identical, then their intersection is an improved bound on the add-multiply. But with fused add-multiply, best-possible bounds are always returned without worrying about such things.

13.5.4 Fused product ratio

Algebra says $(a \times b) \div c = a \times (b \div c)$, so a fused divide-multiply is the same as a fused multiply-divide. With a float-type representation, doing the multiply *first* is always the right thing to do since it always can be done exactly in the scratchpad, whereas the divide usually produces a remainder.

The reason for including fused multiply-divide is that it helps cling to every possibility of getting an exact result. For example, with floats, $(x \div 7) \times 7$ will usually produce a number different from x (caused by rounding error), and with unums, $(xu \odot \hat{7}) \otimes \hat{7}$ will usually produce an inexact ubound that contains xu but is not identical to xu. A fused multiply-divide with those numbers will return xu exactly, with no information loss.

The hardware is really already there for fused multiply-divide, since a multiply in the scratchpad produces an exact integer product. Dividing by the fraction simply means long division with the entire integer product instead of first clipping off bits past the maximum fraction size. When results are inexact, the ubound ULP width of a separate multiply and divide will typically be twice the width of a fused multiply-divide, so the fused form helps retain information about the answer.

It is easy to generalize the fused multiply-divide without violating any rules about how long a calculation sequence can remain in the g-layer. Suppose we want to compute

$$X = \frac{m_1 \times m_2 \times \ldots m_j}{n_1 \times n_2 \times \ldots n_k}$$

The multiplications can be exact in the g-layer; the final long division is only taken to the point where it is at the limit of what can be expressed in the u-layer. Call this a *fused product ratio*. It includes fused multiply-divide and fused divide-multiply as special cases. Notice that j and k are independent; the numerator and denominator lists do not need to be the same length.

The prototype function for fused product ratio with general interval lists *numeratorsg* and *denominatorsg* in the g-layer is `fprodratiog[`*numeratorsg, denominatorsg*`]`. The u-layer version is `fprodratiou[`*numeratorsu, denominatorsu*], which also tallies the bits moved and numbers moved. As an example of how it can help, try evaluating $5 \times (3 \div 10)$ without fusion, where we have purposely grouped the operations unfavorably by doing the divide first:

$$\text{view}\left[\hat{5} \otimes \left(\hat{3} \odot \hat{10}\right)\right]$$

(1.49609375, 1.50390625)

In contrast, the fused operation preserves the exactness of the answer:

$$\text{view}\left[\text{fprodratiou}\left[\left\{\hat{5},\ \hat{3}\right\},\ \{\hat{10}\}\right]\right]$$

1.5

Centuries ago, an English mathematician named John Wallis found a bizarre formula for π (the "Wallis Product"). It provides a way to demonstrate a fused product with lots of multiplications in both numerator and denominator:

$$\pi = 2 \left(\frac{2 \cdot 2}{1 \cdot 3}\right) \left(\frac{4 \cdot 4}{3 \cdot 5}\right) \left(\frac{6 \cdot 6}{5 \cdot 7}\right) \left(\frac{8 \cdot 8}{7 \cdot 9}\right) \left(\frac{10 \cdot 10}{9 \cdot 11}\right) \ldots$$

In other words, the product of infinitely many pairs of even integers in the numerator divided by the product of infinitely many pairs of odd integers in the denominator, but skewed. (Incidentally, Wallis is the person who invented the "∞" symbol, though his mathematical definition of it was more like (*maxreal*, ∞) than exact infinity.)

Here is how the fused product ratio routine computes the first few terms of the Wallis Product, as well as the initial factor of 2 thrown in:

```
setenv[{2, 4}];
view[fprodratiou[{2̂, 2̂, 2̂, 4̂, 4̂, 6̂, 6̂, 8̂, 8̂, 1̂0, 1̂0, 1̂2, 1̂2, 1̂4, 1̂4},
    {1̂, 3̂, 3̂, 5̂, 5̂, 7̂, 7̂, 9̂, 9̂, 1̂1, 1̂1, 1̂3, 1̂3, 1̂5}]]
```

(3.038665771484375, 3.0386962890625)

The environment was set to {2, 4} to demonstrate one of the reasons for using fused operations like this one. In a {2, 4} environment, *maxreal* is about 512, so the dynamic range is enough to express all the numbers shown above but *too small for scratch calculations*. The numerator product is 832 359 628 800 and the denominator product is 273 922 023 375, which if computed with individual unum operations would require a more capable exponent field (an *esizesize* of 3 or more). But the fused calculation worked just fine and returned an answer of the maximum possible accuracy. All the big integers were handled in the *g*-layer, which returned an in-range *u*-layer number as the final ratio. The average unum size in the calculation was 13.5 bits, which is surprisingly cheap for an answer accurate to five decimals!

It takes about *n* terms of the Wallis product to get within about $\frac{1}{n}$ of the correct value of π, so we should not be too disappointed about the lack of convergence. The point is that any long sequence of products of rational numbers and integers can be, and should be, handled by the fused product ratio routine. The task need not require an environment with a large exponent to handle intermediate calculations if we explicitly request help from the *g*-layer. Furthermore, the fused function can produce an exact result if one is expressible in the *u*-layer.

13.5.5 Fused norm, fused root dot product, and fused mean

The last three fused functions are not part of the prototype, but are presented here as suggestions for a more comprehensive unum computing environment.

"Norm" here means "square root of the sums of squares." It could be as simple as the distance coordinates (x, y) are from the origin:

$$r = \sqrt{x^2 + y^2}$$

Or it could be the square root of the sum of millions of squared numbers. It is one of the most common expressions found in technical computing. It is another reason for fearing the fall off the edge of the world of defined numbers, since numbers get squared in the scratch calculations. The final step of taking the square root brings the dynamic range into something close to that of the input values. If the only part of the computer that has to bear the burden of a temporarily too-large or too-small exponent is the *g*-layer, it may be possible to use a smaller *esizesize* in the *u*-layer.

The squares should be computed carefully, with attention to the issues detailed in Section 11.1. The summations are done with the complete accumulator, so everything is exact integer arithmetic up to the point of the square root, which if inexact need only be as accurate as needed for **fsizemax** fraction bits in the *u*-layer.

For the same reason, it would be valuable to extend the EDP dot product with a version that performs a square root at the end *before* the answer is returned to the *u*-layer. For many geometric calculations, we need to know $\sqrt{a \cdot b}$ as accurately as possible, where *a* and *b* are vectors. Fusing the square root saves a small amount of information loss, but it again saves us from having to use overly large exponent sizes for scratch work.

Finally, it is common after summing a set of *n* numbers to want to find their mean, by dividing by *n*. If *n* is large, having a fused mean that divides by *n* in the *g*-layer can similarly eliminate the need for overly large exponent sizes, since a mean will be within the same dynamic range as the input values that are summed.

We now have enough operations to test the viability of the unum computing environment. Some of the questions we need to answer:

- Do ubounds often expand like traditional intervals, giving uselessly large ranges?
- Do they use more or fewer bits than floats do, to get to a satisfactory answer?
- Are they prone to the same pitfalls as floats, as shown by classic test problems?
- Are they easier to use than floats, or more difficult?

Before applying unums to classic computational physics problems (Part 2), we first put them to some "acid tests," problems that mathematicians have invented over the years to expose the flaws of computer numerics.

14 Trial runs: Unums face challenge calculations

Berkeley Professor William Kahan, primary architect of the IEEE 754 Standard.

14.1 Floating point II: The wrath of Kahan

Berkeley professor William Kahan is skilled at finding examples of mathematical questions to which floating point provides grossly incorrect answers... as well as examples where proposed alternatives are *even worse than floats*. Can unums withstand the wrath of Kahan?

Here is one of Kahan's examples: Iterate

$$u_{i+2} = 111 - \frac{1130}{u_{i+1}} + \frac{3000}{u_i \, u_{i+1}}$$

starting with $u_0 = 2$, $u_1 = -4$. With floats, the iteration starts to converge to 6, but then heads for 100 instead even though 6 is a valid steady-state answer. The iteration is stable for numbers above 6 but unstable below 6. If numerical computing were perfect, the u_i values would never fall below 6; however, rounding error leads them to stumble into the unstable region, and then make a beeline for 100. The loop on the next page shows this, using double-precision floats:

```
Module[{i, u0 = 2., u1 = -4., u2},
  For[i = 0, i < 18, i++, {u2 = (111 - (1130 / u1)) + (3000 / (u1 * u0));
    Print[u2]; u0 = u1; u1 = u2;}]]
```

18.5

9.37838

7.80115

7.15441

6.80678

6.59263

6.44947

6.34845

6.27444

6.21868

6.17563

6.13895

6.06036

5.1783

-11.6233

158.376

102.235

100.132

As an example of a proposed alternative to floats, try the iterative loop using double precision *interval* arithmetic, which consumes 128 bits per value instead of 64. They had better be worth their high price:

```
Module[{i, u0 = Interval[{2., 2.}], u1 = Interval[{-4., -4.}], u2},
  For[i = 0, i < 12, i++,
    {u2 = (Interval[{111., 111.}] - (Interval[{1130., 1130.}] / u1)) +
      (Interval[{3000., 3000.}] / (u1 * u0)); Print[u2]; u0 = u1; u1 = u2;}]]
```

Interval[{18.5, 18.5}]

Interval[{9.37838, 9.37838}]

Interval[{7.80115, 7.80115}]

Interval[{7.15441, 7.15441}]

Interval[{6.80678, 6.80678}]

Interval[{6.59263, 6.59263}]

Interval[{6.44941, 6.44952}]

Interval[{6.34642, 6.35048}]

Interval[{6.19348, 6.35542}]

Interval[{2.88101, 9.52231}]

Interval[{-231.651, 160.46}]

Interval[{-∞, ∞}]

They were *not* worth it. This answer is typical of interval arithmetic: A correct but useless result that says the answer is: **Somewhere on the real number line.** Now you can start to see why interval arithmetic has attracted a very limited fan club.

Now try it with unums, turning up the precision (but not the exponent, since these numbers are not particularly large or small). Please excuse the long decimals for the fractions. The `reportdatamotion` call after the loop reports the data motion cost.

```
setenv[{3, 6}]; ubitsmoved = numbersmoved = 0;
Module[{i, u0 = 2̂, u1 = -̂4, u2},
  For[i = 0, i < 12, i++, {u2 = 1î1 ⊖ (1î30 ⊙ u1) ⊕ 3000̂ ⊙ (u1 ⊗ u0);
    Print[view[u2]]; u0 = u1; u1 = u2}]; Print[" "];
  reportdatamotion[numbersmoved, ubitsmoved]]
```

18.5

(9.37837837837837837683119257548014502390287816524505615234375,
  9.37837837837837838030063952743375921272672712802886962890625)

(7.801152737752161357709379529978832579217851161956787109375,
  7.80115273775216141495525423721346669481135904788970947265625)

(7.15441448097524869582930495681694083032198250293731689453125,
  7.1544144809752500922817031181466518319211900234222412109375)

(6.80678473692361268691797260288467441569082438945770263671875,
  6.80678473692365441395646374900252340012229979038238525390625)

(6.59263276870374305242694656925550589221529662609100341796875,
  6.592632768705150414501048317106324248015880584716796875)

(6.44946593376477887715747527863641153089702129364013671875,
  6.449465933816050482174642155541732790879905223846435546875)

(6.3484520556719081785246316940174438059329986572265625,
  6.3484520576407397034202073626829587738029658794403076171875)

(6.2744385589936915448294740826895576901733875274658203125,
  6.274438637500974615124338384930524625815451145145172119140625)

(6.2186941307408676905821298674936770112253725528717041015625,
  6.2186973498528085035896850740755326114594936370849609375)

(6.1757699006420045638454663361471830285154283046722412109375,
  6.175904725095663815037649868600055924616754055023193359375)

(6.139489115844833831692195502682807273231446743011474609375,
  6.1452292795550675086335701280404464341700077056884765625)

Numbers moved	180
Unum bits moved	20 184
Average bits per number	**112.**

So unums head for the correct number, 6. If you keep going, the left endpoint falls below 6, and then diverges toward −∞, but remember that a unum environment *can automatically detect* when relative width gets too high with **needmorefracQ** and either stop (as we have done here) or increase the precision. Unums do not exhibit *false convergence* like the floats did. Furthermore, they do all this using fewer bits than double precision interval arithmetic (128 bits per value).

The following myth is widely held by many scientific programmers as an easy way to guard against the dangers of float errors:

> If you keep increasing the precision of your floats, and the answer is consistent, then the answer is correct to that precision.

Kahan constructed the following fiendishly clever example to prove how fallacious this myth is. With floats, you often get 0, even at very high precision; the correct answer is 1.

> "Define functions with: $E(0) = 1$, $E(z) = \frac{e^z - 1}{z}$. $Q[x] = \left| x - \sqrt{x^2 + 1} \right| - \dfrac{1}{x + \sqrt{x^2 + 1}}$.
>
> $H(x) = E(Q(x))^2)$. Compute $H(x)$ for x = 15.0, 16.0, 17.0, 9999.0. Repeat with more precision, say using BigDecimal."

"BigDecimal" is a data type in Java that allows extended precision arithmetic, one of many historical attempts to deal with rounding errors by using huge amounts of storage to allow more digits in the user-visible layer. Professor Kahan knows exactly how to expose the flaws of floating point math. Using double precision floats, here is the incorrect answer you get for all four suggested values of x:

```
e[z_] := If[z == 0, 1, (Exp[z] - 1.) / z]
q[x_] := Abs[x - Sqrt[x² + 1.]] - 1 / (x + Sqrt[x² + 1.])
h[x_] := e[q[x]²]
{h[15], h[16], h[17], h[9999]}
```

```
{0., 0., 0., 0.}
```

With even higher precision, like IEEE quad precision or the BigDecimal extended precision, it *still comes out wrong almost every time*. A few scattered values of x give the correct answer: {1, 1, 1, 1}, which can be figured out by working through the algebra symbolically instead of numerically. The point is that simply increasing precision and getting the same, perfectly consistent answer can happen when the answer is *dead wrong*.

What about traditional interval arithmetic, which is supposed to be "rigorous"? To give traditional intervals every possible advantage, we can even let *Mathematica* compute the result symbolically and evaluate it with numerical floats only at the end, by redefining the *e*, *q*, and *h* functions to use interval arithmetic. The test of an interval being zero is similar to the way we test it with unums, where `IntervalMemberQ[z,Interval[{0,0}]]` is True if zero is in the interval *z* and False otherwise:

```
e[z_] := If[IntervalMemberQ[z, Interval[{0, 0}]],
   Interval[{1, 1}], (Exp[z] - Interval[{1, 1}]) / z]
q[x_] := Abs[x - Sqrt[x² + 1]] - 1 / (x + Sqrt[x² + 1])
h[x_] := e[q[x]²]
{N[h[Interval[{15, 15}]]], N[h[Interval[{16, 16}]]],
 N[h[Interval[{17, 17}]]], N[h[Interval[{9999, 9999}]]]}
```

```
{Interval[{-∞, ∞}], Interval[{-∞, ∞}],
 Interval[{-∞, ∞}], Interval[{-∞, ∞}]}
```

Another epic fail. Traditional interval arithmetic again gives you the entire real number line as the possible answer, which then forces you to analyze what went wrong and debug where the calculation lost all of its information. Furthermore, the use of symbolic mathematics is extremely expensive and slow compared to numerical computing. This example looks like a very tough nut to crack!

> There are not many moments of "high drama" in a book like this one, but if ever there was one, this is it. *Can unums defeat Professor Kahan's monster?* Perhaps by using some very high environment setting for the precision?

Here is the unum version of the function definitions:

```
e[z_] := If[z ≈ 0̂, 1̂, (expu[z] ⊖ 1̂) ⊙ z]
q[x_] := Module[{v = sqrtu[squareu[x] ⊕ 1̂]}, absu[x ⊖ v] ⊖ (1̂ ⊙ (x ⊕ v))]
h[x_] := e[squareu[q[x]]]
```

Just for fun… trot out the tiniest, most modest unums of them all, the *four-bit* Warlpiri number set. The only reason to have any hope for success in this battle is that all unums, even those with only a *one-bit* fraction and a *one-bit* exponent, are packing a powerful weapon, something no other computational number system has: *Absolute honesty about what they do and do not know* about a value. Here it goes:

```
setenv[{0, 0}]; ubitsmoved = numbersmoved = 0;
{view[h[15̂]], view[h[16̂]], view[h[17̂]], view[h[9999̂]]}
reportdatamotion[numbersmoved, ubitsmoved]
```

$\{1, 1, 1, 1\}$

Numbers moved	92
Unum bits moved	620
Average bits per number	**6.74**

Warlpiri math got it right, even though it cannot count past two! Unums computed the correct answer with fewer than *seven bits* of storage per number.

14.2 Rump's royal pain

Kahan is not the only one to think up ways to expose the shortcomings of computer number systems. Siegfried Rump, an IBM scientist, came up with a great example of why it is *not* enough just to try different float precisions and look for consistency:

Evaluate $333.75\,y^6 + x^2\left(11\,x^2\,y^2 - y^6 - 121\,y^4 - 2\right) + 5.5\,y^8 + \dfrac{x}{2\,y}$,

where $x = 77617$, $y = 33096$.

Here is Rump's formula and input arguments, and what modern IEEE double precision floats tell us:

```
rump[x_, y_] := 333.75 y⁶ + x² (11 x² y² - y⁶ - 121 y⁴ - 2) + (5.5) y⁸ + x / (2 y)
rump[77 617, 33 096]
```

1.18059×10^{21}

With IEEE-style floats, Rump's formula gives the same answer, about 1.18×10^{21}, for single, double, and even quad precision, yet is *massively wrong in all three cases*. The correct answer is $-0.827\cdots$, so floats don't even get the *sign* of the answer right. Not to mention that little problem of being off by *21 orders of magnitude*.

Rump writes that on a 1970s-vintage IBM mainframe, the function evaluates to

1.172603 in single precision,

1.1726039400531 in double precision, and

1.172603940053178 in IBM "extended" (what we now call quad) precision.

The IBM 370 floating point format used a base-16 exponent and had other differences from modern IEEE floats, which is why it produces such a different trio of highly consistent-looking results. Anyone trying to check the soundness of the computation would conclude that the results are valid, and that only single precision is needed. Notice that every number used in the formula can be expressed as an exact float: 333.75, 11, 121, 2, 5.5, and the two integer inputs are all integers divided by powers of 2, so simply entering the problem will not, in this case, introduce any rounding error.

Being off by 21 orders of magnitude and getting the sign wrong might be more forgivable if it were not presented as *an exact result*. Interval arithmetic to the rescue, with guaranteed containment of the correct result? Maybe.

Try a version of `rump` called **rumpint** that instead uses 128-bit intervals (that is, a 64-bit float for each endpoint), and at least you get a bound for the answer:

```
rumpint[x_, y_] :=
  Interval[{333.75, 333.75}] y^6 + x^2 (Interval[{11, 11}] x^2 y^2 -
      y^6 - Interval[{121, 121}] y^4 - Interval[{2, 2}]) +
    Interval[{5.5, 5.5}] y^8 + x / (Interval[{2, 2}] y)
rumpint[Interval[{77 617, 77 617}], Interval[{33 096, 33 096}]]
```

$$\text{Interval}\left[\left\{-2.36118 \times 10^{21}, \ 4.72237 \times 10^{21}\right\}\right]$$

A *gigantic* bound, but a bound. At least traditional intervals suggest that Rump's problem might be prone to wild errors. (Unfortunately, traditional intervals also often suggest this about perfectly stable calculations.) As usual, the huge width of the traditional interval result leaves us baffled about what to do next. Try traditional intervals with *quad-precision* bounds? They fail as well.

Try it with unums. Because one thing being tested about unums is ease-of-use, here is some explanation of what the following code for the unum version looks like. The $\otimes$ operator has precedence over the $\oplus$ and $\ominus$ operators, just as multiplication takes precedence over addition and subtraction in traditional formulas, so we do not have to use parentheses to write, say, $a \otimes b \oplus c \otimes d$. However, to "break the tie" of an expression like $x \oplus y \oplus z$, we use parentheses to group it from left to right: $(x \oplus y) \oplus z$. It also helps readability to put plenty of space around the $\oplus$ and $\ominus$ operators. For clarity, we introduce some temporary variables for x^2, y^2, y^4, y^6, and y^8. The **rumpu** function computes Rump's formula using ubounds:

```
rumpu[xu_, yu_] :=
  Module[{xsqu = squareu[xu], ysqu = squareu[yu], y4u, y6u, y8u},
    y4u = powu[yu, 4̂]; y6u = powu[yu, 6̂]; y8u = powu[yu, 8̂];
    ((333.7̂5 ⊗ y6u ⊕
        xsqu ⊗ ((((1̂1 ⊗ xsqu) ⊗ ysqu ⊖ y6u) ⊖ 1̂21 ⊗ y4u) ⊖ 2̂)) ⊕
      5.̂5 ⊗ y8u) ⊕ xu ⊙ (2̂ ⊗ yu)]
```

Say we want at least two decimals of accuracy in the answer so we set **relwidthtolerance** = 0.005. Start with the Warlpiri environment, {0, 0}, and let the computer automatically find when we have enough bits to get an acceptable answer, as in the flowchart shown in Section 9.3. It turns out that {3, 7} is the environment at which acceptable accuracy is obtained, and it certainly does quite a bit better than two decimals:

$$(-0.8273960599468213681411650954798162919990\cdots,$$
$$-0.8273960599468213681411650954798162919960\cdots)$$

The unum approach bounded the answer to **39** decimals of accuracy, and it did it using an average of **70** bits per number. The main reason unums did so well here is that they dynamically adjust precision as needed. The intermediate results in creating the expression are quite large, requiring 37 decimals of accuracy for exact representation. Those decimals are later canceled out in the calculation, resulting in small integers **that the unum format stores with very few bits**. The overhead of carrying around eleven bits of utag pays for itself many times over. Had we used fused operations, the bits per number would have been even lower.

Fewer bits, better answers.

14.3 The quadratic formula

The previous examples were carefully crafted to expose problems with floats. The reader may be thinking that the examples are too contrived to reflect most computing tasks, for which floats seem to do fairly well. So the next example shows an example from basic, mainstream computing where rounding errors produce unpleasant surprises for casual programmers. There happens to be a trick for dealing with the particular problem of the quadratic formula, and books on numerical analysis eagerly tell you how to invoke the trick, but we are *not* looking for a collection of tricks that programmers must learn; the computer should automate as much of the numerical analysis as possible. Unums should work even when used in a naïve way, the way floats usually are used.

The formula for the roots r_1 and r_2 of a quadratic equation $ax^2 + bx + c = 0$ has numerical hazards we can use to compare floats with unums, both for accuracy and for storage efficiency:

$$r_{1,2} = \frac{-b \pm \sqrt{b^2 - 4ac}}{2a}$$

The square root typically produces an inexact result, of course. The more subtle hazard is that when b^2 is large relative to $4ac$, $\sqrt{b^2 - 4ac}$ will share many leading digits with b. This causes digit cancellation in computing $-b \pm \sqrt{b^2 - 4ac}$, since either the + or the − case of the ± will cause subtraction of similar numbers, leaving only a few significant digits.

For $a = 3$, $b = 100$, and $c = 2$, notice how close one root is to zero:

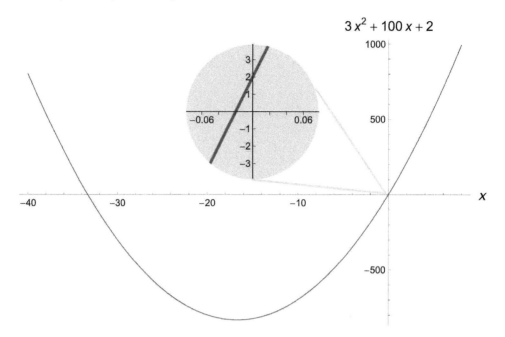

Here are the exact expressions for the two roots, and their values correct to 7 decimals, for reference:

$$r_1 = \frac{1}{6}\left(-100 + \sqrt{9976}\right) = -0.02001201\cdots$$

$$r_2 = \frac{1}{6}\left(-100 - \sqrt{9976}\right) = -33.31332\cdots$$

In IEEE single precision, $b^2 = 100$ can be represented exactly; so can $4\,a\,c = 24$, and $b^2 - 4\,a\,c = 9976$. So far, so good. The square root of 9976 is

$$99.8799279\cdots$$

but single precision floats ask us to instead accept the following incorrect number, where we use orange digits beyond the correct ones and the "↓" notation at the end to emphasize that the decimal expansion shown is *exactly* what the float expresses:

$$99.87992\,85888671875\!\downarrow$$

This is certainly within one ULP of the correct square root. The problem arises in the next step of the calculation, adding $-b = -100$ to the above number. Doing so wipes out two left digits, and the erroneous orange digits are **amplified as a relative error**:

$$-1.20071411328125\!\downarrow$$

In the final step, division by $2a$, we wind up with only four decimals of agreement with the reference answer, and the result is wrong by over 100 ULPs:

$$-0.02001190185546875\!\downarrow$$

If the computer had tracked how the ULP error was amplified, it would know to express the answer as "~−0.02001" or something with only four decimals. Instead, as a single precision answer it is deceptively displayed as "−0.02001190" with nothing to let us know that the last three precise-looking decimals are wrong.

The trick taught in numerical analysis books is to use the identity $r_1 r_2 = \frac{c}{a}$ to find r_1. Since r_2 does not experience the digit cancellation, precision is generally maintained. However, most people forget that particular identity about the quadratic formula soon after they learn it, and in any case it seems like an unnecessary burden to ask programmers to remember and apply such a trick. Instead try a unum environment, the smallest one that can mimic single precision floats:

```
setenv[{3, 5}]
```

Here is the unum version of the quadratic formula for the root near zero, using the previously developed arithmetic operations. Reset bit- and number-motion counters so we can compare the bits-per-value with the 32 bits required for single precision:

```
ubitsmoved = numbersmoved = 0;
Module[{a = 3, b = 100, c = 2},
  r1u = (sqrtu[squareu[b̂] ⊖ (4̂ ⊗ â) ⊗ ĉ] ⊖ b̂) ⊘ (2̂ ⊗ â);
  view[r1u]]
```

(−0.02001201609891722910106182098388671875,
 −0.02001201361417770385742187l875)

Numbers moved	22
Unum bits moved	548
Average bits per number	**24.9**

The bound is accurate to seven decimals: −0.02001201⋯. The {3, 5} unum environment computes seven valid decimals in the answer *and* rigorously bounds the error, yet uses an average of *fewer than 25 bits per number*. So in this case, unums provide almost twice as many digits of precision but use fewer bits than standard IEEE floats, despite the overhead of the utag. Unum storage automatically shrinks when there is cancellation of digits. We can make this even more economical automatically if we have a policy of using the **unify** operation. The final 87-bit ubound answer here unifies to just 41 bits, a single unum:

```
nbits[unify[r1u]]
```

```
view[unify[r1u]]
```

(−0.020012017339468002319335937 5, −0.0200120136141777038574218 75)

The `unify` operation expands the width of the bound by only about 1.5 times, while reducing storage by about 2.1 times, so `unify` increases information-per-bit in this case. The result is still good to seven decimal places.

While we do not show it here, if you compare a {3, 6} environment result with a double-precision IEEE result, unums get 16 significant digits versus 12 digits using doubles, and the average unum size is 33 bits versus 64 bits for doubles. We get over *33% more accuracy with about half as many bits.*

The main point of this example is: **No tricks were needed**. The quadratic formula example supports the claim that besides being more accurate and using less storage than floats, unums are also *easier to use* than floats. It is far more likely with unums than with floats that a calculation will accommodate whatever formula a programmer wants to evaluate, without having to perform sophisticated rearrangement of the calculation using algebraic identities.

> "We have so many mathematical techniques that we never stop telling people how to do things… Physics teachers always show technique, rather than the spirit of how to solve physical problems." —*Richard Feynman*

14.4 Bailey's numerical nightmare

David Bailey, a numerical expert who has built libraries for extended precision arithmetic, came up with a very unstable system of two equations in two unknowns to use as a test:

$$0.25510582\,x + 0.52746197\,y = 0.79981812$$
$$0.80143857\,x + 1.65707065\,y = 2.51270273$$

The equations certainly look innocent enough. Assuming exact decimal inputs, this system is solved exactly by $x = -1$ and $y = 2$. We could try Gaussian elimination or Cramer's rule. For systems this small, Cramer's rule actually looks simpler. Recall Cramer's rule:

To solve the pair of equations $\begin{pmatrix} a\,x + b\,y = u \\ c\,x + d\,y = v \end{pmatrix}$ for x and y, compute the determinant $D = a\,d - b\,c$. If D is nonzero, then $x = \frac{u\,d - v\,b}{D}$ and $y = \frac{a\,v - c\,u}{D}$.

Geometrically, solving two equations in two unknowns means finding the intersection of two lines:

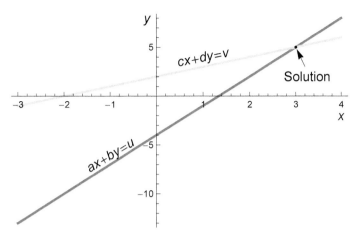

The smaller the determinant D is, *the closer the lines are to being parallel.* If D is exactly zero, then the lines are either separate and never intersect, or they could be the same line and intersect everywhere.

Suppose the lines are *almost* parallel, and there is a little bit of "wobble" in the input values a, b, c, d, u, and v, caused by rounding error of ±0.5 ULP. The intersection point will move all over the place, which means the problem is unstable: Tiny changes to the inputs cause massive changes in the output. In Bailey's nightmarishly unstable problem, the exact determinate D is

$$D = 0.25510582 \times 1.65707065 - 0.52746197 \times 0.80143857$$
$$= 0.4227283669661830 - 0.4227283669661829$$
$$= 0.0000000000000001 = 10^{-16}.$$

The determinant is not zero, but it is so small that even 0.5 ULP wobble in the inputs causes the answer to change by about ±*100 million* from the answer $x = -1$, $y = 2$. There is no problem with a determinant being small if it is exact; but because it is calculated by subtracting very similar float values here, the inaccuracy in the calculation is magnified, including the conversion from decimal input to binary format.

Since IEEE double precision is capable of almost 16 decimal accuracy and the inputs only have 8 or 9 decimals, it seems plausible that it could get this right. **It fails.** When you use double precision and Cramer's Rule, there is massive loss of significant digits through subtraction. The determinant *should* be exactly 10^{-16}:

`det = 0.25510582 x 1.65707065 - 0.52746197 x 0.80143857`

1.66533×10^{-16}

The determinate is not zero, so calculate x and y by Cramer's Rule:

```
{ (0.79981812 × 1.65707065 – 2.51270273 × .52746197) / det,
  (0.25510582 × 2.51270273 – .80143857 × 0.79981812) / det}
```

```
{0., 1.33333}
```

That answer has no resemblance to the correct one, $\{x, y\} = \{-1, 2\}$. By this point, the reader will not be very surprised to learn that double precision *interval* arithmetic fails as well, despite using twice as many bits to represent the numbers. Both x and y evaluate to $[-\infty, \infty]$ with traditional intervals.

Unlike the Rump problem where all input values were exactly expressible with floats, the Bailey problem has two sources of error: Conversion of decimal inputs to binary floats, and inadequate precision for intermediate results. We can separate the two effects. It is not too much to ask a programmer to realize that any linear system can be scaled without changing the answer. Since floats can represent a range of integers exactly, simply scale the problem by 10^8 to make all the values integers, and try again:

```
det = 25 510 582. × 165 707 065. – 52 746 197. × 80 143 857.
```

```
1.
```

That time, the determinant was computed exactly. The problem of decimal-to-float conversion is thus eliminated. So forge ahead:

```
{ (79 981 812. × 165 707 065. – 251 270 273. × 52 746 197.) / det,
  (25 510 582. × 251 270 273. – 80 143 857. × 79 981 812.) / det}
```

```
{0., 2.}
```

Still wrong in x, but at least it got y right.

The error in x is because of the second shortcoming, which is inadequate precision for intermediate results. The problem is that floats only express integers exactly up to the size where an ULP becomes width 2 instead of width 1. The first product $79\,981\,812 \times 165\,707\,065$ should be

$$13\,253\,551\,319\,901\,780$$

and that just happens to be expressible in double precision because it is an even number. The product $251\,270\,273 \times 52\,746\,197$ should be

$$13\,253\,551\,319\,901\,781$$

but that integer is *not* expressible, so it gets rounded down to the nearest even, making the difference between the two products 0 instead of -1.

Bailey has argued that vendors should start supporting quad precision IEEE floats in hardware so that problems like this become easier to solve. Unums point to a different approach, one that does not ask the programmer to schlep 34-decimal values around just to get to answers like −1 and 2: **Rely on the scratchpad.** The values computed in Cramer's Rule like $u\,d - v\,b$ are simply dot products, and the fused dot product only rounds after all the products have been accumulated. Instead of quad precision, go the other way: Use a {3, 5}, environment, the one that can simulate IEEE *single* precision. The input values are exactly representable. By using the fused dot product in the prototype, the programmer does not have to worry about the fact that intermediate calculations take more space.

```
setenv[{3, 5}];
ubitsmoved = numbersmoved = 0;
Module[{au = 25 510 582, bu = 52 746 197, cu = 80 143 857, du = 165 707 065,

   uu = 79 981 812, vu = 251 270 273, det, dxu, dyu},

  det = fdotu[{au, cu}, {du, -1 ⊗ bu}]  (*au⊗du ⊖ bu⊗cu*);

  dxu = fdotu[{uu, vu}, {du, -1 ⊗ bu}]  (*uu⊗du ⊖ vu⊗bu*);

  dyu = fdotu[{au, cu}, {vu, -1 ⊗ uu}]  (*au⊗vu ⊖ cu⊗uu*);

  Print[{view[dxu ⊙ det], view[dyu ⊙ det]}]];
reportdatamotion[numbersmoved, ubitsmoved]
```

{−1, 2}

Numbers moved	30
Unum bits moved	914
Average bits per number	**30.5**

That did the job, computing the *exact* answer with fewer than *half the bits* used by the wildly inaccurate double precision solution. Ironically, the answer to Bailey's demanding problem can now be stored using the most primitive unum environment, since 101 to the left of an all-zero utag always represents −1 and 010 to the left of an all-zero utag always represents 2.

The lesson here is that the selected environment should at least initially match the precision of the input values and the desired output values, with the g-layer handling the need for temporarily more precise scratch calculations. In Part 2, we will explore a technique that automatically finds the complete set of solutions to the problem of two equations in two unknowns when the input values are ranges, not exact values. That technique immediately exposes when problems are ill-posed, like Bailey's.

14.5 Fast Fourier Transforms using unums

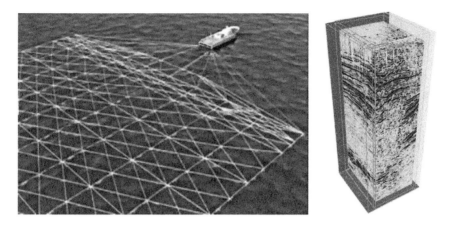

Exploring for oil and gas deposits is done with huge arrays of seismic detectors that record the echoes of intense pulses of sound. This may be the oldest "big data" problem, because the bits that are recorded for processing come back, literally, by the truckload. The exploration industry has long struggled to make data motion and storage more economical. Unums hold the key to doing just that, and in particular can reduce the data motion cost of the most important kernel in signal processing: The Fast Fourier Transform.

This section touches on some fairly high-level math, so consider it optional. Signal processing is one of the most important uses for computers, and it is an area where unums can be of particular benefit. Signal processing hardware is ubiquitous in consumer electronics, where it improves camera and television images, encodes and plays back sound for audio players, and even controls automotive systems.

One of the most important operations in digital signal processing, as well as science and engineering in general, is the *Fourier transform*. If you hear a musical chord and have an ear good enough to pick out the individual notes, you are doing what a Fourier transform does. It transforms a signal, or any function, into sine and cosine functions of different frequencies and amplitudes. The Inverse Fourier transform converts the frequency component representation back into the original signal.

The usual methods for computing the Fourier transform for discrete input data is some variant of what is called the Fast Fourier Transform, or FFT for short. For a signal with n data points, the FFT algorithm produces the Fourier transform with about $5\,n\log_2(n)$ floating point operations. On modern computers, the performance is severely limited not by the amount of arithmetic, but by the need to move the data around. If we can safely reduce the number of bits in the operands, say with unums, we can run proportionately *faster*, plus we get provable bounds on the answer.

A lot of signal information comes from analog-to-digital converters that are only accurate to a few bits. The standard for sound encoding is currently 16-bit accuracy per channel, but for many applications (like oil exploration) the data is only 12-bit accurate (equivalent to about 3.6 decimal digits).

If you look at a graphic equalizer on sound equipment (or displayed by music software), you are seeing Fourier transforms in action. They continuously and in real time calculate the intensities of frequencies from the lowest (bass) to highest (treble):

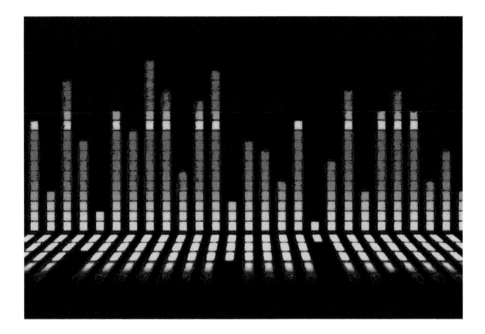

Suppose the signal ranges from −8 to 8 and has only 12 or 16 bits of fraction. In that case, all we need is a small unum tag to represent everything that the analog-to-digital converter can deliver:

```
setenv[{1, 4}]
```

The standard benchmark for signal processing has long been the "1K CFFT," meaning a 1024-point Complex Fast Fourier Transform.

On the next page is a simple general-length version of the algorithm, for the purpose of comparing floats with unums. Double-letter variables like **gg** and **ww** are complex numbers of the form $a + b i$.

```
cfft[rr_List, n_Integer, iflg_Integer] :=
 Module[
  {gg = rr, k = n / 2, th = If[iflg ≥ 0, -π, π], twiddle, ww, t, i, j},
  While[k ≥ 1, ww = N[-2 (Sin[th / (2 k)])² + i Sin[th / k]];
   twiddle = 1. + 0 i;
   For[j = 0, j < k, j++,
    For[i = 1, i ≤ n, i += 2 k,
     t = (gg[[i+j]] - gg[[i+j+k]]);
     gg[[i+j]] = gg[[i+j]] + gg[[i+j+k]];
     gg[[i+j+k]] = twiddle t];
    twiddle = twiddle * ww + twiddle];
   k = k / 2];
  For[i = j = 0, i < n - 1, i++,
   If[i < j, {t0 = gg[[j+1]]; gg[[j+1]] = gg[[i+1]]; gg[[i+1]] = t0}];
   k = n / 2;
   While[k ≤ j, {j = j - k; k = k / 2}]; j = j + k];
  gg]
```

Why not simply use IEEE half-precision floats to save storage, if the input data only has 12 to 16 significant bits? One reason is that IEEE half-precision has 10 bits of fraction, which is not enough. The other reason is the trigonometric functions, which for an n point FFT, sample the unit circle at n locations. Say $n = 1024$; the cosine of $\frac{2\pi}{1024}$ is $0.999981175\cdots$, but with only 10 bits of fraction, that just rounds to the number 1. Similarly for $2\pi \times \frac{2}{1024}$ and $2\pi \times \frac{3}{1024}$. Clearly, something important is getting thrown away by the too-low precision. Hence, signal processing has long been done with single-precision floats despite the low precision of the input data.

One way to think of the Fourier transform is as a sum of waves. Here is a 1024-point FFT that has two wave frequencies, with the real part shown in blue and the imaginary part shown in purple.

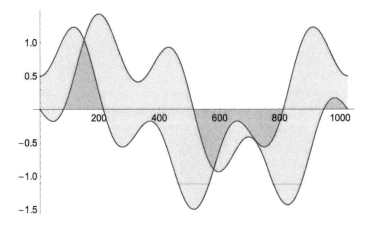

The unum version of the above FFT code is in Appendix C.17. Unlike the previous examples in this chapter, this calculation involves tens of thousands of calculations and provides the best basis yet for comparing the storage efficiency of unums and floats. First, a sanity check on the answer; here is a graph of the same transform, but using unum arithmetic and with tracking of the bit and number motion:

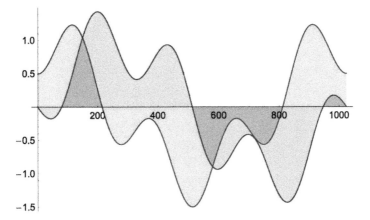

The differences between the two results are well below the resolution of the graphs; the width of the bounds in the unum version are too small to see in the graph as well. Check the data motion costs:

Numbers moved	178 152
Unum bits moved	4 043 801
Average bits per number	**22.7**

The utag is only six bits long, and the maximum unum size is 25 bits. Since many or most of the results of the arithmetic in the FFT are ubounds, and the twiddle factors introduce irrational numbers that typically use the maximum number of fraction bits, you might expect that the average bits per number is pretty close to the maximum for a ubound with two elements, or 51 bits. Why did it do so much better, using fewer than 23 bits per ubound?

Much of the answer can be found in the FFT kernel operation, sometimes called a "butterfly" because of the shape of its data flow diagram:

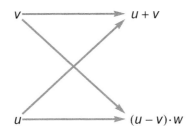

(The "w" is the "twiddle factor" corresponding to `twiddle` in the algorithm above.)

Imagine that *u* and *v* are 12-bit integers, that is, numbers between 0 and 4095. In the top case, adding *u* and *v* risks integer overflow. We might need 13 bits to represent the sum exactly. In the bottom case, we subtract *v* from *u*, which will often require *fewer* than 12 bits. The algorithm practically begs for flexible precision at the bit level, since fixed-sized fractions risk rounding error in one case and wasted storage on the other. The unum environment promotes and demotes the precision as needed, but no more.

The FFT provides a chance to experiment with `smartunify`. The prototype unum function uses `smartunify` with a target ratio of 1, so that a 2-part ubound is only unified when doing so increases information-per-bit. Without it, the cost of the preceding 1K FFT rises to 28.5 bits per value. The very slight loosening of the bounds has negligible effect on the final result, as the graph shows; furthermore, the unified answer is still *bounded*, not a guess as in the case of the float computation.

The oil and gas industry has long relied on 32-bit floats to protect the intermediate calculations on 12-bit data, but the unum method delivers a provably bounded result with fewer bits per number. Much greater economization is possible depending on the application, once the user makes a concession about how much accuracy is needed in the final result.

Part 1 Summary

Unums represent the entire real number line with a limited number of bits, and make it possible to define a computing environment that does not round, overflow, or underflow. Just as floats incorporate scaling information into the number to make it easier to do math on computers, unums take this idea a few steps further by incorporating the exact or inexact status of a number, the number of bits of precision, and the dynamic range. In practice, floats are grossly over-sized as insurance against things that can ruin the result. Because unums automatically scale the bits needed up and down, they usually use *fewer* bits than floats, yet produce more accurate answers that come with provable bounds.

> The biggest objection to unums is likely to come from the
> fact that they are variable in size, at least when they are
> stored in packed form.

> The biggest objection to unums is likely to come from the fact that they are variable in
> size, at least when they are stored in packed form.

What you see above are two identical sentences, one with a monospaced typeface (Courier) and the other with a variable-width typeface (Times) of exactly the same font size. Notice how much less space the variable-width typeface uses. There was a time (before Xerox PARC and Apple's commercialization of PARC's display technology) that computers *always* displayed monospaced font, since system designers thought it inconceivably difficult to manage text display with letters of variable width. The worst-case width of an "M" or a "@" dictated the width of every letter, even an "l" or a ";". With a fixed-width font, legibility suffers and more paper and ink are needed to convey the same message, but things are easier for the engineer who has to figure out where to place the pixels for letter forms on a screen or a printed page. Unums offer the same trade-off versus floats as variable-width versus fixed-width typefaces: Harder for the design engineer and more logic for the computer, but superior for *everyone else* in terms of usability, compactness, and overall cost.

Fewer bits means unums will be *faster* than floats in any computer that is bandwidth-limited, which is just about every computer currently being built. Besides the savings in storage and bandwidth (which results in reduced energy and power consumption), the ability to unpack unums into fixed-size register storage with "summary bits" and the lack of any need for rounding logic means that much of the task of designing hardware for unums is *easier* than for floats, and unums can require fewer gate delays *on chip*.

Some have postulated that any calculation that tracks interval bounds rigorously will always take more storage than the equivalent float routine; actual unum experiments like the FFT in Section 14.5 show this does not need to be the case if *ubounds* are used to track intervals, instead of traditional interval methods.

Programming with unums and ubounds is less treacherous than with floats, for those untrained in the hazards of rounding error. They provide mathematical rigor that even interval arithmetic cannot attain. The information they contain about their own inexactness allows the computer to automatically manage most or all of the numerical analysis burden. Careful definition of what is allowed in the scratchpad calculations guarantees perfect, bitwise identical results from one computer system to another, and even allows unums to obey the associative and distributive laws of algebra. That lets us distinguish between a *programming bug* and a *rounding error*, when doing things that change the order or grouping of calculations, including paral-lelization of software.

There is some unfinished business, however. Since unum bounds resemble interval arithmetic, how do we know they will not suffer the same fate as traditional intervals in producing bounds that are much looser than they should be and need to be? If we already have an algorithm designed for floats, how do we make it work with unums without requiring that the programmer learn about interval arithmetic and its hazards? There is a general solution for this, the *ubox* approach.

Which is the subject of Part 2. ∎

Part 2

A New Way to Solve: The Ubox

15 The other kind of error

15.1 Sampling error

Part 1 showed how to avoid rounding errors, simply by letting numbers track and express their accuracy. However, that is only half the problem with using computers to solve problems that involve real numbers. The other kind of error is variously called *discretization error* or *truncation error* or *sampling error*. "Sampling error" seems the most descriptive, since it is the error caused by sampling something that varies as if it did not vary and was just a constant, representable by just a single value. Whereas rounding error is replacing a correct value with a different value, sampling error is replacing a *range* of values with a single value. Statisticians use the phrase "sampling error" to describe the problem of observing a sample instead of the whole population. That is the way we mean it here as well, where a "population" is all real numbers in the possible variation of a quantity.

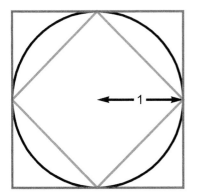

As an example, start with a very ancient problem, one that the ancient Greeks were certainly curious about. If the number we now call π is the area of a disk of radius 1, what is the numerical value of π? The Greeks pursued this question over a thousand years before the invention of calculus, and the tools at their disposal were simply integers and fractions with plus-minus-times-divide arithmetic operations, and elementary geometry.

For example, they could easily prove that the area of the unit disk was more than 2 and less than 4 by using an inner bound (the blue diamond shape) and an outer bound (the red square), The relative width of this bound is $\frac{4-2}{4+2} = 0.33\cdots$ which is low accuracy, but at least it *is* a provable bound and not just an estimate.

One Greek mathematician, Archimedes, used regular polygons with more sides as the inner and outer bounds. That does produce a tighter bound, but it is really the hard way to do the job because it requires the evaluation of square roots that themselves have to be bounded with rational numbers.

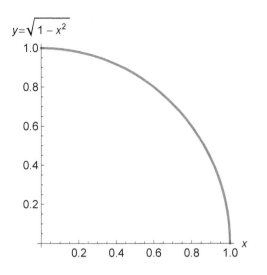

Jump ahead to the way modern textbooks about numerical methods say we should find the area, and watch for the two kinds of error that creep in. Since $x^2 + y^2 = 1$ on the unit circle, the function that describes a quarter circle is

$$y = \sqrt{1 - x^2}$$

for x in $[0, 1]$. The area of the unit circle is four times the area under this curve.

Rounding error shows up when you compute $\sqrt{1 - x^2}$ with floats, and have to round the answer up or down to the nearest number that is representable (as a float).

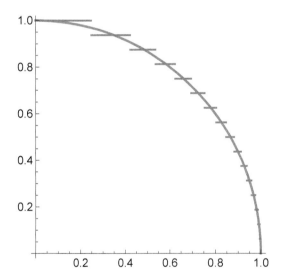

The horizontal magenta lines show what the quarter circle looks like with exaggerated rounding error (only four bits for the fraction), but sampling a large number of points on the x axis. This shows high rounding error; the correct values of the quarter circle (blue) are being replaced by incorrect floats (magenta). However, the *sampling error* is low; the graph on the left samples hundreds of points along the x axis.

It is rounding error that creates an approximation that is very jagged near the top of the semicircle. We *could* use the magenta bounds to get an approximation to π by adding up the areas under each horizontal line segment, but the jagged part at the top does not give us much confidence about the correctness of the result.

Suppose we instead sample the x axis at only, say, 16 equal-spaced points, but compute the function to *double-precision accuracy*.

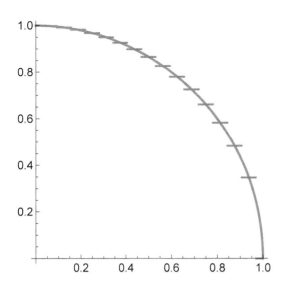

The computation of the circle now looks like this set of horizontal orange lines. The *y* value represented by each orange line segment is a constant, calculated to very high precision but erroneously used as a stand-in for a *range* of values on the blue curve.

This is *sampling* error. Now the error is most obvious when *y* changes quickly as a function of *x*, which is at the right side of the quarter circle.

Both approaches are patently absurd. The first one uses thousands of sample points in *x*, yet can only find *y* to an accuracy of about 6%, by skimping on the numerical accuracy. The second one uses overkill high precision to evaluate the function, but only samples the function at 16 discrete points that grossly misrepresent what the function looks like for nearby values. Would it not make more sense to *balance the two sources of error within whatever storage or computing effort is budgeted?*

15.2 The deeply unsatisfying nature of classical error bounds

The classical numerical analysis way to estimate the area under a curve described by a function $f(x)$ is to approximate the shape with rectangles and add their areas.

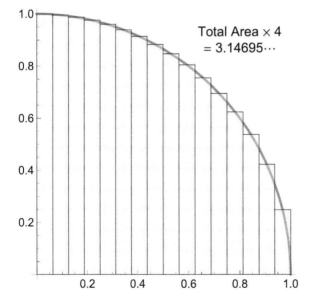

Total Area × 4
= 3.14695···

The "rectangle rule" says to break the horizontal range into subintervals, and then evaluate f at a sample point in each subinterval to get the height of the rectangle. Add up the rectangle areas to estimate the total area. The midpoint looks like the best guess in general. *How close is the midpoint guess?*

The figure on the left shows the quarter circle area, approximated with midpoint-sampled rectangles.

The sum of the areas times four is 3.14695··· which looks close to the value we know for π, 3.14159···, at least to three decimals. But without knowing the value of π, can we put a numerical *bound* on the error?

Pick up any numerical analysis text and it will tell you that the error of using the rectangles to estimate the area under a function $f(x)$, on an interval from a to b with spacing h, is strictly bounded by this formula:

$$\text{error} \leq \frac{(b-a)\,h^2}{24}\left|f''(\xi)\right|$$

Check out this formula. Imagine that you want to actually punch it into a calculator and find out how big the error is, **as a number**. We know a is 0, we know b is 1, we know h is $\frac{1}{16}$… wait a minute, what is this $f''(\xi)$ thing? You have every right to ask,

"What the hell is *that*?" It is the second derivative of the function, in this case $\sqrt{1-x^2}$, at *some unspecified point ξ between a and b*. To compute the second derivative, we first have to know calculus to figure out what the function f'' is, and then we have to somehow find the *maximum possible absolute value* of that derivative over the range from a to b. What has been presented as a nice rigorous bound turns out to be nothing of the sort, because for many functions, including the quarter circle example, $|f''(\xi)|$ can be as large as *infinity*. An infinitely large bound is… **no bound at all**. This is why the classical error bounds that are still taught to this day are deeply, profoundly unsatisfying.

Some people take comfort in the rest of the expression, however, reasoning that "Whatever the error is, if I use twice as many sample points it will make h half as big, and because the error is proportional to h^2, that will make the error one-fourth as big." This is comforting only if the error is finite, of course, since one-fourth of infinity is still an infinite amount of error. If the answer changes with an increased number of sample points, how do you know when the number of sample points is "enough"? *You don't.*

> Refining the sample points is the approach that drives a poorly based reason for demanding ever-faster supercomputers. Computational scientists demand ever *more sample points*, trying to reduce the amount of *sampling error*, but without knowing how big the sampling error *is*.

The values at the coarse sample points are usually calculated to a ridiculously high level of accuracy (single or double precision) in the hope of making h^2 smaller and thus reducing the (unknown) amount of error. It is also possible to take too many sample points, another costly form of over-insurance.

There are other methods besides the Rectangle Rule, of course: Trapezoidal Rule, Simpson's Rule, Gaussian Quadrature, and so on. Every method has an "error bound" formula that relies on taking some order derivative of the function at some unknown point in the region of interest, which could be infinite, as it is for the trivial example of the quarter circle. And many common functions, like ones that have sharp kinks, do not even *have* a derivative.

We can do far better than this, without knowing any calculus at all.

15.3 The ubox approach

> **Definition**: A *ubox* is a shape in *n*-dimensional space that is represented by a list of *n* unums.

Because unums can be exact or inexact, the width of any ubox dimension may be zero (exact), or one ULP (inexact). In three dimensions, for example, a ubox might look like a point, a line segment that does not include its endpoints, a rectangle that does not include its perimeter, or a rectangular box that does not include its surface, depending on how many of the dimensions are exact. Uboxes provide a perfect "tiling" of *n*-dimensional space the way unums perfectly tile the real number line. The unum list is written inside braces `{...}`, like `{xu,yu}`, which in the prototype could look like the way we express a ubound, `{xlowu,xhighu}` if we are using two-dimensional uboxes. We depend on context to show which is which, but in an ideal computing environment there would be a distinct notation for the two concepts.

What does it mean to compute the *solution* to a problem involving real numbers? Not the same thing as a mathematical solution, so we give it a slightly different name, the *c-solution*.

> **Definition**: In computing, a *c-solution* to a problem involving real numbers is the smallest complete set of uboxes that satisfy the problem statement.

That may sound like an empty statement at first reading, but *smallest* and *complete* make it different from the common use of the word "solution" in the context of computation. The examples in sections that follow will make clear how different most computations with floats are from the above definition of a solution, a *c-solution*. Think of the "c" in *c*-solution as either "computational" or "complete."

With uboxes, there are several ways to find the complete yet minimum set of uboxes that satisfy a problem. We frequently begin with trial uboxes, and use unum arithmetic to determine if they are solutions or failures. When the problems are one-dimensional or two-dimensional, we can depict the rounded edges that reflect an open interval, as shown in the table on the next page:

	Exact x, exact y	Exact x, ULP–wide y	ULP–wide x, exact y	Both x and y are ULP–wide
Solutions				
Trials				
Failures				

One strategy when the answer set is known to be *connected* is to find a single ubox that is a solution, examine all of its neighbors, and sort them into solution or non-solution categories. If a problem is computed using the four arithmetic operations and there is no division by zero, a connected input set remains connected through the computation, and by examining neighbors we eventually find every possible solution at a given precision. Another strategy is to try all possible uboxes, and start pruning away those that fail. Both approaches work toward the same goal: Find *all* the uboxes that are part of the solution set, but *exclude* all the uboxes that are not.

If the number of uboxes is too large to compute on a particular computer system, that means the precision is set too high for the amount of computing power available. The ULPs are so small that the solution set will take longer to enumerate than the time we are willing to wait. The ubox approach thus provides a very flexible trade-off between answer quality and computing power, where the computer power is the amount of *parallelism*. Because unums can adjust precision one bit at a time, we can start with very low precision and refine it with smaller and smaller ULP sizes; that means the approach also provides a trade-off between the answer quality and the time we are willing to wait.

> *Uboxes are to computing what atoms are to physics.*

15.4 Walking the line

Just as the reference implementation provides a set of tools for working with unums (like viewing, performing arithmetic, and comparing them), it also provides a set of tools for manipulating uboxes. They are introduced gradually here as needed, but also are collected for easier reference in Appendix D. The first of these will be tools to find the neighbors of a ubox.

First, set the environment to a low precision so the ULPs are large enough to show up on graphs. A {1, 3} environment has unums with at most $2^1 = 2$ bits of exponent and $2^3 = 8$ bits of fraction, and a total size ranging from 8 to 16 bits.

```
setenv[{1, 3}]
```

The first ubox routine we need is a way to find the *nearest neighbors* of a unum. In the prototype, the **nborlo** and **nborhi** functions take a unum as an argument and find an adjacent unum that is lower (closer to negative infinity) and higher (closer to infinity). The code for both functions is in Appendix D.2. Sometimes, the range of exploration includes zero, and it becomes tedious for the ULPs to become extremely small as they scale their way down to the smallest subnormal unum, then back up again. Therefore, the functions take as parameters not just the unum but also a **minpower** value that sets the minimum preferred size of an ULP to $2^{\texttt{minpower}}$. Here are a couple of examples, for the **nborhi** of the exact number zero. The first asks for a minimum ULP size of $2^{-1} = 0.5$ and the second for a minimum ULP size of $2^{-3} = 0.125$. The third asks for a minimum ULP size of 2^{-10}, but ULPs do not go that small in the {1, 2} environment so instead we get the smallest ULP available, which is $2^{-4} = 0.0625$:

```
view[nborhi[0̂, -1]]
view[nborhi[0̂, -3]]
view[nborhi[0̂, -10]]
```

(0, 0.5)

(0, 0.125)

(0, 0.00390625)

The neighbor of an exact unum is an inexact one, and vice versa; the **nborhi** of the unum for (0, 0.125) will always be the unum for exactly 0.125 and its **nborlo** is always exactly 0, no matter what the preferred minimum ULP size is. However, the preferred minimum ULP size is *overridden* if the starting value does not land on an even multiple of that ULP size.

For example, if we ask for the **nborlo** of $\frac{5}{8} = 0.625$ with preferred ULP size $2^{-1} = \frac{1}{2}$, the routine will return finer-grained ULPs until it gets to a multiple of $\frac{1}{2}$, and only then can it take steps of size $\frac{1}{2}$. Here is what happens in "walking the line" three steps to the left, always requesting an ULP of $2^{-1} = \frac{1}{2}$:

$$\texttt{view}\Big[\texttt{nborlo}\Big[0.\hat{6}25, -1\Big]\Big]$$
$$\texttt{view}\Big[\texttt{nborlo}\Big[\texttt{nborlo}\Big[0.\hat{6}25, -1\Big], -1\Big]\Big]$$
$$\texttt{view}\Big[\texttt{nborlo}\Big[\texttt{nborlo}\Big[\texttt{nborlo}\Big[0.\hat{6}25, -1\Big], -1\Big], -1\Big]\Big]$$

(0.5, 0.625)

0.5

(0, 0.5)

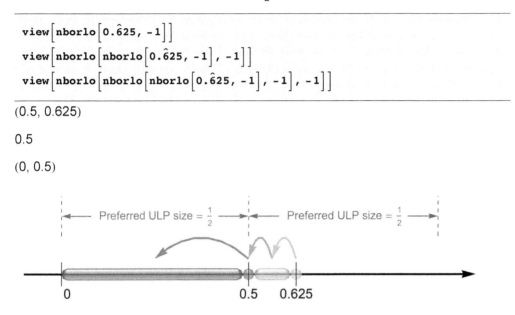

Think of **nborhi** and **nborlo** as the basic ways to walk up and down the real number line, without skipping over any possible values and without ever confusing exact numbers with inexact ones.

The **nborhi** and **nborlo** routines work on "almost infinite" and infinite numbers as well. In the {1,2} environment, the largest positive exact number is 7.5, so its upper neighbor is (7.5, ∞), and the upper neighbor of *that* is exact infinity. And of course, if we try to go "to infinity and beyond," we get nonsense, or NaN. The "preferred ULP size" does not matter in such cases, so we just use 0:

```
view[nborhi[maxrealu, 0]]
view[nborhi[nborhi[maxrealu, 0], 0]]
view[nborhi[nborhi[nborhi[maxrealu, 0], 0], 0]]
```

(7.96875, ∞)

∞

NaN

The only neighbor of NaN is NaN. NaNs are like black holes; once you stumble into one, there is no getting out. We use quiet NaN instead of signaling NaN, since we usually want to continue computing.

15.5 A ubox connected-region example: Unit circle area

One way to use uboxes is the "paint bucket" method, named for the drawing tool invented over thirty years ago. Here is what it looked like in Andy Hertzfeld's brilliant MacPaint program that came with the original 1984 Macintosh:

Even in more modern programs like Photoshop, the tool still does the same thing: It fills in an irregular shape with a color or pattern, starting from a "seed" point and automatically finding the edge of the shape. The only condition is that the irregular shape is a *connected region* with a defined edge.

Frequently, in solving computational problems, we at least know one point in the solution set and we often have the situation that the region is known to be connected. The "edge" of the region is determined by the statement of the conditions of the problem being solved.

That is certainly applicable to the problem of finding the area of the quarter circle. The points inside the quarter circle satisfy the condition that they are distance less than 1 from the origin, and both x and y are greater than or equal to zero. Trivially, $x = 0$ and $y = 0$ is a solution, so $\{\hat{0}, \hat{0}\}$ could be the "seed" ubox, the first set in the solution set we typically call **sols**:

---

sols $= \left\{ \left\{ \hat{0}, \hat{0} \right\} \right\};$

The prototype function **findnbors**$[set, \{minpower_1, ..., minpower_n\}]$ uses the **nborhi** and **nborlo** functions to find the neighbor uboxes for all the uboxes in *set*, with preferred minimum ULP sizes given by a different *minpower* for each of the n dimensions. The routine excludes the original *set* values.

In two dimensions, a ubox has *eight* neighbors. We assign the neighbors of initial seed ubox $\{\hat{0},\hat{0}\}$ to the `trials` list, where we use 2^{-4} and 2^{-4} as the preferred minimum ULP sizes for purposes of illustration. That will make the uboxes square if both dimensions are inexact, but in general they can be rectangular. After all, we might be solving a problem where x goes from 1.0 to 1.01, say, while y ranges from 10^{10} to 10^{12}, so the minimum ULP sizes must be independent for each dimension.

```
minulp = -4;
trials = findnbors[sols, {minulp, minulp}];
```

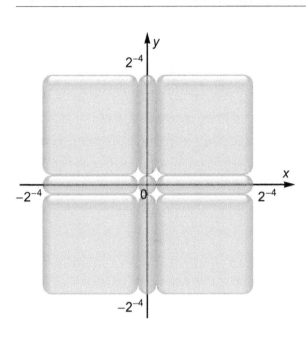

Here is a close-up view of the initial seed solution surrounded by trial uboxes. Four of the neighbors are line segments one ULP wide to the right, left, top, and bottom of the seed solution point; they do not include their endpoints, since ULP-wide unums are open intervals. The rounded endpoints help remind us of that. The other four neighbors are squares one ULP on each side, and those squares do not include their borders. Every real-valued location that is within one ULP of the starting point in x and y is accounted for, exactly once.

Exercise for the reader: How many neighbors does an n-dimensional ubox have?

Each trial element is tested to see if it is *inside the unit circle* and is *in the upper right quadrant*. That is, the problem statement is to find all computer-representable x and y that satisfy three conditions:

$$x \geq 0,$$
$$y \geq 0, \text{ and}$$
$$x^2 + y^2 < 1.$$

The `touchQ[{`*xu,yu*`}]` function returns `True` if these conditions are met and `False` if not, for a ubox $\{xu, yu\}$. The code for the `touchQ` function on the following page matches the problem statement. Every problem statement has a different `touchQ` test. In the prototype, we can create a "greater than or equal" test by putting a logical NOT symbol "¬" in front of a "less than" test, so "$x \geq 0$" translates to "$\neg \left(\mathbf{xu} < \hat{\mathbf{0}} \right)$".

```
touchQ[{xu_, yu_}] :=
```

$$\neg \left(xu < \hat{0} \right) \bigwedge$$

$$\neg \left(yu < \hat{0} \right) \bigwedge$$

```
fdotu[{xu, yu}, {xu, yu}] < 1̂
```

Notice that a fused dot product of $\{x, y\}$ with itself was used to compute $x^2 + y^2$. This assures perfect decision-making in comparing the value with 1, since the result of each of the two multiplies and the add will not always be representable in *u*-layer without some information loss. For example, if $x = \frac{9}{16}$ and $y = \frac{13}{16}$, the $\{1, 2\}$ environment represents x^2 as (0.3125, 0.375) and y^2 as (0.625, 0.6875), which then sum to (0.9375, 1.0625). The range spills outside the unit circle. But a *g*-layer computation computes $x^2 + y^2$ exactly as $\frac{125}{128}$, which then becomes (0.9375, 1) as the *u*-layer value, and is correctly classified as being inside the unit circle.

In a proper unum hardware environment, a test like this **touchQ** is computationally very cheap. In the *g*-layer, it does two multiplications, one addition, two comparisons against zero (which really means testing a single bit), and one comparison against the number 1. And remember, these are extremely small unum bit strings, so the time to do a multiply with 2014-era chip technology is probably better measured in picoseconds than nanoseconds.

The **fails** set holds any ubox that starts in **trials** but when tested with **touchQ**, proves not to touch the solution. Keeping track of failures lets us avoid testing a possible ubox more than once. Failure uboxes will be shown in red. Initially, **fails** is empty, but it will grow to the set of uboxes that *encloses* the solution.

Test each trial ubox with **touchQ**: Does it satisfy $x \geq 0$, $y \geq 0$, and $x^2 + y^2 < 1$? Notice that *every trial can be done independently*, which means perfect parallelism to exploit multiple processors in a computer system. A solution is appended to the **new** set if it was not previously known, and also appended to the known solution set, **sol**. The reason we track which ubox solutions are new is because those are the only ones that can generate neighbors that have not already been looked at. After all of the trial uboxes have been checked, the set of trial uboxes is reset to empty.

Here is the loop that classifies uboxes in the **trials** set for any **touchQ** test:

```
new = {}; fails = {};
For[i = 1, i ≤ Length[trials], i++, Module[{t = trials〚i〛},
  If[touchQ[t],
    If[¬ MemberQ[sols, t], {
      new = AppendTo[new, t]; sols = AppendTo[sols, t]}],
    fails = AppendTo[fails, t]]]]
trials = {};
```

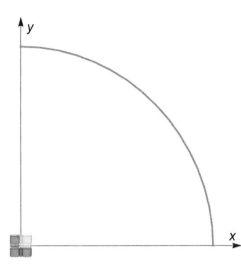

Here is the result of classifying the trial uboxes using the code. The seed box is far from the actual edge of the circle, so the only neighboring uboxes that fail are those not in the upper right quadrant. Three new solution uboxes were found, which are recorded in both the **sols** and the **new** set.

Next, generate a set of trials from all of the new solutions just found. To do this, we find the neighbors of every member of the **new** set, and test if each neighbor even qualifies as a trial ubox.

A neighbor of an element of **new** does not qualify as a trial ubox if

- it is a member of the set of not-new solutions already, or
- it is a member of the set of known failures.

This is succinctly expressed by finding the complement of the neighbor set with **fails** and **sols**. Ubox methods often look more like set theory than arithmetic.

```
trials = Complement[findnbors[new, {minulp, minulp}], fails, sols];
```

That line of code found new trial uboxes to test. They fail unless $x \geq 0$ and $y \geq 0$:

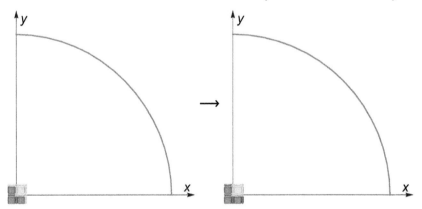

Exercise for the reader: How many amber "trial" boxes will be created in the next pass? How many of them will be added to the solution set?

We have everything we need to make this an automatic loop, except for a condition that says *when to stop*. We stop when *there are no "new" solution uboxes*. That means every viable ubox has been tested and categorized as a success or a failure, and the last pass did not produce a single success.

Here is the complete loop for the paint bucket method, in ten lines of code:

```
While[Length[new] > 0,
  trials = Complement[findnbors[new, {minulp, minulp}], fails, sols];
  Module[{newtemp = {}},
    For[i = 1, i ≤ Length[trials], i++, Module[{t = trials⟦i⟧},
      If[touchQ[t], (* If the problem statement is satisfied, *)
        If[¬ MemberQ[sols, t],
          (* and if it is not already in the solution set, then *)
          newtemp = AppendTo[newtemp, t];
          (* append it to the temporary "new" set; *)
          sols = AppendTo[sols, t]],
          (* also add it to the solution set. Else... *)
        fails = AppendTo[fails, t]]]];
  (* add it to the set of known failure uboxes. *)
  new = newtemp]] (* "new" ubox neighbors form the next trial set. *)
```

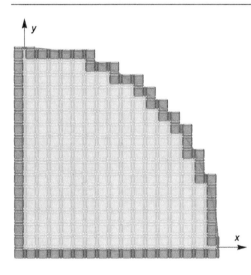

That brief code above produces a rigorous bound for the location of the quarter circle, without rounding errors *or* sampling errors, using a very small set of bits for each ubox.

But the solution set can be made even more concise. The solution uboxes can be *coalesced* to create a lossless compression of the answer set, by repeatedly looking for contiguous trios of inexact-exact-inexact uboxes aligned to the ULP with one less fraction bit, and replacing them with a single inexact ubox.

Here is the compressed solution set (and we compress the **fails** set as well):

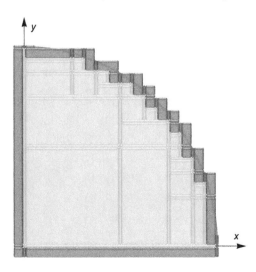

Code for **coalesce**[*set*] is shown in Appendix D.4. The compressed answer set is assigned to **areasols**.

In the prototype, the **volume**[*set*] function finds the total *n*-dimensional volume (in this case, area) of a set of uboxes. If we use it on this solution set and multiply by four, we have a lower bound on the value of π. Notice that the only contribution to the volume is from uboxes that are inexact in both dimensions. Four times **volume**[**areasols**] is 2.890625.

The next step is to find the *upper* bound. Apply the same procedure with a slightly different **touchQ** test, one that requires $x^2 + y^2 = 1$ instead of $x^2 + y^2 < 1$:

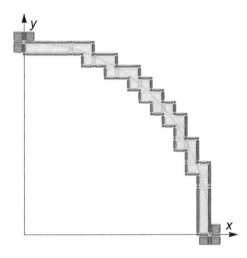

As a seed value, we can start with an obvious solution point like $x = 0$, $y = 1$. That will find the edge set.

An animation of the algorithm looks like a fuse has been lit on the left, and it burns its way around the arc of the circle until it reaches $x = 1$, $y = 0$ on the bottom right, and automatically stops since there are no elements in the **new** set to test.

The volume of *this* set of uboxes is

```
volume[edgeset]
```

$$\frac{31}{256}$$

That value, times four, is the difference between the lower bound and the upper bound. This proves that

$$2.859375 < \pi < 3.34375$$

If all we care about is area, or if we only want to know the *shape* (for graphical display, say), we can ignore the uboxes that have zero area. There are only 20 such uboxes in the area solution set, and 18 in the edge set. The extraordinary thing is the total number of bits required to store both sets: *788.* That is about the same number of bits as just a dozen IEEE double-precision floats. That set of bits is all that would need to be stored in memory at the conclusion of the calculation as the *c*-solution.

> **Exercise for the reader**: How many bits are needed to store the bound on π as computed here? What is the relative width of the bound?

Here are what the interior and edge sets look like after the exact points and line segments have been removed, where we still use the rounded look as a reminder that the rectangles do not include their borders:

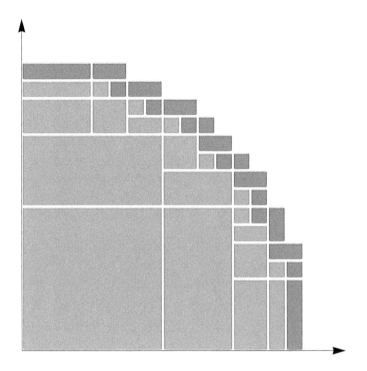

We just computed π and provably bounded the error, using numbers with seven to eleven bits. The more surprising thing is what we did *not* use: Trigonometry, calculus, or infinite series. In fact, we did not even use square root or divide operations.

There was also no need to worry about balancing the sampling error and the rounding error. Both types of error were eliminated and replaced by bounded ULP values, which tracked the shape of the region equally well in the horizontal and vertical directions. All the computer user has to do is specify the **touchQ** function that describes what is considered a solution, a starting point, and perhaps a preferred amount of accuracy in the result, and the computer hammers out a *c*-solution that is *complete* yet *as small as possible* for the computing environment. The brute force of the computer replaces the need for users to be clever.

Far more sophisticated versions of the paint bucket method are possible, of course. The uboxes could be coalesced as an ongoing part of the solution search, to economize the use of temporary storage. The ULP size could start as large as possible and refine wherever a ubox is partly inside and partly outside the solution region. This may remind some readers of a very old idea: *Grid refinement*. Divide squares into smaller squares whenever the need for more detail is detected.

But grid refinement performed with floats has very limited ability to detect *where* more detail is needed, since the float calculations are only sampling what may be a highly variable function. They also take up a lot of space, far more than uboxes do. To store the quarter-circle shape and boundary shown above using the smallest IEEE floats (half precision) takes *2432* bits, which is more than three times as many bits as the ubox solution requires. In a computer system for which the performance is limited by bandwidth (the usual situation), a ubox solution would be three times faster than a grid refinement method using half-precision floats, and *six* times faster than one using single-precision floats.

15.6 A definition of information and computing "speed"

The volume of the edge set is a measure of answer quality; it tells us how much information we have about the answer. In Part 1, we talked about information-per-bit in deciding whether to compress unums. In the context of computational problems involving real numbers, we can more generally define what we mean by "information":

> **Definition**: *Information*, in the context of a particular computing question, is the reciprocal of the total volume of a set that bounds the answer.

For a ubound or a unum, the "volume" is simply the width: The upper bound minus the lower bound of the interval it represents. Information is the reciprocal of the width, the one-dimensional form of "volume." In two dimensions, "volume" is the area. Whatever the number of dimensions, a bound has volume if it is inexact in all of its dimensions.

The definition has a number of shortcomings, such as not dealing with infinite bounds. A more rigorous definition might use the number of possible uboxes at the finest possible ULP size; information would be the fraction that have been classified as solutions or failures to the problem statement. A problem starts with every ubox as a trial ubox (information = 0) and ends up with the *c*-solution at the smallest possible ULP size (information = 1). This definition is usually impractical to calculate, hence the volume-based definition above is the one used in this book.

The Greek bound for π (the diamond lower bound and square upper bound in Section 15.1) showed the circle area was between 2 and 4, so its information about the answer was $\frac{1}{(4-2)} = \frac{1}{2}$. The information produced by traditional numerical methods like the midpoint rule (Section 15.2) is *zero*, since the error "bound" is infinite; they provide a *guess*, but no *information*. For the low-precision ubox computation of the circle area, the volume of the uboxes describing the edge was $\frac{31}{64}$, so the information is $\frac{64}{31} = 2.06\cdots$.

Cutting the ULP size in half will approximately double the information of the quarter-circle *c*-solution. If we really want to nail the value of π via the paint bucket approach, set the environment to something like {1, 5} where we have up to $2^5 = 32$ fraction bits, and apply the preceding procedure. The volume of the edge uboxes gets about 270 million times smaller than possible with only $2^2 = 4$ fraction bits, so the information gets about 270 million times greater. That provably bounds the value of π to eight decimal places. There are only a few billion uboxes in the edge set at that precision, so a single processor core with hardware support for unum arithmetic (and 2014-era chip technology) should take less than one second to compute the edge uboxes and their volume using the paint box method.

Which brings up another capability we get with uboxes: A decent definition of what is meant by "speed" when talking about computations involving real numbers.

> **Definition**: In the context of a particular computing task, *computational speed* is information achieved per second.

That definition may sound peculiar to those accustomed to integer calculations, where nothing is known about the answer until it is completed, and then everything is known. Information about the answer is nil until the end of the algorithm, when it becomes perfect since the answer is known exactly. A problem like "What are the prime numbers between 2 and 1000?" has a set of exact integer values as the result, for example, so computational performance is simply the reciprocal of the time it takes to find the entire set. Calculations involving real numbers typically produce inexact results, so any definition of computing speed needs to first define a measure of what is being accomplished by the calculation.

With the above definition, it becomes possible to rigorously compare the information-per-second or information-per-bit of different computers, no matter how different they are in architecture or what algorithm or language they use to bound the uncertainty.

If we do not have a bound on the error, then the error must be assumed *infinitely large* and the information about the answer is therefore *zero*. This is the case with most floating point calculations, since they only estimate answers and usually do not produce rigorous bounds.

> The traditional measure of computational speed using floats is FLOPS, for FLoating Point Operations Per Second. The FLOPS measure describes only the activity of a computer's float units and *not what the computation is accomplishing*.

212 | The End of Error

The FLOPS measure looks even more ridiculous when you try to use it for anything other than multiplications, additions, and subtractions, because everyone has different ideas about how to score the FLOPS done in a square root or a cosine or even a simple division operation. And how should FLOPS of different precisions be compared? The FLOPS measure is a holdover from the 1970s, when float arithmetic took much longer than memory operations; now that situation is reversed.

Some have tried to rescue the unsound currency of FLOPS by creating elaborate rule books for how to score the computer activity in doing float operations, but FLOPS are not salvageable as a rigorous way to measure or compare computational performance. To see why, consider this: It is easy to create examples of two different algorithms that arrive at the same answer but the one that takes *longer* gets a *higher* FLOPS rating. By defining information as the purpose of a computation, we avoid confusing the ends and the means. Using FLOPS is like judging a foot race by how many footsteps per second were taken by a runner instead of judging by when the runner reaches the finish line.

Recall that we also can measure the number of bits moved in a computation as a first-order approximation to how much *work* the computation does. The cost (time, money, energy, and power) of moving data presently dwarfs the cost of operating logic gates, and the ratio is getting larger with each new generation of computers, so the number of bits moved is a much better measure of work than operation counts. The "best" algorithms are therefore those that produce the highest quality answer for the least bit motion. We can compare different approaches using *information per watt* and *information per joule* or per any other measure of cost, depending on the priority of the application.

Of course, the definition of information is different for every problem, so it does not work to try to describe a piece of computer hardware as having a "peak information per second" rating the way people have done with FLOPS for so many years. The idea of measuring speed with information per second is to allow fair and logical comparison of any two ways to solve the same specific problem.

15.7 Another Kahan booby trap: The "smooth surprise"

Since computational scientists have not had rigorous mathematical bounds on their answers, they often use visualization to see if the answer "looks right," (sometimes called "the eyeball metric"). Professor Kahan shows just how far this can lead us astray with the following example: Suppose we need to know the minimum value of

$$\frac{1}{80} \log(|3(1-x)+1|) + x^2 + 1$$

in the range $0.8 \le x \le 2.0$; so we plot the function and look at it:

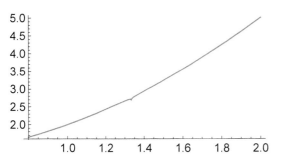

It certainly looks like the minimum is where *x* is 0.8, over on the left, but there is a funny little blip around 1.33···. Most people would ignore it. A diligent computer user *might* ask for higher resolution to tease out what that thing is.

So plot it again but use *500 000* plot points, to try to reduce the sampling error:

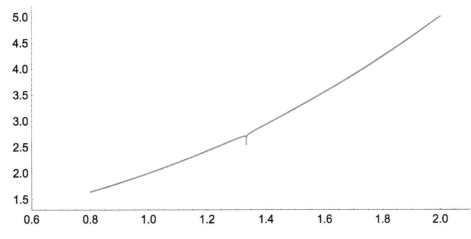

The little blip still looks like nothing to worry about.

Exercise for the reader: Find the two IEEE double-precision numbers closest to $\frac{4}{3}$ (rounded up and rounded down.) What is the value of the above function for those values?

What if we plotted it using unum arithmetic, that is, with low-precision uboxes that completely tile the range of *x* values from 0.8 to 2.0? First, define the function in unum form; call it `spike[u]`:

```
setenv[{2, 3}];
spike[u_] := (1̂ ⊕ squareu[u]) ⊕ logu[absu[1̂ ⊕ 3̂ ⊗ (1̂ ⊖ u)]] ⊙ 8̂0
```

Instead of demanding a huge number of sample points, try plotting the function using a ubox tiling of the range with ULPs of size $\frac{1}{32}$ so that only a few dozen function evaluations are needed. The plot on the following page shows the plot using double-precision floats in orange, and the plot using uboxes in cyan:

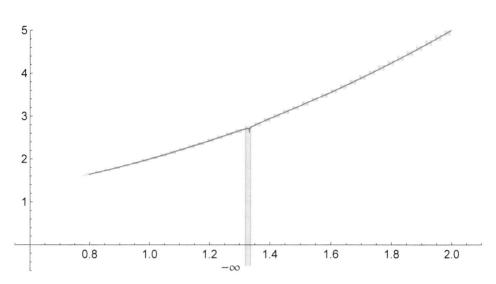

Low-precision uboxes immediately reveal the plummet to negative infinity when $x = \frac{4}{3}$, something even trillions of sample points with IEEE double-precision floats cannot do. This shows how different the ubox method is from "grid refinement" with floats; a grid refinement method would not even be able to detect the *need* for more points near the singularity, but would blissfully pass over it. Unum arithmetic uses an exact value or a one-ULP-wide set of real numbers, and captures the *entire range* of what could happen within that ULP, overlooking nothing. For the ubox that spans the value $\frac{4}{3}$, the **spike** function evaluates to $\left[-\infty, \frac{175}{64}\right)$, and notice that the interval is closed on the left. The function evaluation does not just "approach" $-\infty$, but correctly *includes* $-\infty$ as a value. No amount of sampling will ever produce a binary float that lands exactly on $\frac{4}{3}$ which then leads to evaluating the logarithm of zero. But unum/ubox computation takes into account the entire range without sampling, despite using just a handful of unum values that take up at most 19 bits each.

In preliminary discussions about the relative merits of unum arithmetic, floats were defended with "It all comes down to price-performance." It does indeed, but here, **floats *lose* by a country mile.** The plot using floats tested 500 000 sample points with 52-bit fraction sizes and completely missed the mark. With a few dozen uboxes, where the bounds had at most 8 bits of fraction, the ubox plot easily identified the singularity and the user did not have to be careful to select how many sample points to use. For both programming effort and computer performance, uboxes are vastly more cost-effective than floats in this case. It actually does not make sense to compare price-performance, since unums get the answer and floats do not. If you do not care about getting the right answer, it is possible to compute very quickly indeed!

Unums cannot make rounding errors because they *do not round.*

Uboxes cannot make sampling errors because they *do not sample.*

16 Avoiding interval arithmetic pitfalls

16.1 Useless error bounds

Interval arithmetic seems like a great idea, until you actually try to use it to solve a problem. Several examples of interval arithmetic in Part 1 wound up with an answer bound of "between $-\infty$ and ∞," a correct but completely useless result. The information obtained is $\frac{1}{\infty} = 0$. Unums, ubounds, and uboxes also deal with intervals, suggesting that whatever plagues interval arithmetic and has prevented its widespread adoption will also affect unum arithmetic and send would-be unum users scurrying back to the familiar turf of unknown rounding and sampling errors that produce exact-looking results.

Most of Part 2 is about why uboxes need *not* suffer the drawbacks of interval arithmetic. This chapter is about the two major pitfalls of traditional interval arithmetic and how uboxes avoid them. Later chapters give examples of real problems solved with uboxes, drawn from a wide variety of classic scientific and engineering applications and not contrived as problems-that-fit-the-solution.

There are two major reasons why the bounds produced by interval arithmetic grow so quickly instead of tightly containing the answer: "The wrapping problem" and "the dependency problem."

16.2 The wrapping problem

Interval arithmetic suffers from a drawback called *the wrapping problem*. Here is one of the simplest illustrations of the problem. Suppose we know x and y as intervals, and want to calculate

$$u = x + y$$
$$v = 2x - y$$

Just one addition, one multiplication, and one subtraction; computing doesn't get much simpler than this. Say $2 \le x \le 5$ and $1 \le y \le 2$. With traditional intervals, we get this result, where the "`Interval[{a,b}]`" notation means the set of numbers x such that $a \le x \le b$:

```
x = Interval[{2, 5}];
y = Interval[{1, 2}];
u = x + y
v = 2 x - y
```

```
Interval[{3, 7}]

Interval[{2, 9}]
```

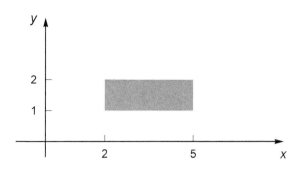

The interval result hides the fact that a great deal of information has been lost, using "information" as defined in the previous chapter. Graphically, we start with the rectangular bound shown in purple in the plot on the left.

The true range of values of *u* and *v* are shown by the tilted blue parallelogram below.

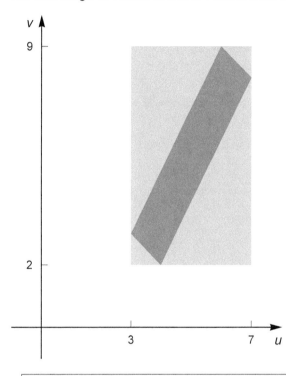

However, what interval arithmetic gives us is the *entire rectangle containing the parallelogram*, which is the "wrapper" for the solution set: $3 \le x \le 7$ and $2 \le y \le 9$. The area of the blue parallelogram is 9, whereas the area of the gray wrapping rectangle is 28, which is more than a *threefold* loss of information about the actual *c*-solution. If the crude description of the *u* and *v* bounds is then passed as input to another simple routine, information about the shape of the solution set is again lost. Without considerable care and expertise, interval arithmetic gives a bound too loose to be useful.

Definition: *The wrapping problem* is excessively large bounds in interval calculations caused by bounding a multidimensional result set with a single axis-aligned box (that is, a simple bound in each dimension).

Some interval arithmetic proponents say the fix for this is to keep track of the complicated shapes using sophisticated mathematical curves and surfaces that cling more closely to the shape, in the hope of keeping the bound as tight as possible. The explosion of complexity that results is unmanageable either by the computer or the programmer. All the arithmetic operations have to be defined not for intervals but for every possible enclosing shape! Each shape requires many numerical values to describe, so instead of economizing bits to save bandwidth and energy, you can easily wind up with ten or a hundred times as much data to move around.

To see why unums and ubounds might not suffer the same problem, imagine the original rectangle in x and y as *a collection of uboxes*. Suppose the environment is set to a very low precision, like {2, 2}:

```
setenv[{2, 2}]
```

That is still enough precision to create ULPs as small as $2^{-4} = \frac{1}{16}$ inside the purple rectangle, but we purposely use larger ULPs so the uboxes are large enough to show graphically. Time for another ubox tool: The $\texttt{uboxlist}\,[\textit{ubound},\textit{minpower}]$ function creates a list of unums of preferred minimum ULP size $2^{\textit{minpower}}$ that perfectly tile *ubound*. For example, $\texttt{uboxlist}\big[\{\hat{2}, \hat{5}\}, \texttt{-2}\big]$ creates a list of unums representing the general intervals 2, (2, 2.25), 2.25, ..., (4.75, 5), 5 with ULP size 2^{-2}. That covers the x values in the purple rectangle. Similarly, $\texttt{uboxlist}\big[\{\hat{1}, \hat{2}\}, \texttt{-1}\big]$ creates a list for the y direction where we pick a minimum ULP size of $2^{-1} = \frac{1}{2}$; ULP sizes can be different in each dimension. The complete set of uboxes for the rectangle set is then formed by taking tuples of the x and y lists:

```
rect = Tuples[{uboxlist[{2̂, 5̂}, -2], uboxlist[{1̂, 2̂}, -1]}];
```

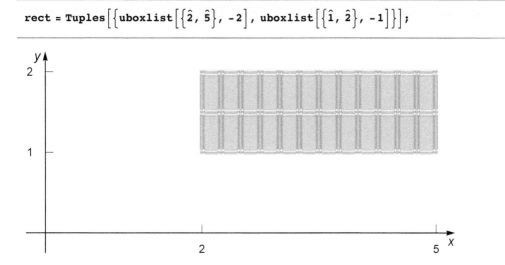

This is a perfect tiling; the open and closed endpoints fit together perfectly with no overlap. Every real number x and y within the rectangle or on its border is represented exactly once.

Traditional interval arithmetic sometimes uses a technique that is sometimes called "mincing" or "paving" that looks very much like the ubox approach shown here. There are several significant differences, however:

- Both exact and inexact dimensions are used for the subintervals.
- The degree of "mincing" is the ULP width, at a selectable level of fraction precision.
- The interval in each dimension is expressed with a *single* number, a unum.
- The results are "self-mincing" in that they always break into ULP-sized quantities.

Here is the computation on *all* the points in the rectangular closed bound:

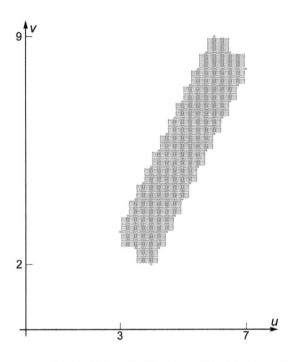

This almost fits our definition of a *c*-solution, because it has *every* ubox that touches the answer set, but it does contain uboxes that fail to touch the true answer set (at the minimum ULP size of 2^{-2}. The area of the answer set is 11, greater than the ideal volume of 9 because of the ragged edge. That is an information loss of only 18%, compared to an information loss of 68% for interval arithmetic. If you want even less information loss, simply reduce the ULP size. Since almost all of the work is data parallel, using a large number of uboxes need not take a lot of time; **it is adjustable to the parallelism of the computing system**.

Exercise for the reader: Find the six uboxes in the preceding result that are outside the *c*-solution. Which uboxes in the *x*-*y* plane led to their occurrence in the result in the *u*-*v* plane?

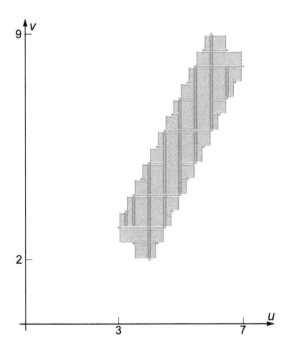

But if this really is the final result to be sent back to main memory, it can be coalesced to save bits without loss of information.

The solution set shown above has 613 uboxes that take up 15069 bits. Here is what it looks like after using **coalesce** on the above solution set. The coalesced set has only 107 uboxes, and a total size of 2281 bits.

Depending on the purpose of the set, you can often delete all the uboxes that have no area as a way to save even more space ("all tiles, no grout").

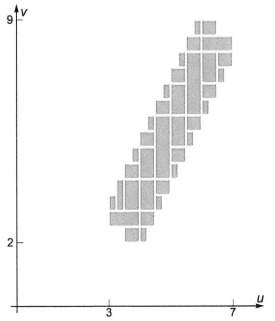

This last set has only 36 uboxes, and a total storage requirement of 803 bits. Imagine, for example, that the purpose of computing the above shape is graphical display of the tilted blue parallelogram, where the pixel scale is the same as the minimum ULP size used for the computation. The shape on the left occupies 176 such pixels, and each pixel might need a pair of 16-bit integers to specify its location on a display. That means $176 \times (16+16) = 5632$ bits that would have to be moved by the graphics processor, instead of 803.

A small amount of logic in the memory chips used as the *frame buffer* would suffice to interpret the ubox format and expand the bit meanings within the chip instead of the far more expensive chip-to-chip transfer. The ubox method is about seven times faster and uses about one-seventh the energy, since it moves about one-seventh the amount of data over the expensive path.

This is the essence of the ubox approach: Mindless, brute-force application of large-scale parallel computing using ULP-wide multidimensional elements to track the answer.

Floats try to conceal the fact that answers are selected from an ULP-wide set, misrepresenting them as exact. In contrast, uboxes embrace the fact that computing with real numbers means computing with *sets,* and track what happens to each ULP-wide set of real numbers.

16.3 The dependency problem

Consider the calculation of a simple expression like

$$y = \frac{x+1}{x+3}$$

which looks like the curve on the following page, for positive x:

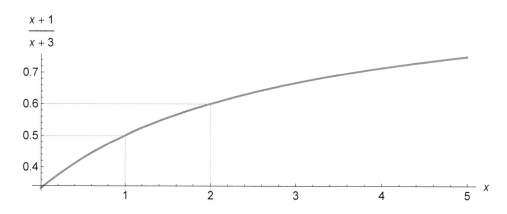

Suppose x is an interval, say $x = [1, 2]$. The numerator evaluates to $[2, 3]$ and the denominator to $[4, 5]$. The plot above makes it clear that the result *should* be the interval $\left[\frac{1+1}{1+3}, \frac{2+1}{2+3}\right] = [0.5, 0.6]$. Instead, here is what traditional interval arithmetic gives as the bound:

```
x = Interval[{1, 2}]; x+1.
                       ____
                       x+3.
```

```
Interval[{0.4, 0.75}]
```

Why did the bound become unnecessarily loose? Because the computer makes no use of the fact that the values of the numerator and denominator are *dependent* on each other. The computer evaluates the ratio of the two intervals $\frac{[2,3]}{[4,5]}$ using every independent combination of the endpoints: $\frac{2}{4}, \frac{2}{5}, \frac{3}{4}, \frac{3}{5}$, and then picks the minimum and maximum from those four possibilities: $\left[\frac{2}{5}, \frac{3}{4}\right] = [0.4, 0.75]$. That bound is 3.5 times larger than it needs to be, so about 71% of the information was lost.

> **Definition**: *The dependency problem* is excessively large bounds in interval calculations caused by treating multiple occurrences of the same variable as if they were independent, when they are not.

The dimensions of inexact uboxes are intervals; **will ubox computations also suffer from the dependency problem?**

Suppose we are using a {2, 3} environment, and for some reason we are willing to accept the low accuracy that results from a preferred minimum ULP size of $2^{-4} = \frac{1}{16}$. The following line of code generates the ubox list describing $x = [1, 2]$ as the list of unums representing 1, (1, 1.0625), 1.0625, ... , 1.9375, (1.9375, 2), 2:

```
xub = uboxlist[{1̂, 2̂}, -4];
```

If we put those through a unum computation and store the entire resulting shape (and also coalesce the result, for clarity), we get this set of uboxes:

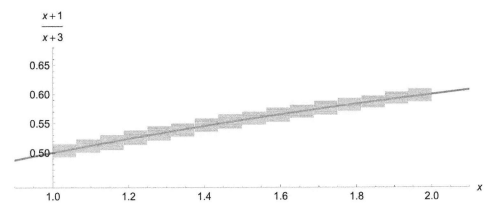

Of course, if what we wanted was simply the range as a ubound, there would be no need to compute the entire shape of the function like this; the computer would simply keep track of the smallest and largest result of calculating using elements of **xub**. The smallest computed value of $\frac{x+1}{x+3}$ was the open interval (0.4921875, 0.5), and the largest was (0.59375, 0.609375), so the final result (stored internally as a ubound) is (0.4921875, 0.609375). The width of that bound, which is the "volume" in one dimension, is 0.1171875, just slightly larger than the ideal volume of 0.1.

So uboxes are not immune to accuracy loss, but **the accuracy loss is generally limited to whatever you sign up for when you select the ULP size** (or perhaps the relative width, set as an environment value that allows the computer to automatically decide the right ULP size). The use of ULP-width sets creates work for the computer compared to what traditional interval bounds do, but it is almost all parallel work that need not take significantly more time (or significantly more electricity to run, since the work is on-chip).

The programmer does not have to think. At most, the programmer might have to declare whether a result is a final answer or an intermediate value, so that the computer does not waste time coalescing the answer (or summarizing it as a ubound) only to have to break it back up into ULP-size sets for some following step.

16.4 Intelligent standard library routines

The preceding two sections show that uboxes can reduce the wrapping problem and the dependency problem down to the limit of whatever ULP size we are working with. The trade-off is that smaller ULPs create more (parallel) work to do and take up more storage space. There is another approach that makes more sense for commonly-used routines: Whenever we start depending on a particular operation on real numbers and using it often, we may want to add enough logic to it so it becomes as perfect as the basic arithmetic operations.

A simple example of this in Part 1 was the routine for *squaring a number*. Why not simply write x^2 as x times x? Because it invites the dependency problem; the two occurrences of x are treated as independent by the multiplication routine. The interval $(-1, 1)$ squared is $[0, 1)$. But $(-1, 1) \times (-1, 1)$ is $(-1, 1)$, a clumsy expansion of the bound that loses half the information about the answer. Breaking up the interval into small uboxes would help, as shown in the previous two sections; the result of the multiplication would be $(-\text{ULP}, 1)$. However, that still contains negative values that squaring a real value cannot produce, which could cause a catastrophic error if the program needs the square guaranteed to be nonnegative. So with a few extra lines of logic, routines for **squareg** and **squareu** produce bounds that are as tight as possible.

The approach in general is to use knowledge of minimum and maximum points. If the function $\frac{x+1}{x+3}$ was actually called repeatedly in a program, it might be worth the effort to exploit the fact that the function is monotone increasing for x greater than -3. That means if x ranges from x_{low} to x_{high}, the bound on $\frac{x+1}{x+3}$ is simply $\frac{x_{\text{low}}+1}{x_{\text{low}}+3}$ to $\frac{x_{\text{high}}+1}{x_{\text{high}}+3}$, with the appropriate open-closed qualities for each endpoint. The function only needs evaluation twice if the function is monotone increasing or monotone decreasing, and sometimes it is very easy to figure out that there are no minimum or maximum points except at the endpoints and get rid of the dependency problem that way. If it is *not* easy, and there is no library function for the task, then let the computer do the work via uboxes.

Programmers already depend on libraries of software. Just about every programming language uses libraries for input/output and mathematical operations. Library routines are the ideal place to put human effort into making bounds as tight as possible. While much of the unum/ubox approach is intended to make programming much easier, it increases the work of the hardware designer, and also makes more work for the engineers who create standard software libraries. Because their efforts then make life easier for millions of computer *users*, this shifting of the burden makes economic sense and assures a net productivity gain. But do not expect hardware and software library designers to go down without a fight when asked to support the new paradigm described here!

16.5 Polynomial evaluation without the dependency problem

One of the simplest cases of a dependency problem is evaluating a quadratic, $a + bx + cx^2$. Since x occurs twice, there are cases where applying the formula directly gives a loose bound, that is, causes unnecessary information loss. That is also true when the equation is written as $a + x(b + xc)$, Horner's rule, which at least saves a multiplication. Having a loss-free routine that evaluates x^2 does not save us, nor does simply evaluating everything in the *g*-layer.

To illustrate with a very simple example, look at the plot for $y = 1 + 2x + x^2$:

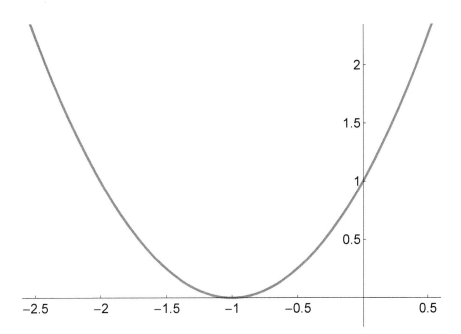

If we used a set of uboxes that are exact or width $\frac{1}{2}$, this is what we would *like* to see as the approximation (in an environment of {1, 2} or better, where the endpoints can be represented as exact numbers):

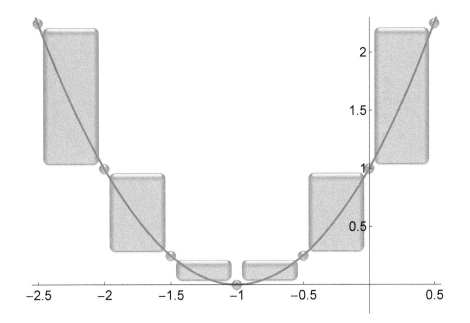

Instead, two of the ubounds will produce unnecessarily loose results, shown in amber below:

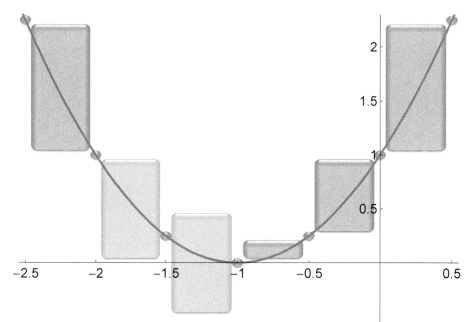

This is particularly annoying when looking for the roots, solutions to $x^2 + 2x + 1 = 0$. The answer should be exact −1, but according to the above graph, it will be the ubound representing (−1.5, 1]. Using finer detail (more uboxes) and higher precision *reduces* the amount of information loss and moves the sloppy calculation closer to the root at −1, but does not *eliminate* it. The problem occurs in all polynomials of degree two and higher, since x occurs multiple times in the expression. Generally speaking, the higher the degree of the polynomial, the worse the information loss.

This is not merely annoying but downright exasperating to engineers trying to solve control problems. Control theory says that a linear system is *stable* if all the roots of a certain polynomial are strictly in the left half of the complex plane. That is, their real parts are all strictly less than zero. However, the values of x are never known **precisely** in real world systems, but using an interval for x instead of a point value smears the roots out so badly that you cannot tell whether the answer to the question "Is the system I just designed stable?" is yes or no. The coefficients are generally not exact numbers either, but they only occur once in each evaluation of the polynomial, so they cause no more bound expansion than they are supposed to. The big problem is all the x occurrences being treated as independent ranges, when they are not independent at all.

Polynomials are far too useful for us to settle for this information loss. A unum environment should include a fused polynomial routine that executes in the *g*-layer and provides a *tightest-possible* evaluation, and the user simply submits the coefficients instead of having to figure out how or why the routine works.

The author was surprised to discover how long the problem of ideal polynomial evaluation has been in existence, unsolved: At least sixty years, since the introduction of traditional interval arithmetic by Ramon Moore. IBM tried to deal with the problem using an iterative package called ACRITH that used Kulisch's complete accumulator to alleviate the error, but it does not eliminate the error. Occasionally a paper or PhD thesis comes out that shows a way to reduce the error by 25% or so, through various combinations of operation reorderings. Like much of the literature on numerical analysis, this is a sort of "our method sucks less" approach, chipping away at the problem instead of solving it. Perhaps the reason the problem has not been previously cracked is that people are using floats or traditional intervals, instead of a representation like unums that is careful about open-closed endpoints and correctly represents ranges of real numbers instead of sample points.

In any case, **a general approach for evaluating polynomials with interval arguments without any information loss is presented here for the first time**. The code for the `polyg[`*coeffsg, xg*`]` function is in Appendix D.7. The list of coefficients, from constant to the coefficient of x^n, is the set of general intervals *coeffsg*. The input argument, a general interval, is *xg*. *A ubox approach allows automatic avoidance of the dependency problem.*

Start with the general quadratic, so the algebra is easier to read. The equation $a + bx + cx^2$ can always be rearranged as follows, for a real number x_0, where the powers of arguments are shown in blue to make it a bit easier to see that the right-hand side is also a quadratic equation:

$$a + bx + cx^2 = \left(a + bx_0 + cx_0^2\right) + (b + 2cx_0)(x - x_0) + c(x - x_0)^2$$

In other words, a quadratic equation in x can always be written as a quadratic equation in $x - x_0$. This can be done for *any degree polynomial*. One way to do it is to replace x on the left with $((x - x_0) + x_0)$ and expand the result, and then gather similar powers of $(x - x_0)$. (An effort is being made to steer clear of calculus and stick to high school algebra here, but calculus offers a different way to think about this: A Taylor series about x_0 with a finite number of terms.) Nothing new here; the approach was used by Horner, hundreds of years ago. To help see the pattern, here is how to express the fifth-degree polynomial $a + bx + cx^2 + dx^3 + ex^4 + fx^5$ as a polynomial in $(x - x_0)$, with some English words "times" and "plus" thrown in for readability:

$$(x - x_0)^0 \text{ times } \left(a + bx_0 + cx_0^2 + dx_0^3 + ex_0^4 + fx_0^5\right) \text{ plus}$$
$$(x - x_0)^1 \text{ times } \left(b + 2cx_0 + 3dx_0^2 + 4ex_0^3 + 5fx_0^4\right) \text{ plus}$$
$$(x - x_0)^2 \text{ times } \left(c + 3dx_0 + 6ex_0^2 + 10fx_0^3\right) \text{ plus}$$
$$(x - x_0)^3 \text{ times } \left(d + 4ex_0 + 10fx_0^2\right) \text{ plus}$$
$$(x - x_0)^4 \text{ times } (e + 5fx_0) \text{ plus}$$
$$(x - x_0)^5 \text{ times } f$$

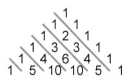

Perhaps the reader is familiar with *Pascal's triangle*, where every number is the sum of the two above it. The blue diagonal lines show where the coefficients in the rewritten polynomial come from.

Notice that the new coefficients each use a, b, c,... at most once. That means they can be computed without information loss. The new coefficients can be computed exactly in the g-layer of the computer.

So what use is this type of transformation?

The transformation allows us to move the "origin" of the polynomial. About the origin, here is how each power of x behaves when its coefficient is a positive number:

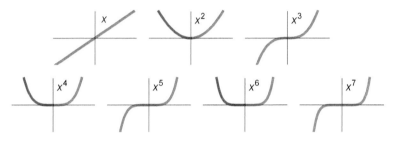

Blue: Power function preserves min–max ordering
Red: Power function reverses min–max ordering

If the coefficient is a negative number, the above graphs turn upside-down:

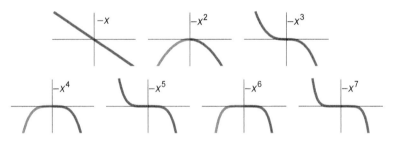

Blue: Power function preserves min–max ordering
Red: Power function reverses min–max ordering

If all the terms in a polynomial slope the same way, then evaluating the interval is simply a matter of evaluating its endpoints. When they do not slope the same way, the dependency problem can appear as the result of adding slope-up to slope-down terms. Far from the origin, the higher powers dictate the slope of the polynomial and the interval evaluation is again simply the evaluation at the endpoints. The influence of the higher powers is small near the origin, which means that the closer the function is to the origin, the more its behavior is dominated by the lowest-order term.

> **Exercise for the reader**: For the quadratic $y = 1 + 2x + x^2$ shown in the beginning of this section, find the slope of each term for the open intervals $x = (-2, -1.5)$ and $(-1.5, -1)$ that produced the amber-colored bounds. Then show why the dependency problem does not affect the interval $x = (-2.5, -2)$.

The trick of rewriting the polynomial in terms of powers of $(x - x_0)$ lets us use the left or right endpoint of any interval as the origin. Then, by subdividing the interval in the *g*-layer, the interval always can be made either small enough that the linear term determines the slope of the polynomial, or small enough that the dependency problem is *less than the smallest ULP*. The last possibility is crucial: It can happen that the maximum or minimum of a polynomial occurs at an irrational number, which would lead to an infinite loop trying to home in on it. The algorithm knows to stop when the finite-precision *u*-layer result is determined. Here is the general scheme of the procedure for evaluating the polynomial $p(x)$ where x is a general interval.

Set the *trials* to the single interval x, bounded by exact values x_{lo} and x_{hi}.
Evaluate $p(x_{lo})$ and $p(x_{hi})$ exactly and use them to initialize *min* and *max* values.
Find the ubound value of the general interval $\dfrac{min \quad max}{minQ \quad maxQ}$.

While *trials* is nonempty, for the first entry in *trials*:
  Evaluate the polynomial, rewritten in terms of $x - x_{lo}$.
  Evaluate the polynomial, rewritten in terms of $x - x_{hi}$.
  Find the intersection of those two evaluations.
  If the intersection is within the ubound value of $\dfrac{min \quad max}{minQ \quad maxQ}$,

    Then remove the first entry from trials.
    Else subdivide the interval along coarsest-possible ULP boundaries;
      Prepend the two new subintervals to the trials list, as open intervals;
      Evaluate the polynomial exactly at the point of subdivision;
      Use that evaluation to update $\dfrac{min \quad max}{minQ \quad maxQ}$ and its ubound version.

The ubound version of $\dfrac{min \quad max}{minQ \quad maxQ}$ is the optimal *u*-layer polynomial result.

Perhaps one reason this approach does not seem to appear in the literature for interval arithmetic is that it makes heavy use of the ubound ability to express whether an endpoint is open or closed. Traditional closed intervals start out with a significant misrepresentation of the polynomial even in their ideal case. Here is what the quadratic from the beginning of the section would look like if it could be evaluated without the dependency problem. The amber regions in the plot on the next page should not be there, but with closed intervals, there is no choice.

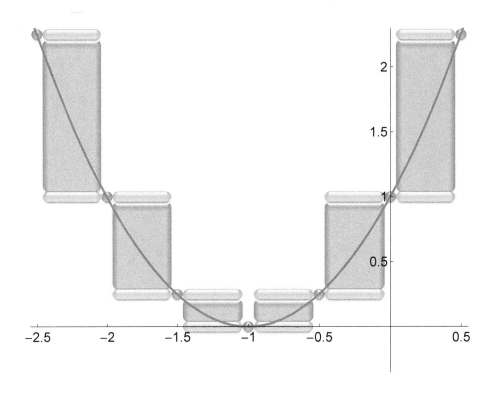

> The distinction between "greater than" and "greater than or equal" is ignored by traditional intervals, so the *entire range from −1.5 to −0.5 looks like a root*.

Of course, traditional intervals use single or double precision IEEE floats as endpoints, but that means a problem in double precision like

$$\text{"Where is } 10^{-300} x^2 \text{ equal to zero?"}$$

gets the spectacularly useless answer of a huge range, from about -10^{138} to 10^{138}, where the calculation "equals" zero. **No, it doesn't**. In a unum environment, no matter how low the precision, the correct answer "At the point $x = 0$, exactly" is quickly returned since the arithmetic knows that a strictly positive number times a strictly positive number is a strictly positive number, even in Warlpiri math.

As a test case, the cubic polynomial $-\frac{1}{2} - x + x^2 - \frac{1}{8}x^3$ has irrational roots and irrational local minima and maxima. Set the precision low to exaggerate the dependency problem, and create a ubox list covering $[-2, 7]$ in steps of 2^{-1}:

```
setenv[{2, 2}];
set = uboxlist[{-2̂, 7̂}, -1];
```

Here is the sloppiest approach: Horner's rule applied to unfused unum arithmetic.

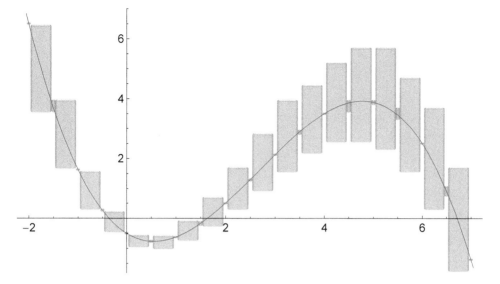

On the left, the cubic terms dominate so there is no dependency problem. For the rest of the plot shown, there is conflict over slope-up and slope-down terms in the polynomial for the inexact intervals, producing quite a bit of information loss. The loose bounds alternate with much tighter evaluations at the exact points, where the only information loss is expressing the exact value with low precision. In this case, the loose bound actually did a good job of locating the roots of the cubic as well as possible at such low precision. However, it did a lousy job of finding the local minimum and local maximum of the function, where we would like a value only one ULP wide (at the smallest possible ULP width). Here is what the same cubic looks like if evaluated with the `polyg` function:

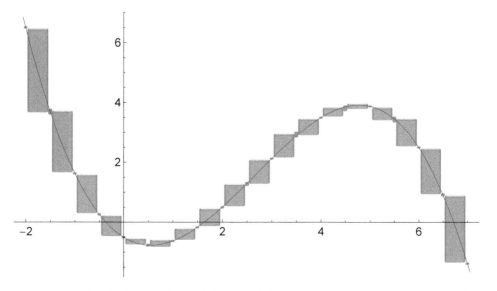

That's more like it. The polynomial cannot be expressed more precisely using a {2, 2} environment.

Where other polynomial interval algorithms have the most trouble is near a root, especially a *multiple* root. For example, $8(x-1)^3$ has a triple root at $x = 1$ and the function is so flat and close to the x axis that it is difficult for conventional methods to find the crossing point. The function expands to

$$8(x-1)^3 = -8 + 24x - 24x^2 + 8x^3$$

and if it is presented in the expanded form on the right-hand side above, the triple root is certainly not obvious. A computer evaluating it with Horner's rule would evaluate it as $-8 + x(24 + x(-24 + x(8)))$. With unums, simply using Horner's rule gives the following graph for x in the range -1 to 3:

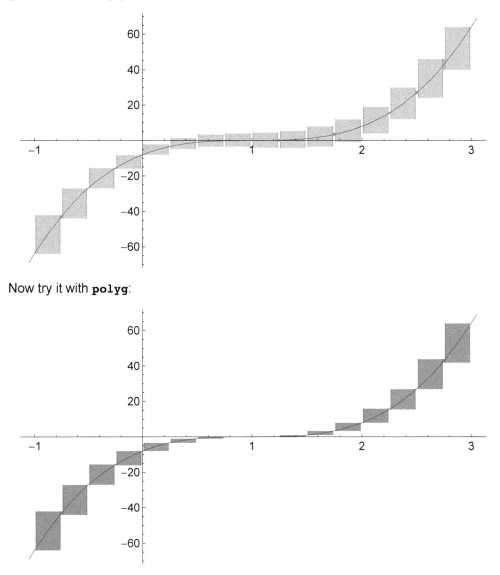

Now try it with **polyg**:

That is clearly more precise near the root. Here is an expanded view of the ubox evaluation of the polynomial, right near the triple root $x = 1$:

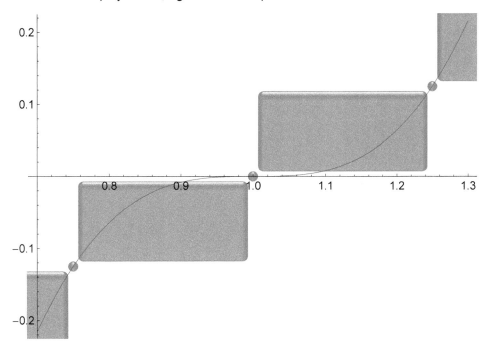

The **polyg** function can be trusted to find the exact root when that value can be expressed exactly in the unum environment. This is another case where unum arithmetic solves problems that cannot be solved by other methods; engineers can finally get answers to questions like "is this system stable?" when the answer depends on tightest-possible evaluation of a polynomial.

16.6 Other fused multiple-use expressions

There are other common expressions in which the argument occurs more than once, and deciding what should be part of a standard unum environment is the sort of thing that standards committees should decide, after enough use cases have been accumulated that a "ten most wanted" list forms. There are several ways to extend the polynomial evaluator developed in the previous section:

- Polynomials of more than one variable, like $x^3 + 3xy^2 + y - 9$
- Rational functions (the ratio of two polynomials), like $\frac{1+x^3+x^7}{2-x+5x^6}$
- Polynomials with terms of negative degree, like $\frac{3}{x^2} + \frac{1}{x} + 2 + x^4$

Each of these is an "exercise for the reader," but more like a term project than an exercise. The general approach is similar: Evaluate the endpoints of an interval exactly in the *g*-layer, then evaluate the open interval between the endpoints. If the latter result is between the evaluations at the endpoints, then we're done. If it is not, it could be either because of the dependency problem causing a loose bound, or a sign that the function has a local minimum or maximum inside the interval. The interval is subdivided along the coarsest-possible ULP boundary and prepended to the `trials` list, and the evaluation of the function at the bisection point is incorporated into the known minimum-maximum range *at the precision of the* u-*layer.*

The process repeats until all subintervals have been bounded. As long as the function is continuous (or has a finite number of discontinuities, like the rational functions or the polynomials with terms of negative degree), the procedure will terminate and return a tightest-possible result.

3. Find *x*.

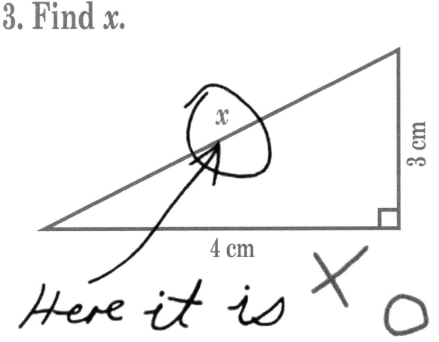

A test question answered in desperation. This chapter is also about finding *x*. Like the
example above, we may "find *x*" in a way other than what your math teacher had in mind

17.1 Another break from traditional numerical methods

Before attempting full-scale applications that can be solved with uboxes, we will
need ubox versions of some basic equation-solving tools like you might find in a
software library. There is a mountain of literature on sophisticated methods for
solving equations, but the ubox approach is quite unlike any of the classical
methods. For one thing, ubox methods are *not* sophisticated, and thus easy-to-use.

Classical methods try to minimize rounding errors. Since unums and uboxes do not
make rounding errors but instead track exact numbers or ULP-wide ranges, we need
to reexamine what it really means to "solve" an equation on a computer. As usual,
we will be looking for an approach that does *not* require programmer knowledge of
numerical methods, and instead puts most of the burden on the computer to do the
right thing.

Even the simple case of *one linear equation in one unknown* shows the need for a
new approach.

17.2 A linear equation in one unknown, solved by inversion

In elementary school, everyone sees problems like "$3x + 6 = 0$. What is x?" The implicit meaning of a math teacher who asks you to "solve for x" or "find x" is "rearrange things so the unknown quantity is all by itself on one side of the equation, and simplify the other side of the equation as much as possible."

Algebra teaches the approach of repeatedly "undoing" the things done to x in the equation, by applying inverse operations until x is left by itself on one side of the equation and the other side has operations on numbers with known values. So to undo the "$+6$" on the left-hand side we subtract 6 from both sides to get "$3x = -6$". Then we undo the multiplication by 3 with the inverse operation, division by 3, applied to both sides: "$x = -6/3$". The quantities on the right are all known, and simplify to -2. With that meaning, the solution to the question "What values of x satisfy $bx + c = 0$?" is $x = -c/b$. That certainly looks like a reasonable way to program a computer to solve $bx + c = 0$. Negate c, divide that by b, done.

But what if b is zero? Does that mean the computer should divide $-c$ by zero and get a NaN? No; re-examine the original question. If b is zero (and we assume x is finite), the question becomes "What values of x satisfy $c = 0$?", which seems like a silly question since x does not even appear in the equation. However, the question has a perfectly logical answer. If c is not equal to zero, then the answer is "None. *No* values of x satisfy $c = 0$." But if c happens to be zero, then the answer is "*All* values of x satisfy the equation $0 = 0$."

What if b is infinity, or negative infinity? If c is finite, the $\frac{-c}{b}$ formula says zero is a valid answer. But it does not work, since $\infty \cdot 0 = \text{NaN}$. What if c is infinity or negative infinity? There is no value you can add to infinity or negative infinity that produces a finite result. So the more careful answer has three cases:

$$\text{If } bx + c = 0, \text{ then } x = \begin{cases} \dfrac{-c}{b} & \text{if } b \text{ and } c \text{ are nonzero and finite,} \\ \text{The empty set} & \text{if } b \text{ is zero and } c \text{ is nonzero,} \\ & \text{or if } |b| = \infty \text{ or if } |c| = \infty \\ \text{All real numbers} & \text{if } b \text{ and } c \text{ are zero.} \end{cases}$$

The general principle is this: "Solving" an equation computationally means something different from "solving" an equation mathematically. You may have been taught to first solve mathematically by inverting the operations and getting x on one side of the equation, and then enter the expression on the other side into a computer. As shown in the previous paragraphs, doing so invites the mistake of forgetting special cases.

More importantly, conventional computer arithmetic makes rounding-underflow-overflow errors, so it cannot be trusted to match the mathematical answer, *even for something this simple.*

> To "solve an equation" on a computer really means this: *Find the minimum but complete set of representable values for which the computer version of the math equation is true.* In other words, find the *c*-solution to the **translated** equation.

Oddly enough, very few computer programs that claim to "solve equations" actually do this! They might give one value and omit other values that also work, and thus the solution is incomplete. Or, they include values that do *not* give equality when tested in the original equation. Even *Mathematica* gets confused about this, and says that the numerical values of *x* that satisfy "0 = 0" is the *empty set*:

```
Solve[0 == 0 ⋀ NumericQ[x], x]
```

```
{}
```

If you use floating point arithmetic, the solution to $b\,x + c = 0$ is almost always *wrong*. For instance, if you try to solve $7\,x - 2 = 0$ with floats and the $\frac{-c}{b}$ formula, you get

$$x = \frac{2}{7} \text{ is } \frac{585}{2048} \text{ half precision; but } 7\,x - 2 \text{ is } -\frac{1}{2048}, \text{ not 0.}$$

$$x = \frac{2}{7} \text{ is } \frac{9\,586\,981}{33\,554\,432} \text{ single precision; but } 7\,x - 2 \text{ is } \frac{3}{33\,554\,432}, \text{ not 0.}$$

$$x = \frac{2}{7} \text{ is } \frac{10\,293\,942\,005\,418\,277}{36\,028\,797\,018\,963\,968} \text{ double precision; but } 7\,x - 2 \text{ is } \frac{3}{252\,201\,579\,132\,747\,776}, \text{ not 0.}$$

Therefore, if a program solves for *x* using floats and later arrives at a test like

```
if (7 * x - 2 == 0) then ...
```

it will *always* do the wrong thing no matter what precision is used. The current workaround with floats is to test if a result is *close*, where "close" has to be defined using human judgment on a case-by-case basis. And should "close" be defined using absolute error or relative error? More decisions that humans have to make. The need for such workarounds makes code hard to write and even harder to read.

17.2.1 Inversion with unums and intervals

Since unums follow the rules of mathematics, it is safe to use them with the inverse operation approach. The result will always *contain* the solution as defined above, but it will not always be the *minimum* set of unums that satisfy the original equation, that is, the *c*-solution. If the answer is more than one ULP wide using the smallest ULP available, then one approach is to create the ubox set for the result of the inverse operation and test each one in the original equation, discarding any that fail. Only then do you have the *c*-solution to the equation.

For linear equations, the $\frac{-c}{b}$ result is always of minimum width and no refinement is needed. For example: With unums and a modest {1, 3} environment, the unum version of the equation $7x - 2 = 0$ is $\hat{7} \otimes xu \ominus \hat{2} \approx \hat{0}$, and the unum version of the $\frac{-c}{b}$ answer is

$$\texttt{negateu[-}\hat{2}\texttt{]} \odot \hat{7} = \left(\frac{73}{256}, \frac{74}{256} \right),$$

that is, $0.28515625 \le x \le 0.2890625$; a correct statement, and the unum is the minimum possible ULP width, so there is no need to split it further. The solution is concise, too; the unum for x is the binary string $0\,00\,0100100\,1\,1\,111$, which takes only 16 bits. Plugging x into $7x - 2$ gives the open interval $\left(-\frac{1}{256}, \frac{3}{128} \right)$, which contains zero.

A unum computer program could later test that value of x with something like

---

```
If 7̂ ⊗ xu ⊖ 2̂ ≈ 0̂ then...
```

---

and the `True` or `False` result would be correct no matter what the environment settings are, and without the need to insert a "closeness" tolerance as a workaround.

Is this any different from interval arithmetic? Interval arithmetic, say with half-precision floats as endpoints, would conclude that $\frac{585}{2048} \le x \le \frac{586}{2048}$, a true statement but one that needlessly includes the endpoints as possibilities. Half-precision interval arithmetic also consumes a wasteful 32 bits to describe x.

The bigger problem with traditional intervals is that they are susceptible to underflow and overflow problems, even on a problem as simple as one linear equation in one unknown. Suppose we use traditional interval arithmetic, again with half-precision floats, to answer the question

"What values of x satisfy $\frac{1}{64}x - 10\,000 = 0$?"

Integer arithmetic claims mathematical rigor, so we *should* be able to use the formula $x = \frac{-c}{b}$. After all, the values $b = \frac{1}{64}$ and $c = -10\,000$ are representable *exactly* in half precision, and neither b nor c is zero so the special cases are not an issue. However, traditional interval arithmetic overflows for the upper bound of the division operation, and says x could be exactly *infinity*.

Even worse things can happen to an equation like this one:

"What values of x satisfy $10^{-8}x + 10^{-8} = 0$?"

The obvious answer is $x = -1$, but those b and c values underflow half-precision float arithmetic. When both b and c underflow to zero, a float method stumbles incorrectly into the third case of the formula as the answer:

> All real numbers   if b and c are zero

If we try solving the equation with traditional interval bounds, b and c become the closed intervals $\left[0, \frac{1}{16\,777\,216}\right]$. Good luck applying the formula:

$$\begin{cases} \frac{-c}{b} & \text{if } b \text{ and } c \text{ are nonzero,} \\ \text{The empty set} & \text{if } b \text{ is zero and } c \text{ is nonzero,} \\ \text{All real numbers} & \text{if } b \text{ and } c \text{ are zero.} \end{cases}$$

So, **which case is it?** The interval versions of b and c could be considered nonzero, or possibly zero, or possibly *both* zero, presenting us with the ugly choice between $\frac{-c}{b}$ (which evaluates to the interval "$[-\infty, 0]$"), "The empty set," or "All real numbers."

The examples so far use exact values for b and c. What happens when the coefficients are general intervals, representable by ubounds?

17.2.2 Ubounds as coefficients

So far, we have shown what happens when b and c can be represented with single unums, so they are either exact or one ULP wide, and cannot span zero. For linear equations where b and c are not just real values but are general intervals, the situation is more complicated. There are *eighteen* possible cases that can arise, some of them quite weird. The equation $y = bx + c$ describes a line with slope b that intercepts the y axis at c. But when b and c are ranges of values, then $y = bx + c$ describes a shape that reminds us of a bow tie, with a "wobble" in its slope and a nonzero width at the pinch point where it intersects the y axis.

For example, here is the plot of $bx + c$ when the slope b ranges from −0.5 to 0.25 and the intercept c ranges from 0.25 to 0.375:

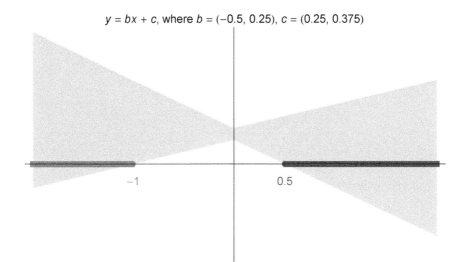

$y = bx + c$, where $b = (−0.5, 0.25)$, $c = (0.25, 0.375)$

The graph shows that the solution to $bx + c = 0$ is *two* disjoint ranges, one from −∞ to −1 and the other from 0.5 to ∞. The closed-open status of boundaries at −1 and 0.5 depends on the closed-open status of the endpoints of the ranges for b and c.

Suppose the slope b is the general interval (−0.5, 0.25) but the intercept slope is (0, 0.375). The solution set consists of every number *other than zero*:

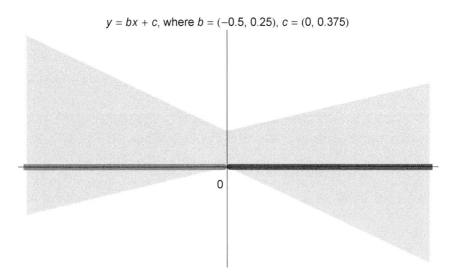

$y = bx + c$, where $b = (−0.5, 0.25)$, $c = (0, 0.375)$

Exercise for the reader: The equation $bx + c = 0$ can also be expressed as a fused multiply-add: $\mathbf{fmau}\left[\hat{b}, \hat{x}, \hat{c}\right] \approx \hat{0}$. Are there cases where using the fused multiply-add results in a tighter solution than the solution to $\hat{b} \otimes \hat{x} \oplus \hat{c} \approx \hat{0}$?

Remember what it means to "solve an equation": Find *all* the values that satisfy the equation, not just one value computed by inverse operations. Inverse operations tend to have lots of special cases to deal with, so they are usually *not* the best approach. To write a program that directly solves $bx + c = 0$, we have to consider six possible cases for the value of b:

1. Exactly zero
2. From exact zero to a positive number
3. From a negative number to exact zero
4. Strictly greater than zero
5. Strictly less than zero
6. From a negative number to a positive number

There are three cases to consider for c:

a. Contains zero
b. Is strictly greater than zero
c. Is strictly less than zero

This is why there are eighteen situations to consider, not counting the special cases of infinite and NaN input values. It *might* make sense to put such complicated code into a math library, but the casual programmer should *not* have to deal with such complexity. The table on the following page shows what the cases look like as *x-y* plots, with $y = bx + c$.

One of the more subtle aspects is in the first column of the table. When $b = 0$ and c includes 0, the solution set is $-\infty < x < \infty$. Why does the solution set not include the endpoints, $\pm\infty$? The graph suggests the solution set should be $-\infty \le x \le \infty$. That *would* be the solution set if we had been careful to define the original problem as

$$\text{solve} \begin{cases} bx + c = 0 & \text{if } b \text{ is nonzero,} \\ c = 0 & \text{if } b \text{ is zero} \end{cases}$$

Without that definition, it is not valid to say that $x = \infty$ solves $0x + c = 0$, since the multiplication produces a NaN. On the other hand, the case on the bottom left of the table, where b includes 0 and c includes 0, **does** produce $-\infty \le x \le \infty$ as the result. An infinity times any general interval that spans zero produces the entire range $-\infty$ to ∞, inclusive. This is an example of how a mathematical statement is *translated* when expressed to a computer, often without it being obvious that it is not identical.

The discussion to this point suggests that it is complicated to produce a *c*-solution for even the simplest math problems. The next section will reverse this; uboxes provide a surprisingly easy and robust approach that works for a huge range of problems and requires *next to nothing in the way of programming expertise*.

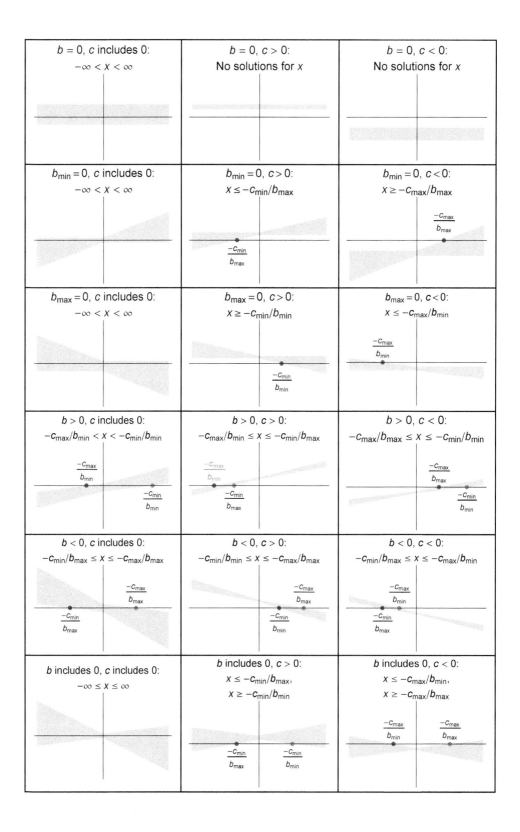

17.3 "Try everything!" Exhaustive search of the number line

17.3.1 The quadratic formula revisited

A cartoon by the inimitable Saul Steinberg.

In Part 1, we saw how the naive use of the quadratic formula to solve the quadratic equation $a x^2 + b x + c = 0$ with unums still could produce safe, usable results. The quadratic formula is a good example of an inverse operation approach for which someone had to sweat bullets to derive it in the first place, in this case by applying clever rearrangements to the quadratic equation $a x^2 + b x + c = 0$:

$$r_1, \ r_2 = \frac{-b \pm \sqrt{b^2 - 4\,a\,c}}{2\,a}$$

Unless you use math frequently, you probably forgot the equation soon after learning it in grade school, and now would have to look it up if you needed it.

The last section showed how complicated solving linear equations can be if you do it with inverse operations. If you actually wanted to create a unum library routine for solving *quadratic* equations using case after case of the quadratic formula, the graph on the next page shows an example of the kind of bizarre case you would have to deal with:

$$y = a\,x^2 + b\,x + c;\ a \text{ spans } 0,\ b^2 - 4ac > 0$$

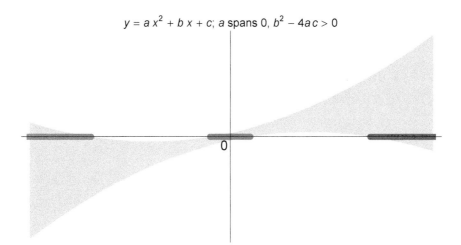

You were always taught there were at most two roots to the quadratic equation, but if the coefficients are inexact, there can be *three* separate regions of the real number line for which $a\,x^2 + b\,x + c$ equals zero. At least we know that if $a = 0$ and x is finite, everything reduces to the eighteen cases for $b\,x + c = 0$ detailed in the previous section. If a is not exactly zero, there are *dozens* of additional cases to wrestle with, like the weird one shown above. The quantity inside the square root can be zero or negative, so the cases $b^2 - 4\,a\,c = 0$ and $b^2 - 4\,a\,c < 0$ require separate cases in the code. Programming all the separate cases does not seem like the most productive approach. There is a far better way.

17.3.2 To boldly split infinities: Try everything

Unum math offers something so simple and painless that it may seem too good to be true. It can quickly test the *entire real number line* to screen out unnecessary work. For example, suppose the quadratic equation to solve is

$$x^2 + 1 = 0$$

If the answer has to be a real number, there *are* no solutions. You could compute the value inside the square root, $b^2 - 4\,a\,c$ and test if it is negative, but that is a lot of work compared to what we could do far more easily: Try $[-\infty, \infty]$. If there is a solution to $x^2 + 1 = 0$ anywhere, then the unum version will yield `True`. Try it:

```
squareu[{-∞, ∞}] ⊕ 1̂ ≈ 0̂
```

```
False
```

With just one multiplication, one addition, and one comparison, unum math screened out the need for any further work. The `False` result does not mean the method simply could not *find* a solution; it proves there *are* no solutions anywhere in the entire real number line, and that $-\infty$ or $+\infty$ also do not solve the equation. The use of "≈" instead of "≡" as the equality test is key, since "$\approx \hat{0}$" means *intersects* zero.

Armed with this idea, we can revisit the same quadratic equation used as an example in Part 1: $3x^2 + 100x + 2 = 0$. In unum form, define it as **quadu[ub]**, where *ub* is any ubound, and this time use **polyu** to prevent any dependency problems from making the function evaluation loose:

$$\texttt{quadu[ub\_] := polyu}\left[\left\{\hat{2},\ 10\hat{0},\ \hat{3}\right\},\ \texttt{ub}\right]$$

The first step is to see if there are *any* solutions. Finding that out takes just one call of **quad**, using the entire real number line $(-\infty, \infty)$:

$$\texttt{quadu}\left[\texttt{g2u}\left[\begin{matrix} -\infty & \infty \\ \textbf{open} & \textbf{open} \end{matrix}\right]\right] \approx \hat{0}$$

True

The **True** result means it is worth doing some more checking. In Part 1, the environment used was the one closest to IEEE single precision, a {3, 5} environment; direct use of the quadratic formula gave the following result for the root closest to zero, the one that had information loss from digit cancellation:

$$(-0.0200120160\cdots, -0.0200120136\cdots)$$

It is hard to read so many decimals, so the digits that differ are shown in orange. The width of the above bound is $2.4\cdots\times 10^{-9}$, which is hundreds of times larger than the minimum ULP size. That means it might contain some unums that do *not* satisfy the original equation, and thus might not meet the definition of a *c*-solution.

The **solveforub[***domain***]** function in the prototype takes a set of ubounds *domain* that define the regions of the real number line in which to search, and it assumes that a condition of truth **conditionQ[***ub***]** has been defined. The code for **solveforub** is in Appendix D.8. A description of the algorithm is as follows:

Set the *trials* set to *domain* and the *solutions* set to the empty set, {}.
While the *trials* set is nonempty:
    Set *new* to the empty set, {}.
    For each ubound *ub* in *trials*:
        Test if *ub* satisfies the truth condition.
        If it does, attempt to split it into smaller ubounds.
            If *ub* is the smallest ULP (unsplittable), add it to the *solutions* set;
            else, add the split to *new* and *trials;* delete *ub* from *trials*.
    Set the *trials* set to *new*.
Return *solutions* as the *c*-solution to the truth condition.

In adding a ubound to the solutions set, it is merged with any of its neighbors that are already there. The prototype function that does this is **ubinsert[***set, ub***]** where *set* contains ubounds and *ub* is the ubound to be added to *set*, either as an independent range of real numbers or (if possible) joined to an existing ubound in *set*.

The "split it into smaller ubounds" task can be done in various ways, and the prototype function that does the splitting is **splitub[***ub***]** for an input ubound *ub*. If you give **splitub** the entire real number as well as the infinity endpoints, it first will split it into −∞, (−∞, ∞), and ∞. The priority is to get the region of exploration to be finite, so (−∞, ∞) is split as (−∞, −*maxreal*), −*maxreal*, (−*maxreal*, ∞). Then the right-hand range (−*maxreal*, ∞) is divided at *maxreal*. Priority then goes to splitting positive and negative reals, so the next point of division is at zero. After that, the coarsest-possible ULP sizes are used. The figure below shows this initial splitting sequence when starting with the infinite range [−∞, ∞]:

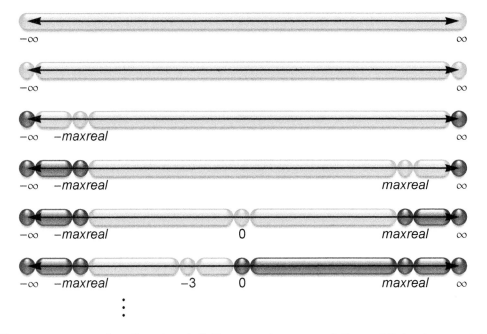

The **solveforub** function, by definition, produces *c*-solutions: Minimal, complete, tight-as-possible sets of expressible values that make a condition true. So try it on the quadratic equation and see if it can automatically find a better set. We do not need a large exponent; a {2, 5} environment should suffice. The solution ranges found by **solveforub** are assigned to **sols**, and then displayed in full decimal form:

```
setenv[{2, 5}]
conditionQ[u_] := (quadu[u] ≈ 0̂)
sols = solveforub[{{-∞̂, ∞̂}}];
Table[view[sols₍ᵢ₎], {i, 1, Length[sols]}]
```

$\{(-33.31332132220268249511718750, -33.31332131475210189819335937\mathit{5}),$
$(-0.0200120144218089990317821502685546875, -0.02001201441817102022469043731689453125)\}$

These two open ranges bound the two roots to one ULP of minimum size, thereby producing *hundreds of times more information about the answer* than did the inversion approach of applying the quadratic formula.

17.3.3 Highly adjustable parallelism

Each region split creates three independent ubounds to test. Those tests can be done in parallel, using the easiest form of parallelism: Identical operations on different data. The number of sub-regions in a split *could* be far more than three, if we so choose. Instead of a simple `splitub`, any ubound can be split into a multitude of smaller ubounds using `uboxlist`. If you have, say, 1024 processors handy, why not give them each something to do that progresses the answer? This is a choice the computer system can make, not the user, since the operating system knows which processors are idle and which ones are busy. Perhaps using 1024 processors at once would accelerate the path to the *c*-solution by a factor of 30; is it worth it?

That depends on the trade-off between speed and cost. In this case, the highly parallel version will use more electricity than doing the minimum split of an open ubound into three ubounds. But this is quite likely *on-chip* parallel computation that is far more economical than off-chip communication, so it may add very little to the total power bill. The reduction in human time waiting for an answer has quantifiable financial value that may very well justify the extra work for the processor cores.

Unlike many parallel methods, using extra-fine ubound splitting produces an identical final result no matter how many processors are used. There is only one *c*-solution, and the try-everything approach will find it no matter how the possibilities are tested.

The main point is this: The *option* is there to use parallel processing when using a `solveforub` approach. It is a selectable level of parallelism, and therefore an easy way to make the speed-power trade-off. There are times when it is worth it to do much more total work but increase the rate at which information is acquired.

17.4 The universal equation solver

17.4.1 Methods that USUALLY work

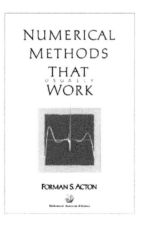

A classic numerical analysis book is *Numerical Methods that Work* by Forman S. Acton. If you get a copy of the original, you can see the word USUALLY embossed, without ink, just before "Work" in the title. In reprints, USUALLY is shown in fine print, but is not part of the official title.

Acton's tongue-in-cheek "watch out for the fine print" warning recognizes the unsatisfactory state of existing numerical techniques. Just as with integration routines, traditional numerical analysis texts offer a smorgasbord of approaches for solving equations in one variable (root finding) with advice like "Use this in situation A, but be careful, because such-and-such can go wrong and then you should try a different method."

The established methods for finding roots of $f(x) = 0$ numerically have drawbacks like the following:

- Some require you to pick a starting guess. If you do not guess well, you will completely miss the root you were seeking, or you will not find *all* of the roots.
- A method requires knowing or computing the derivative of the function f, which may not even exist everywhere.
- Some methods work when the function *crosses* the x axis, but not when it is *tangent* to the x axis; or they work in theory, but run drastically slower.
- Methods that look for zero crossings completely overlook places where a function crosses the x-axis but crosses back *within the space of one ULP*. (Recall the "Smooth Surprise" function described in Section 15.7.)
- Methods that look for zero crossings for *discontinuous* functions also can iterate forever, or mistakenly decide there must be a root at the discontinuity where it jumps across the x axis.
- Divide-and-conquer methods require artistic selection of a "recursion limit" at which to give up, and even when the limit is set to an extremely high level, they will still miss some of the roots. (Again, the "Smooth Surprise" in Section 15.7 is an example for which this happens.)
- Whenever a method uses floats, including traditional interval arithmetic, underflow can be mistaken for a root. If $f(x)$ is the formula for a bell curve (Gaussian distribution), for example, most root finders will mistakenly declare $f(x)$ to be exactly zero for x values not very far from the origin.

> The ubox approach to equation solving/root finding misses *nothing* because it does not sample the real number line; it examines the entire thing, every value. It is a numerical method that (ALWAYS, not embossed) works.

Moreover, it can be very fast because it can start with ultra-low precision and work its way up to any desired level of information about the answer set.

Some examples of root finding with uboxes illustrate how difficult it is to break the "Try everything" approach. The range $(0, \infty]$ can be written `{ubitmask, posinfu}`.

`î ⊙ {ubitmask, posinfu} ≈ ô`

`True`

So $1/x$ has a root somewhere in $(0, \infty]$. Try Warlpiri unums; the only positive value that works is ∞:

```
setenv[{0, 0}]
conditionQ[u_] := (î ⊙ u ≈ ô)
sols = solveforub[{g2u[ 0      ∞      ]}];
              open  closed
Table[view[sols[[i]]], {i, 1, Length[sols]}]
```

$\{\infty\}$

Here is another example: Set the environment fairly low, and ask "For what values of *x* does sin(*x*) = 1?" as the equation condition. Effectively, this computes the arc sine of 1. If you wanted to restrict the answer to the principal value of the arc sine, you could restrict the angle range explicitly as part of the **conditionQ** statement. The following code finds all the *x* values that satisfy the condition sin(*x*) = 1 and that are expressible in a {2, 3} environment:

```
setenv[{2, 3}]
conditionQ[u_] := (sinu[u] ≈ 1̂)
sols = solveforub[{{-∞̂, ∞̂}}];
Table[view[sols⟦i⟧], {i, 1, Length[sols]}]
```

$\{[-\infty, -510), -270, 90, 450, (510, \infty]\}$

For programmers used to working with radians, it may be surprising to see such a succinct, exact answer. The region (−*maxreal*, *maxreal*) has three exact values for *x*, in degrees, for which sin(*x*) = 1. There are infinitely many places where sin(*x*) = 1 outside that region, and the **solveforub** function found them as two infinite ranges.

> **Exercise for the reader**: The function $\sin\left(\frac{1}{x}\right)$ has an infinite number of roots near *x* = 0. What would the Warlpiri unums produce as the ubound set satisfying $\sin\left(\frac{1}{x}\right) = 0$, starting with a domain of [−∞, ∞]?

17.4.2 The general failure of the inversion method

The beginning of this section showed what happens when you use the inversion approach on linear and quadratic equations. There are closed forms for the solution to cubic and quartic equations as well; they are very complicated expressions. No one could figure out the inversion formula for *fifth*-degree equations for a long time and the assumption was that it was because of the challenge of doing that many pages of horrible-looking algebra.

About two hundred years ago, mathematicians found a proof that *no closed-form solution exists* for polynomials of degree five or higher. "That's fine," you might say, "because we can always just use a standard numerical method to find the roots."

Not really. Take a look at the plot of a fifth-degree polynomial on the next page:

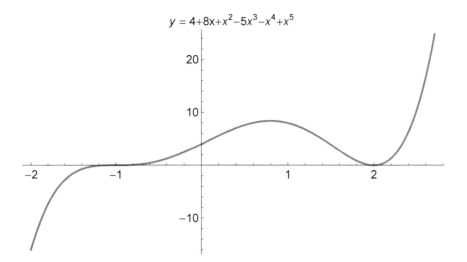

$$y = 4 + 8x + x^2 - 5x^3 - x^4 + x^5$$

Numerical root-finding methods based on floats have a rough time with a problem like this. Even if they are sophisticated enough to deal with multiple-order roots (there is a triple root at −1 and a double root at 2 in the polynomial), they will incorrectly declare many *x* values that are *near* −1 and *near* 2 to be roots, because of underflow to zero.

Exercise for the reader: With IEEE single precision, what range of values around −1 and 2 in the above polynomial will be rounded to zero (underflow) even if the polynomial is calculated to perfect float accuracy?

So there is no inversion formula, and traditional numerical methods give poor answers at best. Can the ubox approach solve this? Check it out:

```
conditionQ[ub_] := polyu[{4̂, 8̂, 1̂, -5̂, -1̂, 1̂}, ub] ≈ 0̂;
sols = solveforub[{{-∞̂, ∞̂}}];
Table[view[sols〚i〛], {i, 1, Length[sols]}]
```

$$\{-1, 2\}$$

The ubox approach works for even higher-order polynomials; there is nothing special about fifth-order except that they are the lowest order polynomial for which there is *no* inversion formula. If the roots are not expressible exactly in the computing environment, the unum with the smallest ULP that contains the root is returned.

It feels like being a kid in a candy store, having a solver that works so well for so many situations and that requires so little thought to use. There are no algebra equations to invert or formulas to memorize.

Even if it were possible to invert $f(x) = 0$ for a general degree polynomial *f*, there are many functions that do not even remotely resemble polynomials and that cannot be inverted in terms of elementary functions.

Like this one:

$$\log(x) - \frac{1}{x} = 0$$

How would you ever rearrange that so *x* is entirely on one side of the equation, using standard mathematical operations? There is no way. But the "try everything" approach requires no rearrangement since it tests the truth condition directly. Before diving into the above example with `solveforub`, consider this interesting way to use NaN. The `solveforub` solver can figure out what the *domain* of a function is, automatically. The condition to test is, "Does evaluating *f* produce a NaN?," and initially we try the entire range $-\infty$ to ∞:

```
setenv[{0, 0}]
conditionQ[u_] := (logu[u] ⊖ (1̂ ⊙ u) == {qNaNu})
sols = solveforub[{{-∞̂, ∞̂}}];
Table[view[sols〚i〛], {i, 1, Length[sols]}]
```

$$\{[-\infty, 0]\}$$

That shows $\log(x) - \frac{1}{x}$ is a NaN unless *x* is strictly positive. The solver caught the fact that exact 0 is not allowed; the log function allows $x = 0$, but the $\frac{1}{x}$ term does not.

Note: Be careful what you ask for! The `solveforub` routine refines regions where it detects `True` for the condition, but it does *not* refine regions for which the condition statement is `False`. This is important when using the above trick. If the test condition was "*f* is not NaN?", and you try $\{-\infty, \infty\}$, the solver will immediately return the empty set, {}, since the condition fails when the entire real number line is tested. It concludes, "Nothing to see here, folks. Move along." This is why the condition test has to be posed as "*f* is NaN?" instead of the inverse question.

Using the domain of the function $(0, \infty]$ and an environment accurate to 32 bits, here is the solver setup:

```
setenv[{2, 5}]
conditionQ[u_] := logu[u] ⊖ (1̂ ⊙ u) ≈ 0̂

sols = solveforub[{g2u[ 0      ∞   ]}];
                     open  closed
```

It produces the following ULP-wide bound on the result, where different digits are shown in orange as was done for the roots of the quadratic equation:

1.763222834095248838471777734375 < *x* < 1.763222834561020135879516601562

There are 64-bit float algorithms that can get this close to the root quickly, but they do not bound the answer. It is possible for interval arithmetic with IEEE floats to provide a bound this accurate, but it would require 128 bits to store the bound. The above bound fits in **43 bits** as an inexact unum. Fewer bits, better answers.

17.4.3 Inequality testing; finding extrema

The examples given so far are tests of equality (or of being NaN), but inequalities can be tested as well. Again, the condition question must be asked carefully. If we use a condition like

---

```
conditionQ[u] := u > 3̂
```

---

then the **solveforub** method will not work, since it stops looking as soon as it finds a region for which the condition fails. The way to pose the question so that the solver explores regions where the condition returns True is to pose the inequality with 3 as **an equality with (3, ∞] instead**:

$$\text{conditionQ[u] := } \left(u \approx \text{g2u} \left[\begin{matrix} 3 & \infty \\ \text{open} & \text{closed} \end{matrix} \right] \right)$$

In other words, ask if there is any *intersection* with the interval $3 < x \le \infty$. If given $\{-\infty, \infty\}$ as the starting set, **solveforub** will see the intersection and begin refining the real line into subsets as it should.

The ability to test for inequalities means it is possible to use **solveforub** *iteratively* to look for the minimum or maximum of a function in a region. (If the region is the entire real number line, it finds the global maximum or minimum, but it can also find local extrema.) If looking for a maximum of a function $f(x)$, begin by setting the condition to see if $f(x) > maxreal$. If any regions are found then we are done, since that is the largest expressible maximum. Otherwise, start bisecting exactly the same as with **splitub**, looking for the largest value *fmax* such that the condition

---

$$\text{conditionQ[u] := } \left(u \approx \text{g2u} \left[\begin{matrix} fmax & \infty \\ \text{open} & \text{closed} \end{matrix} \right] \right)$$

---

has an answer that is not the empty set. Finding the maximum takes no more than $\log_2(n)$ such splits. If a unum environment has n expressible values, then finding the maximum cannot take more than order $n \log_2(n)$ function evaluations, via this approach. It may surprise the reader that finding the maxima and minima of general continuous functions is considered *extremely* difficult in theoretical computer science, one more of those "No one knows how to do it quickly" assertions.

17.4.4 The "intractable" fallacy

A theorem exists that says finding equation roots and extrema in general is "NP-hard," a category computer scientists use to describe problems for which *no algorithm is known that takes polynomial time*; that is, the answer takes exponential time for the search, or worse. The theorem holds even if the function in question is merely a polynomial! How can this be true, if the ubox method makes it so easy?

The preceding section showed a way to find the extrema of a function *f* in (at most) $n \log_2(n)$ function evaluations when the function is expressed with unum math and there are *n* possible unum values. That is certainly not exponential time.

Examination of the theorems shows that they are all about the mathematical definition of "solve," not the computational one. The fallacy is a combination of two things: The assumption that whatever is difficult for pure symbolic mathematics must also be difficult for numerical computation, and a failure to realize that computing with real numbers requires representation of *ranges* (open intervals) as well as exact points. The ubox method does not produce a "solution" in the form of an algebraic formula, but so what? It always produces a bound on the answer, and that bound can be made as tight as you want for whatever practical purpose.

> Ubox methods convert algebra problems into finite *set* problems, and then produce mathematically correct set statements about where the answer must lie.

Vast amounts of work has been done on the "P = NP?" question, the question of whether NP-hard problems are in a separate class from the P problems that can be done in polynomial time. It is well established that if you do not require a *perfect* answer to some NP-hard question, but just something guaranteed to be within a few percent of the answer, then yes, some NP-hard problems can be done in polynomial time. But that is what actual computation with real numbers is all about: Being within some distance of a value, because most reals do not have exact representations.

A similar fallacy occurs in estimating the time required to sort *n* elements into order. Many texts assert that the time is order n^2 for an inefficient sort (like bubble sort), and order $n \log(n)$ for a more sophisticated method like HeapSort. Such analyses ignore the simple fact that there are a *limited number of different values the numbers can represent*; just histogram them, and the time is order *n* (also known as a bucket sort)! Uboxes take advantage of the simplifications offered by discrete values.

17.5 Solvers in more than one dimension

The approach discussed so far generalizes to multiple equations (or inequalities) in multiple unknowns. The example of two linear equations in two unknowns was touched on briefly in Part 1. Now that we are armed with the paint bucket method and the try-everything method, we can try them on a simple system.

Like this one:

$$x + \frac{7}{16}y = 3 \text{ and}$$
$$\frac{7}{8}x + y = 4.$$

Here is the same problem expressed as a matrix-vector equation:

$$\begin{pmatrix} 1 & \frac{7}{16} \\ \frac{7}{8} & 1 \end{pmatrix}\begin{pmatrix} x \\ y \end{pmatrix} = \begin{pmatrix} 3 \\ 4 \end{pmatrix}$$

Any system of linear equations can be thought of as a set of dot product equalities. In this case, we can write $\left(1, \frac{7}{16}\right)\cdot\begin{pmatrix} x \\ y \end{pmatrix} = 3$ and $\left(\frac{7}{8}, 1\right)\cdot\begin{pmatrix} x \\ y \end{pmatrix} = 4.$

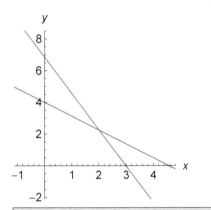

That is important because dot products can be done to maximum accuracy as fused operations in a unum computing environment.

On the left is the two-equation system as a geometry problem, the intersection of two lines. To make this more realistic and more challenging, assume the coefficients of the equations are not exact, but have at least a minimal ULP-wide uncertainty. For example:

$$\begin{pmatrix} (1,\ 1.0625) & (0.4375,\ 0.5) \\ (0.8125,\ 0.875) & (1,\ 1.0625) \end{pmatrix}\begin{pmatrix} x \\ y \end{pmatrix} = \begin{pmatrix} (3,\ 3.125) \\ (4,\ 4.25) \end{pmatrix}$$

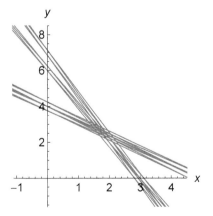

What this does to the solution set is broaden it to a shape similar to a parallelogram, because now the solution set is the intersection of banded regions. It is not a true parallelogram, because opposite sides are not quite parallel.

Plotting *every possible combination* of high value and low value for the coefficients above creates a set of boundaries that look like someone dropped some uncooked spaghetti on the *x-y* plane.

Some advocates of traditional interval arithmetic recommend figuring out which of the 16 lines form the boundary of the quadrilateral, a very tedious task and definitely the "NP-hard" way to go about things.

While the prototype does not include a two-dimensional version of the `solveforub` search function, it is not difficult to tile the *x-y* plane with uboxes and test them for compliance with the following condition:

```
conditionQ[xu_] :=
    fdotu[au⟦1⟧, xu] ≈ bu⟦1⟧ ∧
    fdotu[au⟦2⟧, xu] ≈ bu⟦2⟧
```

where **au** is the unum matrix representing $\begin{pmatrix} (1,\,1.0625) & (0.4375,\,0.5) \\ (0.8125,\,0.875) & (1,\,1.0625) \end{pmatrix}$, **xu** represents the unknown values $(x,\,y)$, and **bu** represents the right-hand side $\begin{pmatrix} (3,\,3.125) \\ (4,\,4.25) \end{pmatrix}$. A $\{1,\,2\}$ environment suffices to express all of these ranges without loss. An easy way to generate a two-dimensional "try everything" ubox test set is to create tuples of the set `uboxlist[{-∞̂, ∞̂},0]`. Surprisingly, the resulting set of fails and solves has *four* separate solves regions.

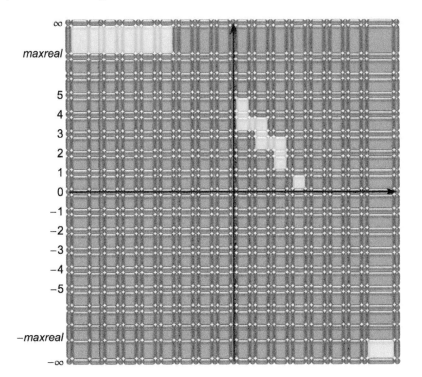

The upper left and lower right solves regions are where the ubox extends across (*maxreal*, ∞) and (−∞, −*maxreal*), so both slanted bands intersect those infinitely wide swaths. This is the two-dimensional form of the effect shown in Section 17.2.2. Notice that exact ∞ and −∞ do *not* satisfy the condition statement, however.

The isolated ubox where $x = (3, 4)$ and $y = (0, 1)$ intersects both bands, but the regions of that ubox that satisfy each of the two equations are *separate*:

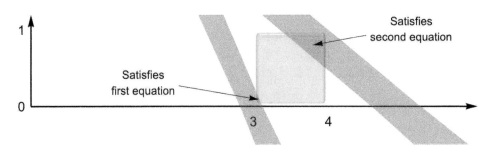

Should we consider that ubox a "spurious" result, one to be discarded? In this case, yes, because refining the ULP size to $\frac{1}{2}$ instead of 1 will cross this ubox off the solves list. If the environment were even lower precision, like {0, 1}, then this ubox *would* be a legitimate part of the c-solution, but in the present environment it is really a trial ubox. We are probably most interested in the zig-zag set of uboxes just above and to the left of this one, where there are point values that do satisfy both equations.

Notice that there is nothing special about the equations being *linear*. As with the single-variable solver, the method tolerates any complicated function you throw at it, since there is no need to find an inversion formula. You simply have to be able to state the conditions you want satisfied using unum arithmetic, and the computer can (in parallel) search all the candidate uboxes.

Now consider the paint bucket method. One way to get a starting seed is to use the try-everything method until a ubox is found that is exact in all dimensions. Another way is to try a traditional numerical method like Gaussian elimination, with exact values picked from the inexact ranges:

$\texttt{NSolve}\left[\texttt{x} + \frac{7}{16}\, \texttt{y} == 3 \bigwedge \frac{7}{8}\, \texttt{x} + \texttt{y} == 4,\ \{\texttt{x},\ \texttt{y}\} \right]$

$\{\{\texttt{x} \to 2.02532,\ \texttt{y} \to 2.22785\}\}$

To make sure we do not have a case like the one above where the solution grazes separate corners, pick an *exact* pair of unums near the result of the traditional solver, like the unums representing 2 and 2.25. Since floats are only guesses, one still has to verify that the unum pair satisfies the condition:

$\texttt{conditionQ}\left[\left\{ \hat{2},\ \hat{2.25} \right\} \right]$

$\texttt{True}$

We have our seed. Before invoking the paint bucket algorithm, consider what would happen if we tried using traditional interval arithmetic with Gaussian elimination.

The result is, as usual, very loose:

$$x = \left[\frac{112}{99}, \frac{67}{24}\right] = [1.13\cdots, 2.79\cdots],$$

$$y = \left[\frac{18}{11}, \frac{133}{36}\right] = [1.63\cdots, 3.69\cdots]$$

The formulas for inverting systems of linear equations, like Cramer's rule or Gaussian elimination, use the input values multiple times and suffer the dependency problem. In contrast, the equations in the `conditionQ` test use each input value *once*, so there is no looseness in the answer.

The ubox results have been coalesced on the plot, below. The black dot shows a single value, like one that might be produced by a traditional numerical "solver." But as you can see, it is just one value in a much larger set when the coefficients are inexact. It is like asking "Find x and y such that $x^2 + y^2 = 1$" and getting the answer "$x = 0$, $y = 1$," period. We need *all* the values that work, not just one. Single-value float solutions hide the instability of some problems; ubox solutions expose instability as eccentric-shaped solution sets, effectively performing "sensitivity analysis" that shows how inexact inputs affect the answer.

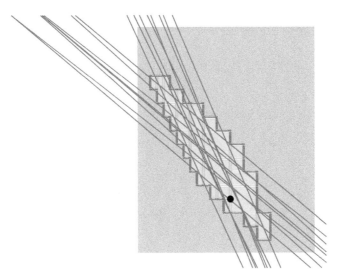

The gray rectangle is the loose bound produced by naive use of traditional interval arithmetic with an inverse method, Gaussian elimination. Though loose, it is not completely useless since it *could* be used as the starting point for a try-everything search, which would go faster than starting the search with the entire *x-y* plane.

The amber lines show every combination of interval endpoints in the coefficients and the right-hand side, where the exact shape of overlap is an exponentially difficult problem in computational geometry that defines the ultimate *mathematical* answer to the linear system.

The ubox solutions and fails values do not take exponential time, and they are not mathematically perfect but they approximate the mathematical answer as accurately as we require and they make heavy use of brute force parallel processing instead of algorithmic sophistication. The paint bucket method has an advantage over the try-everything method in that it finds the connected set of points that satisfy both equations, and does not include uboxes for which only separate parts of the ubox satisfy each equation.

17.6 Summary of the ubox solver approach

A common saying is "If you cannot solve a problem, try redefining the problem." Or something like that. This chapter takes that advice one step further:

If you cannot solve a problem, try redefining what "solve" means.

It is liberating not to have to *invert* equations,
but to ask instead the much simpler direct question:

Where are these equations true?

If there is anything tricky about using the try-everything ubox solver approach, it is making sure that the condition statement asks the right question, one that leads to refining regions that return `True`.

Asking questions of rigorous logicians can be exasperating. Imagine this questioning of one serving as a witness in a court trial:

> Could you tell if the victim was a man or a woman?
>> *Yes.*

> So, was it a man or a woman?
>> *Yes.*

> Can you tell me the victim was a man?
>> (pause) *Yes.*

> So the victim was a man.
>> *No.*

Questions have to be asked very carefully when a lawyer is questioning an extremely logical witness. So it is with trying to find the truth or untruth of a mathematical equation. Once you get the *question* right, the truth will set you free from having to learn numerical "recipes." And the computation itself will only cost you a finite number of ubox operations.

18 Permission to guess

If you are lucky, floats can sometimes guess an answer faster than unums can find it with rigorous methods. Unums face two more challenge problems from Professor Kahan in this chapter. Are you feeling lucky?

18.1 Algorithms that work for floats also work for unums

Unums do *not* discard the vast accumulated literature about how to compute with floats. Unums are a *superset* of floats. Floats are simply unums for which the ubit is zero. Up to now, we have ignored this because of the importance of showing examples where rigorous answers are possible for the first time, using set computation with ranges of real values instead of point guesses.

Guessing is a lot like rounding, **but in a unum environment, guessing must be done explicitly**. Whereas rounding is an invisible error made automatically by float hardware, a guess is something that has to be requested in the code, with a command like "`y = guess[x]`".

> **Definition**: A *guess* is an exact representable value selected from (or adjacent to) a range of real values.

Equipped with a `guess` function, which we will define more specifically, **there is absolutely nothing that a float environment can do that a unum environment cannot do**.

While rigorous mathematics is a powerful tool, there is also some power in guessing when a rigorous approach is elusive. The key is *to admit when you are doing it.* There is nothing mathematically dishonest about a computation that is declared to be a guess; it does not commit "the original sin" of numerical computing. In a well-designed computing environment, a guessed numerical answer should always be marked as such on output, for example by putting a raised asterisk after it:

`Result of pi estimation program : 3.125`*

That notation can be added to the input/output grammar for communicating results to humans described in "Demands and Concessions," Section 5.4.2. Notice it is quite different from writing an *inexact* result, such as

`Result of pi estimation program : 3.1 ⋯`

because the "⋯" notation means the result is *provably* between 3.1 and 3.2. If the guess is to find a value that solves an equation, the guess can be verified by evaluating the equation by rigorous unum arithmetic with no guessing, per Chapter 17, **which then removes the stigma of being only a guess**.

If we are going to guess, we had best do it carefully. We want guessing to resemble IEEE float rounding when a bound is the size of the smallest ULP. Here we will use the round-to-nearest-even mode, but other guessing methods can be crafted to suit your taste. If a unum is one ULP wide but *not* the smallest possible ULP, then it is always possible to express the midpoint as an exact unum.

Exercise for the reader: For a unum u that has its ubit set to 1 but a fraction that is less than the maximum length $2^{fsizesize}$, describe the procedure in terms of bit operations that creates the bit string for the exact unum v that represents the midpoint of u.

When the bounds a and b are more widely spaced, it is tempting to think about more sophisticated methods, like the geometric mean $\sqrt{a\,b}$ instead of the arithmetic mean $(a+b)/2$. That way, if the range is something like $(1, 10^{30})$, the guess would be 10^{15} instead of about 0.5×10^{30}. But sophistication leads to surprises, and it would probably be better if a guess does the simplest thing so a user knows what to expect. Therefore, we use the arithmetic average of the endpoints, $(a+b)/2$.

Here is one way to construct a robust guessing algorithm to create a single exact value from a range of real numbers from a to b. We ignore whether the endpoints of the range are open or closed.

Special case handling:
If either endpoint is NaN, return NaN.
If $a = b$, return a. (Notice that this means the returned value is exact, *not* a guess.)
If $a = -\infty$ and $b = \infty$, return 0.

The guess for distinct endpoints:
Compute the average of a and b, $(a + b)/2$. (Notice that infinite endpoints produce
  infinite averages.)
Find the nearest even unum to that average.
If the nearest even unum has only one fraction bit, return it;
  else compress away the trailing zero.

Notice that the guess can violate containment. If the input real number range is
$(a, a + \text{ULP})$ and the ULP is as small as possible, then the only exact values
available are a or $a + \text{ULP}$, which are both *outside* the bound. (By round-to-nearest-
even rules, the endpoint with the shorter fraction bit string wins.) As with floats, it is
not the least bit difficult to come up with cases where the guess is not merely
inaccurate, but dead wrong.

> **Exercise for the reader**: What will **guessu** do with the open intervals
> $(-\infty, -maxreal)$, $(-smallsubnormal, 0)$, $(0, smallsubnormal)$, and $(maxreal, \infty)$?
> Does that match what IEEE floats do, or is it different?

Also notice that guessing forces abandonment of the "knowledge" metric described
earlier, at least until the guess is checked for validity. It may be possible to guess,
then use rigorous methods to validate or invalidate the guess, and if valid, extend the
set of possible answers that form the *c*-solution, thereby recovering information
about the complete, minimal answer.

In the prototype, the **guessu** function accepts a unum, ubound, or general interval
input and produces an exact unum output guess, selected by the above algorithm.
The code for **guessu** is shown in Appendix D.9. Some examples follow.

If we have enough fraction bits to express the exact midpoint, then we do so:

```
setenv[{2, 3}]
view[guessu[{1̂, 4̂}]]
```

2.5

Notice that this increased the apparent precision. The endpoints 1 and 4 only require
one fraction bit to express, but the average 2.5 requires two fraction bits. An alternative form of guessing would be to round to the same number of bits as the
endpoint with the least precision, which in this case would produce the guess 2
instead of 2.5.

The average of any finite number with $\pm\infty$ is $\pm\infty$.

$$\text{view}\left[\text{guessu}\left[\left\{-\hat{\infty},\ \text{max}\hat{\text{real}}\right\}\right]\right]$$

$-\infty$

The **guessu** function also accepts a general interval as its input argument. In the following example, it is an open interval representing a smallest-possible ULP range; therefore, **guessu** returns the nearest unum with the shorter fraction bit string, adjacent to the open interval. We call it "nearest even" because if it did not compress away the trailing zeros in the fraction, the fraction string would end in zero and therefore look like the bit string for an even integer. When unum bounds are expressed with exact decimals, you can usually spot the "nearest even" endpoint as the one with the shortest decimal expansion:

$$\text{view}\left[\text{guessu}\left[\begin{array}{cc} 3.5 & 3.5078125 \\ \text{open} & \text{open} \end{array}\right]\right]$$

3.5

Guessing can be the most efficient thing to do **when you are certain there is something correcting the guess that keeps pushing it closer to the true answer**. That is, when you can prove that a numerical algorithm is "stable." This is the way biological systems work, after all; when you reach for a coffee cup, the signals the brain sends to your arm are guesses about how much muscle contraction is needed, rapidly corrected by visual and tactile feedback. When a housefly lands on a wall, it certainly does not first calculate solutions to the Navier–Stokes equations that govern flight dynamics. (Which is too bad, because if they did it would make them *much* easier to swat.)

For an example where guessing seems to work well, we turn to another challenge that came directly from Dr. Kahan in response to early drafts of this book:

Suppose $f(x)$ has a fixed-point $L = f(L)$ and for values of x in a given range X we wish to know whether iterating $x := f(x)$ will converge to L and how fast. The function $f(x)$ is provided in the form $f(x) := g(x) - h(x)$ for two expressions g and h that are going to be given as part of the problem. A simple example has $g(x) := 2x$ and $h(x) := x^2/L$ in which $L > 0$ will be supplied later along with $X := [(1 - d)L, (1 + d)L]$ for some small positive $d < 1$.

Let's represent X by a unum, and perform the iteration $x := f(x)$ in unum arithmetic programmed to evaluate g and h as separate subprograms. What will happen? What would happen if we did not use unums nor interval arithmetic but replaced X by an array of samples of floating-point numbers drawn from X?

We are not allowed to use the fact that $f(x) = (2 - x/L)x = L - (x - L)^2/L$ because, in general, f could be far more complicated than that and still exhibit the same kind of behavior. After a lengthy computation of many iterations, the disparity between the behavior of unums or interval arithmetic and the true behavior of the trivial dynamical system f must raise some suspicion that unums may handle other dynamical systems and lengthy computations badly.

18.2 A fixed-point problem

The general idea is this: Some iterative procedures with floats converge to an exact float value as a fixed point. They "trammel" into the right answer, the way the ball in a roulette wheel always lands in one of its integer numerical compartments. However, if you use any kind of interval in the procedure as if it were a point value, the dependency problem might smear each iteration and actually cause *divergence*. It sounds like Kahan is equating unums with traditional intervals and assuming they must suffer the same shortcomings, especially the dependency problem.

The following demonstrates the effect Kahan is referring to, using his variable names. Since we are leaving mathematics and entering the casino of floating point, let's pick a lucky-looking number for L, like 77. For the X interval, use the ubound representing [73, 81]. The evaluation of x^2/L is a good place to use the fused product ratio function, **fprodratiou**, to reduce information loss.

```
setenv[{2, 4}]

L = 7̂7;

g[x_] := 2̂ ⊗ x
h[x_] := fprodratiou[{x, x}, {L}]  (* better than squareu[x]⊙L *)
f[x_] := g[x] ⊖ h[x]
X = {7̂3, 8̂1};
```

A plot of $f(x) = 2x - x^2/77$ shows why f has a fixed point at 77. Exact values selected from the interval X evaluate to values in a much smaller interval [4.9875, 5], and exact values selected from that smaller interval then evaluate to values in [4.99996875, 5], and it converges quickly to the fixed point 5.

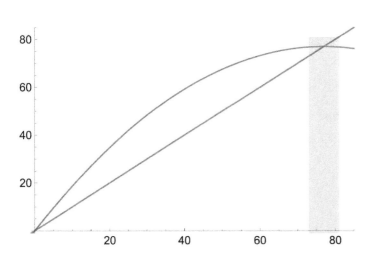

In the plot on the left, the light blue rectangle on the right is the range X=[73, 81], and the very thin light blue rectangle at the top is the range [f(73), f(81)]. The **polyu** polynomial evaluator applied to the entire interval X produces the range [4.9875, 5] without information loss.

But wait; the rules say **we are not allowed** to evaluate the polynomial $f(x) = 2x - x^2/77$ correctly. As the problem is posed, the computer is not allowed to detect the fact that f is a polynomial, since it might *not* be a polynomial for different definitions of g and h. The problem definition is designed to make the dependency problem unavoidable by treating the two functions g and h as unknown to the solution method. And sure enough, if all you do is use unums to mimic traditional interval arithmetic, the bound X expands instead of contracts:

`view[f[X]]`

(60.7919921875, 92.79296875)

On the other hand, if we plug in any specific float in that range, even one from the minimum and the maximum of X, the unum version of the function f does a nice job of landing a rigorous bound *near* the value of the fixed point, L.

$\text{view}\left[\text{f}[\hat{73}]\right]$

$\text{view}\left[\text{f}[\hat{79}]\right]$

(76.7919921875, 76.79296875)

(76.947265625, 76.9482421875)

Notice the unum result does not include the exact value 77, like the general interval (76.5, 77] does, since **that would be an error**. For any exact starting value other than 77, the mathematical function can only *approach* 77 after a finite number of iterations but never actually *reach* it.

By the rules of this game, we are not allowed to fuse g and h. A plot of f would reveal the pattern shown by loosely evaluated polynomials: Precise values for exact unums inputs alternated with wide bounds for inexact unum inputs. Any compiler would immediately detect the use of x in two expressions and invoke techniques for controlling dependency error.

But let's play this game, and see if the problem can be solved *mindlessly*. Just use uboxes. The function **fub** defined below accepts an input ubound **ub** and produces a bound for f. The "−5" in the first line would in general be supplied by the operating system according to how many processors are available to run in parallel; the routine automatically adjusts the unum widths based on the interval width.

```
fub[ub_] := Module[{fset = {}, i, minexp = ⌊Log[2, width[ub]]⌋ - 5},
  xset = uboxlist[ub, minexp];
  For[i = 1, i ≤ Length[xset],
    i++, fset = coalesce1D[fset, f[xset[[i]]]]];
  fset[[1]]]
```

Start with the same interval as before, $X = [73, 81]$:

```
xub = fub[X];
view[xub]
```

(76.2919921875, 77.5)

That iteration looks reasonably well-behaved. Do it again:

```
xub = fub[xub];
view[xub]
```

(76.931640625, 77.0625)

The small amount of dependency error is still causing the result to include 77 and even spill slightly over. Will this cause trouble? Iterate and see:

```
xub = fub[xub];
view[xub]
```

(76.9912109375, 77.0078125)

So far, so good. The location of the fixed point is being pinned down to a smaller and smaller range.

```
xub = fub[xub];
view[xub]
```

(76.9970703125, 77.001953125)

One last time:

```
xub = fub[xub];
view[xub]
```

(76.9970703125, 77.001953125)

Well, well. The iteration method works for uboxes as well, but produces a bound that correctly reminds us that the *computer* version of the problem is not quite the same as the *mathematical* problem. At this low precision, the ubox method captured both the exact fixed point and the fixed point for starting values other than L. The perfect answer would have been $(77 - \text{ULP}, 77]$, the smallest possible ULP-wide range, closed on the right. Because of the rules of the game, there was a small loss of information, and the range (77, 77.001953125) was also needlessly included. The calculation was *not* unstable like it is for traditional interval arithmetic. It proved the location of the fixed point with a bound five ULPs wide, even with rules designed to amplify the dependency problem.

Now let's try it with 16-bit IEEE floats. Pick a float from the interval X about the fixed point. The rules say "$X := [(1 - d)L, (1 + d)L]$ for some small positive $d < 1$." In other words, the largest range X can be is $(0, 2L)$. Wait a minute, why do floats need helpful coaching about *where to start*?

The rules say the starting guess *must* be picked from the open interval $(0, 2\,L)$. Start outside that range, and the *float method diverges*. How the heck would anyone know in advance that this is the magical range where things work, unless they had advance knowledge of g and h and had used calculus to figure out where the slope of f is between -1 and 1? If the problem allows you to do *that*, then of course you have a stable point and might as well use guessing methods; it's like dropping a marble into a big funnel where the rules say you have to drop it from anywhere, so long as you start directly over the funnel.

But ignore that for now, and use the same "small positive $d < 1$" as we did for unums. Pick, say, $x = 81$, representable exactly with the float bit string 0 10101 0100010000.

$g(x) = 2\,x = 162$. So far, so good. Doubling a number never causes rounding error, though it can cause overflow.

$h(x) = x^2/77$. That will round, of course. To what? *It depends on the computer!* Some computer systems will perform the x^2, round, then divide by 77, and round again, while others may do both operations in higher precision and *then* round, creating a different guess. The IEEE Standard says **it can be done either way, without any indication to the user**. The closest float to $81^2 = 6561$ is 6560. The closest float to $6560 \div 77$ is 85.1875, but had we done both operations with 32-bit floats and *then* rounded, the answer would be 85.25 instead! Assume the rounding is done after every operation, so the correct exact value $\frac{6561}{77}$ is replaced with the incorrect exact value $\frac{1363}{16} = 85.1875$.

$f(x) = g(x) - h(x)$ evaluates to $162 - 85.1875 = 76.8125$. That is over twenty times closer to the fixed point, 77, than was our initial value of 81. Repeat the process with this new value, $x = 76.8125$: $g(x) = 2\,x = 153.625$. $h(x) = x^2/77$; the closest float to x^2 is 5900, and the closest float to $5900 \div 77$ is 76.625. $f(x) = g(x) - h(x) = 153.625 - 76.625 = 77$ exactly. The method landed on an exact fixed point that matches the fixed point of the mathematical version of the problem, so it certainly looks like guessing with floats is the big winner.

But... again, wait a minute. Getting *exactly* 77 is **the wrong answer**. Mathematically, a finite number of iterations can never reach 77, but only *approach it from below*. A correct way to express the answer numerically is that the stable point of $f(x)$ lies in the open interval $(76.9375, 77)$ where both bounds are expressible floats. Too bad there is no way to express such a range with floats. To express the **correct answer** for any starting guess other than exactly 77, you need an inexact unum, like this one:

0	1101	0011001111111111	1	11	1111	$(78847/1024, 77)$
+	$2^6 \times$	1+ 13311/65536	$\cdots$	4	16	$= (76.9990234375, 77)$

Apparently the rules allow us to know that the starting interval X is a stable region, a "funnel" that makes guessing stable. So let's play this same game, using the `guessu` function to imitate the reckless behavior of floats. And while we are at it, invoke the tracking of bits per number so we can measure the cost of using unums instead of floats. Since the problem is designed to reward inaccuracy and guessing, turn the environment all the way down to {2, 3}:

```
numbersmoved = ubitsmoved = 0;
setenv[{2, 3}];

L = 77̂;

g[x_] := guessu[2̂ ⊗ x];

h[x_] := guessu[fmdu[x, x, L]];
f[x_] := guessu[g[x] ⊖ h[x]];
```

With the functions defined, try iterating twice with a starting exact unum for 81:

```
view[f[81̂]]

view[f[f[81̂]]]

reportdatamotion[numbersmoved, ubitsmoved]
```

77

77

Numbers moved	30
Unum bits moved	522
Average bits per number	**17.4**

That certainly did not take long. It guessed the mathematical fixed point in *one* iteration. It did it with an average of 17.4 bits per number, just slightly more than the 16 bits used by IEEE floats. If false convergence is what you want, `guessu` can do it.

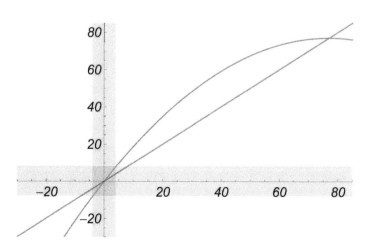

Did it strike the reader as slightly odd that the problem did not search for the *other* fixed point for $f(x) = x$, the one at $x = 0$? It should. Here again are the plots of $f(x)$ and x, but now the light blue rectangles represent ranges for x and $f(x)$ surrounding the *other* fixed point.

Pick a real between −4 and 4 and start iterating $f(x)$, using conventional float math:

```
f[x_] := 2 x - x² / 77
f[-3.]
f[f[-3.]]
f[f[f[-3.]]]
f[f[f[f[-3.]]]]
```

-6.11688

-12.7197

-27.5406

-64.9315

Oops. The guessing method of floats does not work so well at *this* fixed point because it happens to be *unstable*. (Actually, the iterations are heading for −∞, which *is* a stable fixed point if you think about it. And since $f(\text{NaN})$ is NaN, perhaps NaN should be regarded as a fixed point as well.) If you start the guess at a positive value, it heads for the wrong fixed point at 77. It could have been stable if the $g(x)$ and $h(x)$ functions were such that the function flattened out near the origin, but in this case that did not happen.

So the only way to "solve" for the fixed points with floats is to know *in advance* where the function is guaranteed to have the property $-1 < \text{slope} < 1$. That seems to violate the rule about treating $g(x)$ and $h(x)$ as unknown inputs to the problem.

Another drawback to the iteration technique of finding fixed points this way: *It is completely serial.* You cannot do anything with parallel hardware, because you certainly need to know $f(x)$ before you can find $f(f(x))$, and so on. The structure of the algorithm reduces the speed of the world's fastest supercomputer to that of a desktop personal computer.

Should we be impressed by floats in this cherry-picked example? Or with all due respect to Dr. Kahan, *should we be suspicious that this casino game is rigged?*

> The actual question being asked is: *Where are the fixed points?*

Let's try unums and uboxes on that larger, harder question, this time using the **solveforub** method described in Chapter 17. The condition question to ask is, "Where is $f(x)$ **not** identical to x?" and then take the complement of that set:

```
conditionQ[x_] := ¬ (f[x] ≡ x);
set = solveforub[{{73̂, 81̂}}]
Table[view[set〚i〛], {i, 1, Length[set]}]
```

$\{[73, 77), (77, 81]\}$

In other words, starting with the trial interval [73, 81], the solver correctly identified every real number *except* 77 as being a non-fixed point. A supercomputer could do the search very quickly, using as many processors as are available, in parallel.

The beauty of `solveforub` is that it does not need to be coached about where to look, and it has no problem finding *unstable* fixed points just as easily as fixed ones:

```
set = solveforub[{{-4̂, 4̂}}]
Table[view[set⟦i⟧], {i, 1, Length[set]}]
```

$\{[-4, 0), (0, 4]\}$

Everything other than 0 in the starting search interval is a non-fixed point, so 0 is fixed. In contrast, the iterative method has no prayer of locating that fixed point unless it is handed $x = 0$ to start, which is obviously cheating. We could have used the try-everything method starting with the entire domain of f (which is $[-\infty, \infty)$ because $f(\infty)$ is a NaN), and it would have produced the following fixed points:

$$-\infty, (-\infty, -maxreal), 0, 77.$$

The inclusion of $(-\infty, -maxreal)$ is the only difference between the correct solution to the computer problem and the correct solution to the mathematical problem.

To sum up what we have seen so far in this contest between floats and unums:

- Floats needed the head start of being told where to look for a fixed point that had been rigged to be stable. Unums need no such head start.

- By deviating from the mathematical evaluation of the functions, floats got lucky and landed on the stable point L exactly instead of correctly showing an asymptotic approach that cannot be exactly L after any finite number of iterations. Unums can perform the same trick, but converge faster using even *less* precision.

- The float-guess approach can only find one of the fixed points, since it is unstable at the other one. Unums locate both the stable and unstable fixed points.

- The actual values of the guesses using IEEE floats depend on what computer you use. The guesses for unums are bitwise identical across computer systems.

- The float approach is as serial as algorithms get; there is no way to access the speed of a massively parallel computer. Both of the unum methods (ubox tiling and try-everything) are massively parallel, able to use as many processors as the number of representable values in the search range.

Floats are inferior to unums on all counts listed above.

And now, a counter-challenge. An important function in physics is the "Sinc" function, a continuous function defined as $\frac{\sin(x)}{x}$ if $x \neq 0$, and 1 if $x = 0$. It describes things like the intensity of light that has diffracted through a slit opening.

```
sinc[x_] := If[x ≠ 0, (*then*) Sin[x]/x, (*else*) 1]
```

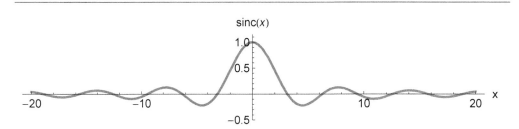

Suppose we make a tiny alteration to the g function. Instead of $g(x) = 2x$, add a very slight bit of Sinc function wobble near $x = L$:

$$g(x) = 2x + \frac{\text{sinc}(16\pi(x-77))}{25}.$$

If the function f is plotted on the same scale as before, this alteration is too small to see. The plot below zooms in to where both x and $f(x)$ are close to L.

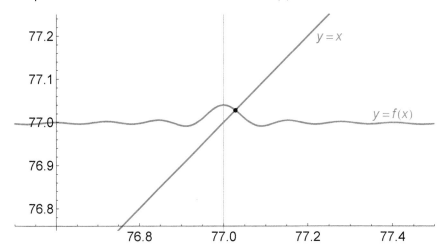

The fixed point is stable, and is located at 77.028···. The slope of the orange curve is between −1 and 1, so this is a stable point. It conforms to the rules regarding the input functions. What will happen when we try the guessing game of float arithmetic on this slight variation?

Start at $x = 81$ again. If the float math library for trig functions is *very* accurate, it will get the following result of zero for the Sinc term:

```
x = 81;
Sin[16 π (x - 77)]/(16 π (x - 77))/25
```

0

That means the computation of $h(x)$ is no different than before. Again, $f(x) = g(x) - h(x)$ evaluates to $162 - 85.1875 = 76.8125$, which is only 3 ULPs away from 77. What is the Sinc function there? The sine has an argument of 3π, for which a good math library should again return zero as the closest float.

The guess hopped over many ULPs worth of bumpy road in going from 84 to 76.8125, blissfully unaware of the oscillations in f. So once again, $f(x) = g(x) - h(x) = 153.625 - 76.625 = 77$ exactly.

Just one problem: 77 is **not** a fixed point of f; the Sinc function is $\frac{1}{25}$ there, so f evaluates to the closest float to 77.04, which is 77.0625. So is *that* the fixed point, at this precision?

```
x = 77.0625;
16 π (x - 77)
```

3.14159

No. Floats can only land on *exact multiples of the ULP size*, which is $\frac{1}{16}$ in the vicinity of 77. Multiply by 16π and the sine part of the Sinc function is zero, which points the next iteration back to 77.

The float algorithm will alternate between 77 and 77.0625 forever, never converging. The *correct* result at this precision is the open interval $(77, 77.0625)$, which is expressible with unums but not with floats.

Kahan's example was constructed to disguise the rounding errors of floats, by providing a situation where the rounding is attenuated with each iteration. But floats have "the other kind of error" described at the beginning of Part 2: *Sampling error*. By sampling the sine function and hitting only values where the sine is zero, an iteration with floats has no way to account for the fact that the true value of f really **is** a range of values in any ULP-wide region between floats. Ignore that at your peril.

The sinc(x) example was a soft toss for the guessing method, and it still messed up. It could have been *much* worse, like a higher-amplitude oscillation that makes all the fixed points unstable near 77. Or it could have had something like "The Smooth Surprise" of Section 15.7. If we added the absolute value of $|\log(x - 77)|/80$ to $g(x)$, the iteration would proceed rapidly toward lucky number 77 and discover that it had stepped on a land mine, sending the next iteration to infinity.

One last remark about not being allowed to notice the dependency of $g(x)$ and $h(x)$ as part of the rules: Compilers have been able to detect common subexpressions since the Jimmy Carter administration; this is not difficult technology. Why force a computer to solve a problem with willfully imposed stupidity?

Even if *g* and *h* are presented to the computer dynamically, any modern compiler (or interpreter) can spot the fact that both depend on *x*, and the dependency problem can be dealt with automatically by any number of methods like uboxes, or a math library call like `polyu` that knows how to avoid information loss when there is dependency of number ranges. There are even compilers that automatically find the derivatives of functions symbolically to discover where the functions have minima and maxima, which makes it easy to eliminate the dependency problem.

> The ground rules for any scientific comparison of two computing environments should be that automatable methods are allowed, but methods requiring human cleverness or special knowledge about a particular input *are not*.

We leave it to the reader to decide the validity of Kahan's indictment, repeated here:

> After a lengthy computation of many iterations, the disparity between the behavior of unums or interval arithmetic and the true behavior of the trivial dynamical system *f* must raise some suspicion that unums may handle other dynamical systems and lengthy computations badly.

If your life depended on a computer calculation, would you prefer unums or floats?

18.3 Large systems of linear equations

One of the proudest achievements of the numerical analysts of the mid-twentieth century (notably James H. Wilkinson) was to show that automatic computation with floats could be made safe for solving systems of linear equations and a number of other important operations. This is the type of special problem where guessing genuinely pays off, because under certain conditions, the algorithm can *attenuate* rounding errors as it proceeds. Here is another challenge from Professor Kahan:

> Here is a second example, now about unums versus rounding errors. Choose a large integer *N* and generate at random an *N*-by-*N* matrix *H*. The choice of random number generator does not matter much so long as it is roughly symmetrical around zero; Gaussian or rectangular distributions are O.K. Then the condition number $\text{norm}(H)\,\text{norm}(H^{-1})$ is very likely to be near *N*, which is pretty small for a matrix of that dimension; we would call *H* "well conditioned." Next generate at random an *N*-column *b* and solve "$Hx = b$" for *x* using Gaussian elimination with pivoting as usual. If the computation is carried out in 8-byte floating-point, and again in unums starting with about 53 sig.bits, how will the accuracies of the two results compare?
>
> Of course, the computation can be repeated starting with unums of much higher precision; how much higher is needed to get a result about as accurate as the result from ordinary floating-point? And how accurate is that? Rather than compute an error bound, I prefer to compute a correct *x* by using iterative refinement that costs a small additional fraction (about 20/*N*) of the cost of computing the first solution *x* of unknown accuracy using just ordinary floating-point. See
>
> www.eecs.berkeley.edu/~wkahan/p325-demmel.pdf
>
> Since there is nothing pathological about the computation of *x*, we have to wonder whether unums cope well with all high-dimensional computations generally. Some are O.K.

There need be no difference between unums and floats for solving a system of equations $Hx = b$, because the `guessu` function can be applied throughout the algorithm to do the same thing as any classic method with floats. The second challenge stems again from the misimpression that unums are simply a form of traditional interval arithmetic (or significance arithmetic). If you have to use intervals, then Gaussian elimination creates a *terrible* case of the dependency problem; that fact was already noted in the previous chapter, where even two equations in two unknowns created a very loose bound (too big an answer), whereas uboxes found a perfect *c*-solution, and floats simply picked out *one* of the complete set of answers and ignored the others (too incomplete an answer).

If you look carefully at the algorithm for Gaussian elimination with a dense matrix, it contains a triply-nested set of loops; one is for the row, one is for the column, and one is for the iteration number where each iteration eliminates off-diagonal entries. That is why solving an N by N dense system by Gaussian elimination takes order N^3 operations. Depending on how the loops are nested, the innermost operation looks like a dot product:

```
For[k = m1, k <= m1 + m, i++,
  H[i, j] := H[i, j] - H[i, k]*H[k, j]]
```

Recall that unums support the fused dot product. So if each entry in H is treated as exact, the unum call `fdotu` with vectors of length m will be accurate to a single ULP, which is far fewer than the approximately $\sqrt{m}$ ULPs of rounding error for conventional Gaussian elimination with rounding after every multiply-add. (The idea that the roundings are statistically independent is true when the entries are random and centered about zero, so the random walk model of Section 9.2 is not a myth in this case. Had they *not* been centered about zero, the error can be biased and severe, as that section showed.)

Kahan's preference for computing a correct x with "ordinary floating-point" (presumably IEEE single precision) and then performing iterative refinement is very much in keeping with the unum philosophy of what it means to "solve an equation": Test if $Hx = b$ is true. Computing Hx only takes N^2 multiply-adds, and once again those are all dot products that can be done as a highly accurate fused operation. With unums, the precision of the initial solve can be **very** low, like a {2, 3} environment where the unums range from 9 to 19 bits long. We do not need a large exponent even for N as large as 10 000 equations, because the sum of N numbers randomly chosen from $[-1, 1]$ will be a bell curve of width of about $\sqrt{10\,000} = 100$, and those sums are all performed with fused dot products in the g-layer.

In other words, unums can be *much* faster than floats in getting an initial guess for the refinement, and then can be *much* more accurate than floats at equivalent precision for the final iterations, through the use of fused dot products. The prototype is too slow to permit an experiment with a large matrix like $N = 10\,000$, but at some point a hardware-to-hardware comparison of unums and floats will be possible.

18.4 The last resort

When it looks like floats are losing on every count, one defense that comes up as a last resort goes something like this:

> Yes, your format looks better for some problems, but not all; sometimes floats do better. Since each format has problems it can solve that the other cannot, let's just stick with floats.

This argument does not hold water if a format contains floats as a *subset* of what it can do. Unum math can make guesses exactly the way float environments do and get the same answers (often with fewer bits and less energy). Unums also can perform naive closed interval arithmetic. The argument against any *alternative* numerical format has to take a very different form than the argument against a *superset* numerical format that can mime the way you presently compute, but also offers something much better for anyone tired of the gambling casino risks offered by floats or the uselessly loose bounds of interval arithmetic.

Technical debates about the merits of floats disguise the *actual* underlying reason for defending their continued use: The enormous amount of legacy and investment associated with floats. Decades of algorithm tweaking, countless hours spent in arguments about how many exponent bits there should be and whether the reciprocal of the square root of negative zero should be negative infinity, trillions of dollars in chip hardware tuned to the IEEE definition and in developing software that tolerates calculations making errors inconsistently from system to system. The desire to preserve the legacy of floats is very reminiscent of the desire in the 1990s to cling to serial computing, with *one* memory and *one* processor, years after parallel computing had antiquated that 1940s model of computation.

At some point, everyone computing with real numbers will recognize that the validity of using a computer as a tool depends on its ability to represent sets of numbers with open and closed endpoints. This is something to embrace and start working on, not resist because it seems daunting to "boil the ocean."

19 Pendulums done correctly

When physicists analyze pendulums, they prefer to talk about "small oscillations."
Have you ever met a child who didn't prefer the large kind?

19.1 The introductory physics approach

The reader may be getting impatient to see some examples of *real* problems that can be solved without error, not just fiendishly clever challenges thought up by mathematicians. The behavior of a pendulum is one of the first dynamics problems introduced in any physics course, so start with that.

> Every introductory physics course uses a pendulum as an example of a *harmonic oscillator.* The greater the displacement from equilibrium, the greater the restoring force, and that means the solution is the *wave equation.* That is, the displacement angle is a sinusoidal function of the time.

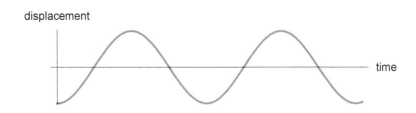

The analysis also shows that the period of oscillation does not depend on the amplitude, a counterintuitive result first pointed out by Galileo. If the length of the pendulum is L and the gravitational acceleration is g (about 9.8 meters per second per second), then the period of the pendulum is $2\pi\sqrt{L/g}$.

Unfortunately, it simply is not true. The restoring force is *not* proportional to the displacement. A pendulum is *not* a harmonic oscillator. Larger swings take *longer* than small ones, as anyone who has ever used a playground swing knows. Applying the wave equation to a pendulum is one of the "little lies" that sneak into the teaching of physics. The myth stems from the powerful desire to use the elegant analytical tools we have at our disposal and get results that look like elementary functions, even when... they do not quite fit the problem. A pendulum is approximately a harmonic oscillator the same way the earth is approximately flat, not round.

There *is* a theoretical solution for the true behavior of a pendulum, but it involves mathematics so ghastly that only a few undergraduate physics majors ever are exposed to it, and when you look at the "solution" it provides no insight into what is going on. There are many ways to express the exact description of pendulum motion, but all of them require combining an *infinite number* of elementary functions.

19.2 The usual numerical approach

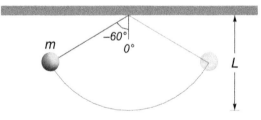

Suppose that the pendulum swing starts with the pendulum held at an angle –60° from vertical, for a mass m on a string of length L.

The usual approach to simulating physical dynamics with a computer is to march through time steps, by performing the following steps repeatedly:

- Use the position to estimate the force on the mass.
- Use the force to estimate the acceleration.
- Use the acceleration and the present velocity to estimate the future velocity.
- Use the velocity to estimate the future position.

Each estimate is subject to sampling error because position, force, and velocity all **vary** within a time step. There are many sophisticated methods that try to be clever, say, by using an interpolation between past, present, and future accelerations and velocities to get closer to the truth, but they all commit sampling errors.

When performed with floating point numbers, the calculations accumulate rounding errors as well. Because of both sampling and rounding errors, the computed physics simulation drifts farther and farther from reality. And you have no idea *how far* the answer has drifted because all the errors are silent and invisible. Traditional numerical analysis also states the sampling error for time step methods with another deeply unsatisfying error bound like the ones for numerically computing the area under a curve. The amount of sampling error is written in terms of unknown quantities that could be infinite.

> Sampling and rounding errors are not the only drawbacks to the time step method: *It also cannot be done with parallel computing.* Each time step depends on the previous one, so the method leaves no choice but to do the time steps sequentially. This leads people who run simulations to demand very fast *single-processor* computers so they can take finer time steps (to try to reduce sampling error)

The ubox method is fundamentally different, and treats the time dimension simply as one more ubox dimension. Each ubox in the two-dimensional space of position and time must

- conserve energy,
- preserve continuity (no teleportation to a new point in space), and
- have no instantaneous changes in the velocity, since that would require an infinite amount of force.

Say the starting position is angle $x = -60°$ as shown in the figure. Envision it as a child weighing 25 kilograms on a swing 2 meters long. Suppose for the moment that all values are exact, including the force of gravity as 9.8 meters per second squared.

The potential energy is mass m times gravity g times relative height h, or mgh. Here, the relative height is the distance to the bottom of the arc of the swing, which for a displacement angle θ is $L(1 - \cos(\theta))$. At the starting position, *all* the energy is potential energy because the velocity is zero. For this swing, the total energy is always 245 joules. (We can ignore air resistance, and assume someone is giving the child a slight push at the end of each cycle to keep the amplitude at $\pm 60°$.)

$$\text{potential energy}(\theta) = mgL(1 - \cos(\theta))$$
$$\text{total energy} = \text{potential energy}(-60°) = 245 \text{ joules}$$

Kinetic energy is $\frac{1}{2}mv^2$, where v is velocity. It is the total energy minus the potential energy, since the total amount of energy is conserved.

$$\text{kinetic energy}(\theta) = \text{total energy} - \text{potential energy}(\theta)$$
$$= \frac{1}{2}mv^2.$$

That means velocity *can be derived from the position*. The velocity is in meters per second along the arc of motion, so the angular velocity in degrees must be scaled by $\frac{\pi L}{180°}$ to find the velocity of the suspended mass M. The following formula infers the velocity from the kinetic energy; notice that this requires kinetic energy to be nonnegative, or else the velocity will be an imaginary number!

$$v(\theta) = \sqrt{2 \text{ kinetic energy}(\theta)/m} \; .$$

We can also determine the *acceleration* from the position. It is simply gravity times the negative sine of the deflection angle. (First-year physics approximates $\sin(\theta)$ with θ to make things easier, albeit wrong.)

$$a(\theta) = -g \sin(\theta).$$

Problems like pendulum dynamics are called *nonlinear ordinary differential equations*. The convenient situation where position, velocity, and acceleration are related by simple linear equations does not happen nearly as often in real-world problems as it does in physics courses. Traditional numerical methods offer *guesses* about the solution to such nonlinear problems, but no guaranteed bounds for the answer to the simple question, "Where is the mass at any point in time?"

19.3 Space stepping: A new source of massive parallelism

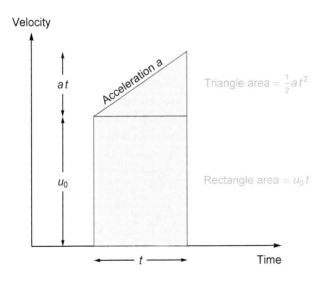

This diagram is a reminder that much of physics does *not* require calculus. Distance is velocity times time. If speed varies with time, then distance is the area under the function that describes the velocity. It does not take integral calculus to find the area if acceleration is constant; we only need elementary geometry for the area of a triangle and a rectangle.

$$x_{\text{new}} = x_{\text{old}} + u_0\, t + \frac{1}{2} a\, t^2.$$

Why not simply use $x_{new} = x_{old} + u_0 t + \frac{1}{2} a t^2$ with traditional interval arithmetic, using bounds on velocity u_0 and acceleration a to bound the value of x_{new}? Because doing so creates a chicken-and-egg problem: You do not know the range of possible accelerations and velocities until you know the range of possible positions in the time step. And you do not know the possible positions in the time step until you find out the range of possible accelerations and velocities. In the case of the pendulum, during the first quarter-period the velocity goes as low as zero and the acceleration also goes as low as zero, which means interval arithmetic predicts that the pendulum *might not move at all*, but just hang there having a Wile E. Coyote moment.

Instead, we treat *time as a function of location*. Within a location range, there is a rigorous bound on the velocity and the acceleration. (This happens for many types of dynamics problems, not just the pendulum.) From the bounds on velocity and acceleration, we can derive rigorous bounds on the time interval spent traversing each location range using elementary geometry, as we will show in a bit. This might make you uncomfortable because space stepping is backwards from the usual way we think about motion, where the location is a function of time. But consider this fact:

> *All the traversal time bounds can be computed in parallel.*

That is a stunning result, because it means we not only get rigorous bounds on the physical behavior, but we can use *as many processors as we have* in a computer system to get any desired answer quality. Nature may operate a pendulum as a serial function of time, but that does not mean we have to compute its behavior serially as well.

When people seek excuses not to use parallel computing to solve problems, they often point to problems that seem *inherently* serial, where things apparently have to be computed in sequential order (each step depends on the previous one). Pendulum simulation is a perfect example, and so is the problem of calculating an orbital trajectory. For many years, computer users have misused Amdahl's law as an excuse not to apply parallel programming, asserting that doing so would have very limited benefit. Similarly, the time dependency of physical simulations has been misused as an excuse not to change existing serial software to run in parallel. It is now time to retire that excuse, also.

Each calculated traversal time will be off by some number of ULPs, and when we line all the times up to create the graph of where the mass is at any range of time, the ULPs accumulate. The unum environment should be selected so that the accumulated ULP error is about the same size as the accumulated theoretical bound between the slowest and fastest time, which can be made as small as we like by using more space steps. This is very different from rounding error and sampling error; it eliminates *both* sources of error in the numerical calculation, replacing them with guaranteed bounds.

Here is a simple example of how to bound the amount of time spent in an angle range. Start with the entire range from the start time at –60° to the point where the pendulum passes through vertical, 0°. (After that the behavior is symmetrical, so calculating a quarter-period suffices to calculate the entire period of the swing.) The angle subscripts denote physical values at particular angles, so $a_{-60°}$ means the acceleration when the pendulum is at –60°, for example:

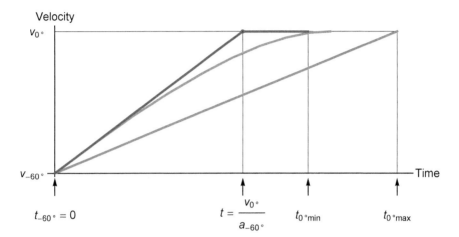

The purple curve shows the correct velocity as a function of time. We do not know what that velocity function is, but we do know what its physical limits can be. The *maximum* it can be is if it increases with the maximum acceleration $a_{-60°}$ up to the time when it reaches the maximum velocity $v_{0°}$, which is the curve shown in red. Notice the red dot where the maximum acceleration changes; that will help with visualization when we use the velocity to find the location.

The *minimum*, shown in blue, would be if the velocity accelerated only enough to get it to the maximum. Why can it not be even lower than that, and perhaps curve up toward the right? Because the acceleration is *decreasing* in this angle range. It cannot curve up anywhere, since that would mean acceleration is increasing.

The corner where the top curve hits the maximum velocity is when time is equal to $v_{0°}/a_{-60°}$. With that, we have the coordinates of every point needed to find the formula for the area under the red curve and under the blue curve, both of which represent the distance traveled. But remember, we are *not* computing the distance traveled. We **know** the distance traveled; what we do not know is the *time* it takes to go through that part of the pendulum swing.

Adding up the area of the triangle and the rectangle under the red curve gives the **minimum time** to traverse the angle range, since that is the **maximum velocity**.

$$\text{Distance traveled} = \text{Area} = \frac{1}{2}\left(\frac{v_{0°}}{a_{-60°}}\right)v_{0°} + \left(t_{min} - \frac{v_{0°}}{a_{-60°}}\right)v_{0°} = L \times \left(\frac{60°\pi}{180°}\right).$$

Solving this equation for t_{min} tells us the following:

$$t_{min} = \frac{L\pi}{3\,v_{0°}} + \frac{v_{0°}}{2\,a_{-60°}}.$$

For this particular pendulum, t_{min} evaluates to 0.733896··· seconds as the *minimum possible* time, which is interesting because the harmonic oscillator model says the period is 0.709613··· seconds! **We just proved that the usual model derived from calculus is false, and did it with elementary geometry and algebra.**

Now for the lower bound, the blue curve, which tells us the *maximum* time to traverse the angle range since it is the slowest possible speed. This one is simpler because the distance traveled is simply the area of a triangle:

$$\text{Distance traveled} = \text{Area} = \frac{1}{2}\,t_{max}\,v_{0°} = L\times\left(\frac{60°\pi}{180°}\right).$$

Solving this for t_{max} tells us this:

$$t_{max} = \frac{2L\pi}{3\,v_{0°}}.$$

That gives an upper bound on the time of 0.94632··· seconds. If the ubox bounds in time and space are calculated with a {2, 3} environment (at most eight bits of precision in the fraction), the bounded solution is still tight enough to exclude the first-year physics solution, shown in amber on the graph below.

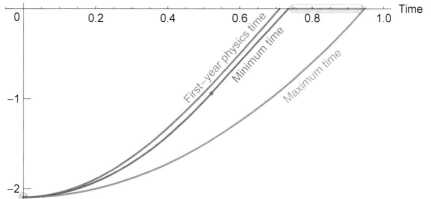

The dot on the red curve corresponds to the dot on the velocity curve in the previous plot. It shows when the acceleration changes, which is hard to see otherwise because the curve is smooth there. The plot shows a cyan ubound at the start and one at the end, but information about the bounds on the acceleration and velocity mean we have parabolic trajectories bounding the time and position in between. Notice that the starting ubound is exact in both dimensions, because we know the time is exactly zero and the location is exactly –60° (though the angle is plotted as position on the arc of the swing).

The final ubound is exact in the position because we know the position is exactly zero, and inexact for time. The unum bounds for the two red acceleration-velocity pairs and the single blue acceleration-velocity pair yield a rigorous bound for the trajectory using finite-precision arithmetic.

You can guess what comes next: Split the space region in two. Evaluate the acceleration and velocity at −30°, $a_{-30°}$ and $v_{-30°}$, and "pin" the bounding regions and slopes to those values. The shape bounds get more interesting:

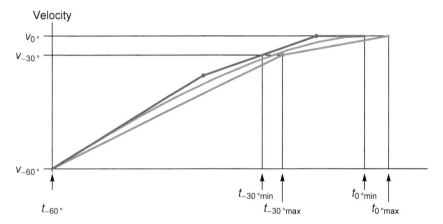

A glance at the graph shows that this is already a much tighter bound on the unknown purple velocity curve. Plus, two processors can calculate the bounds on the two time periods *in parallel*. For this new subdivision of angle space, the area bounds work out to the following:

Time spent going from −60° to −30°: 0.50486⋯ to 0.55291⋯ seconds.
Time spent going from −30° to −0°:   0.24595⋯ to 0.25494⋯ seconds.

That makes the total time from −60° to 0° somewhere in the interval 0.7508⋯ to 0.8079⋯ seconds, which is almost four times as tight as the bound we computed before, 0.7338⋯ to 0.9463⋯ seconds:

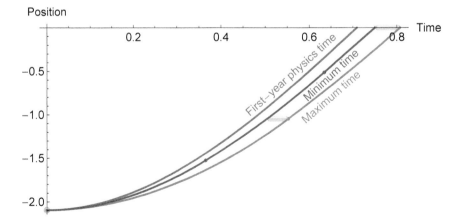

The computation at the midpoint angle adds a new ubound at exactly −30°. With the general technique defined for computing changes in time, try bounding the pendulum time using space steps of 10° arcs. The graphs are so precise that we have to use thinner lines to make the minimum time distinguishable from the maximum time. The first-year physics answer is so far off that it is easy to distinguish.

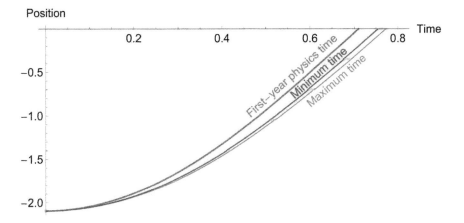

The cyan ubox bounds are almost too small to see, but they are exact in the vertical dimension and have a tiny range in the horizontal dimension that accumulates from the leftmost ubox (where we know the starting time exactly) to the rightmost one. The range is the result of both unum ULP widths and the variation in the velocity in each position range. If we had used higher-precision unums to reduce ULP widths, the final time range would be 0.759494⋯ to 0.768758⋯ seconds, a width of about 0.009 seconds. Recall that we used a {2, 3} unum environment, so the largest possible fraction is only $2^3 = 8$ bits. This causes an additional information loss at each step, so the resulting bound is between 0.773438⋯ to 0.753906⋯ seconds, a range of about 0.02 seconds. This indicates a good match between the space step discretization and the unum precision, since neither one is using "overkill" to get the answer more accurate. If we had used, say, a {2, 4} unum environment that could have up to $2^4 = 16$ bits in the fraction, that would have merited decomposing the problem into more space steps, staying in balance.

With a total of only six independent space steps, we calculated the pendulum behavior to almost three decimals of accuracy and with a provable bound. Furthermore, the most complicated math we used was grade school algebra. Without deep human thought but with brute force computing, we can bound the solution to any desired accuracy. In other words, we just *"solved" a nonlinear differential equation without using any calculus.*

This shows why it may be time to overthrow a century of numerical analysis. The classical way of solving a dynamics problem is either to substitute a different problem that has a tidy-looking solution involving only elementary functions, or to perform an error-filled time-stepping simulation that cannot run in parallel and has no rigorous bound on the numerical guesses it makes.

By combining space stepping with uboxes and unum arithmetic, we can obtain a bound that can be *as tight as we want* and can be computed using as many processor cores as we have at our disposal.

Method	Pros	Cons
High school and freshman physics	Answer looks like a simple elementary trig function	Answer is wrong
Advanced physics	Looks like a "closed form" solution	Closed form is not a finite combination of elementary functions
Traditional numerical methods (time steps)	It's... traditional	Forces serial execution, accumulates rounding error, accumulates sampling error, provides no answer bounds
Ubox method (space steps)	Bounds answer rigorously; massively parallel	Computer has to do more work

19.4 It's not just for pendulums

The approach works whenever the kinetic and potential energy of a mass depend only on its position along a known path, allowing us to compute the velocity and the force as a function of position in space instead of time. It can, for example, bound the solution for a true harmonic oscillator (like a mass on an ideal spring). Many simulations of electric circuits that compute current and voltage in time steps can be rewritten to compute the time from the change in the circuit state, again applying energy conservation.

The pendulum example shows the promise of a general technique, one worth building a set of tools for. The technique requires that the force on a mass, and its velocity, are determined by the position of the mass. That would cover, say, the motion of a ball rolling down a slide with a complicated shape. It would not apply to something like a rocket, which decreases its mass as it burns fuel. (The popular expression "It's rocket science" stems from the challenge of solving equations where the physics depends on both the position and the *history* of the rocket.)

Here is a general type of dynamics problem: A mass moves from point x_1 to point x_2. Its velocity at x_1 is v_1; its velocity at x_2 is v_2, and between x_1 and x_2 the velocity never goes outside that range. Similarly, the acceleration is a_1 at point x_1 and a_2 and point x_2, and never goes outside that range. What range of time t is required *for the mass to make the trip?* Notice we do not require that $v_1 \le v_2$ or $a_1 \le a_2$.

The graph on the next page indicates subtleties about velocity ranges when $a_1 > a_2$, with red indicating the maximum velocity and blue indicating the minimum velocity:

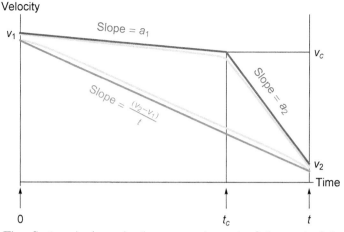

First, notice that the v_1 and v_2 are shown as inexact to remind us that they are computed values that will generally be represented with a ubound; that is, there is a gap between the red and blue lines at the left and right endpoints.

The first red slope is the upper bound of the calculated acceleration a_1 and the second red slope is the lower bound a_2. Actual physical behavior is mathematically bounded. Also, notice the two green curves. The bottom one starts out with a slope of a_1 as required, but almost immediately takes on a slope close to that of the line connecting the start and end points. Just before it gets to the end time t, it curves down just enough to match the slope of a_2, the acceleration at the end time. The lower bound shown as a blue line bounds the lowest possible velocity at any time. The upper green line takes on the higher acceleration a_1 for as long as it can, but at some time t_c it has to switch to acceleration a_2 or it cannot reach the final velocity v_2. Since instantaneous changes in velocity are impossible (it would take an infinite amount of force), the upper green line shows a rounded elbow at t_c.

The minimum travel time occurs when the velocity is maximum; the maximum time occurs when the velocity is minimum. The area under the blue line is the distance traveled, $x = x_2 - x_1$. If we were lazy, we could just say the minimum velocity is v_2 and the maximum velocity is v_1, so

$$v_1 t_{min} = x = v_2 t_{max}, \text{ hence } \frac{x}{v_1} \le t \le \frac{x}{v_2}.$$

The simple bound made it easy to invert the problem to solve for t as a function of x. The problem is that the minimum velocity could be *zero*, in which case the bound on the travel time is infinity, and that does not provide a useful answer.

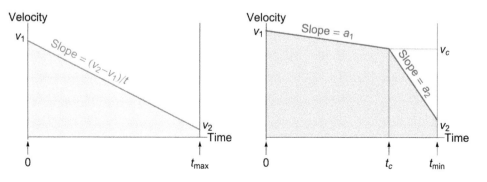

The area under the blue line is easy to compute: $x = \frac{1}{2}(v_1 + v_2) t_{max}$, which inverts to $t_{max} = \frac{2x}{v_1 + v_2}$. The area under the two red line segments is much more complicated, because we first have to find the point of intersection and the area is a second-degree polynomial in the time t, which means that inverting to solve for t requires solving a quadratic equation. Brace for algebra.

The two red lines intersect when

$$v_1 + a_1 t_c = v_c = v_2 + a_2(t_c - t), \text{ so}$$

$$t_c = \frac{v_1 - v_2 + a_2 t}{a_2 - a_1} \text{ and } v_c = \frac{a_1 v_2 - a_2 v_1 - a_1 a_2 t}{a_1 - a_2}.$$

The area under the left red line segment is $\frac{1}{2}(v_1 + v_c) t_c$ and the area under the right red line segment is $\frac{1}{2}(v_c + v_2)(t - t_c)$. After expanding those products and collecting terms, we get the total area:

$$\text{Total area} = x = \frac{1}{2}(v_1 t_c + v_c t + v_2 t - v_2 t_c)$$

Solving for t gives the minimum time, a complicated expression but nothing a computer cannot handle. With expressions for t_{min} and t_{max}, we can directly compute the range of possible times:

$$t_{min} = \frac{v_2}{a_2} - \frac{v_1}{a_1} - \frac{\sqrt{(a_2 - a_1)\left(a_2\left(v1^2 + 2 a_1 x\right) - a_1 v_2^2\right)}}{a_1 a_2}.$$

The minimum can also be computed using the **solveforub** function described in the previous chapter instead of using the quadratic equation, if a tightest-possible bound is needed. The situation is different if $a_1 < a_2$:

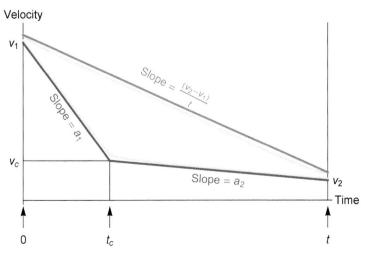

In this case, the equations for t_{min} and t_{max} are interchanged and we need the other ± case of the quadratic formula. We also need to handle the case where a_1 or a_2 is zero, and the trivial case where both are zero (that is, velocity is constant).

$$V_1 + a_1\, t_c = V_c = V_2 + a_2(t_c - t),\ \text{so}$$
$$t_c = \frac{V_1 - V_2 + a_2 t}{a_2 - a_1} \ \text{and}\ V_c = \frac{a_1\, V_2 - a_2\, V_1 - a_1\, a_2\, t}{a_1 - a_2}.$$

The implementation function **delt** below derives elapsed time from acceleration bounds, velocity bounds, and distance and is careful to account for the cases where either acceleration is zero. It was used to produce the preceding plots and is shown below. Its arguments are the acceleration bounds **a1** and **a2**, the velocity bounds **v1** and **v2**, and the distance to be traversed, **x**. A unum version of **delt** was used to do the ubox calculations of the pendulum behavior.

```
delt[a1_, a2_, v1_, v2_, x_] := Which[

  a1 == 0 ∧ a2 == 0, {x / v1, x / v2},

  a1 == 0 ∧ a2 < 0, { (v2-v1/2)/a2 + (x - 1/2 v2²/a2)/v1 , 2x/(v1+v2) },

  a1 == 0 ∧ a2 > 0, { 2x/(v1+v2) , (v2-v1/2)/a2 + (x - 1/2 v2²/a2)/v1 },

  a1 < 0 ∧ a2 == 0, { 2x/(v1+v2) , (v2/2-v1)/a1 + (x + 1/2 v1²/a1)/v2 },

  a1 > 0 ∧ a2 == 0, { (v2/2-v1)/a1 + (x + 1/2 v1²/a1)/v2 , 2x/(v1+v2) },

  a2 > a1, { 2x/(v1+v2) , v2/a2 - v1/a1 + √((a2-a1)(a2(v1²+2 a1 x)-a1 v2²))/(a1 a2) },

  True, { v2/a2 - v1/a1 - √((a2-a1)(a2(v1²+2 a1 x)-a1 v2²))/(a1 a2) , 2x/(v1+v2) }]
```

If our computer resources involve processors that can work in parallel, **delt** is the task that each processor can perform independently to solve this type of space-stepping problem. The only limit to the parallelism is how much precision is justified in the final answer. For instance, if we only know $g = 9.80$ to three decimals, it makes little sense to compute pendulum behavior to ten decimals of accuracy. The value of g ranges from 9.78036 at the equator to 9.83208 at the poles, and it also varies with altitude. If you use the following formula for g in meters per second squared, where p is the latitude and h is the altitude in kilometers, and the pendulum has little air resistance, **then** you might be justified calculating to seven decimals:

$$g = 9.780356 \left(1 + 0.0052885\,\sin^2(p) - 0.0000059\,\sin^2(2\,p) - 0.003086\,h\right)$$

The adding of the times to reach any desired accumulated time can be done in parallel with a reduction operation like a binary tree sum collapse. Summing n values takes $\log_2(n)$ passes with that method, a big improvement over doing it on one processor in n steps. Besides parallel summation being fast, the availability of a *fused* sum means there is no difference in the answer no matter how the time intervals are added.

With just a little added complexity to the sum collapse, the *prefix sum* can be calcu-lated in about the same time, producing the partial sums t_1, $t_1 + t_2$, $t_1 + t_2 + t_3$, as well as complete sum. That restores the space-parallel answer to the usual way we think about simulations: Position (or some other quantity) as a function of time.

This chapter started with the sinusoidal plot of the classical solution to the pendulum problem. Here is the plot again, along with a plot of the *rigorous ubox time bounds* for the coarse 10° angle spacing. It may initially look like a small difference between the two results in the horizontal (time) dimension, but look at the difference in the *displacement* value for times toward the right part of the graph. For instance, when the pendulum is actually passing through 0° for the fourth time, the classical sinusoidal solution shows the pendulum around –40°, a huge error.

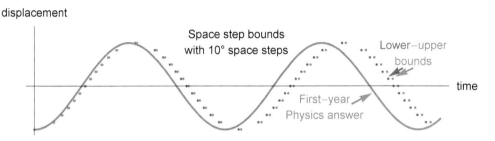

Since the uncertainty in the ubox method accumulates as the ranges of elapsed time accumulate from left to right, it is possible to see the spread of minimum time (red) and maximum time (blue). Decreasing the space step to $\frac{1}{16}$° not only makes the spread too small to show on the graph, it also fills in all the gaps in the plot. Since the maximum time is plotted after the minimum time, the blue plot covers the red plot.

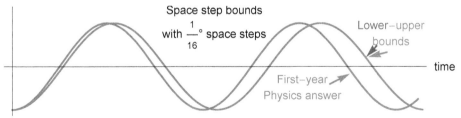

What are the limits of applying ubounds and space-stepping to numerical physics? The chapters that follow show that the technique is certainly *not* limited to nonlinear ordinary differential equations.

20 The two-body problem (and beyond)

Stargazers spotted a 60-meter asteroid heading toward Earth eleven months before a possible collision in February 2013. NASA expert David Dunham said, *"The Earth's gravitational field will alter the asteroid's path significantly. Further scrupulous calculation is required to estimate the threat of collision."* Computing this two-body problem with error-prone floats and traditional time step methods provided only *guesses*, not scrupulous assurance, that the asteroid would *probably* not strike Earth with the destructive power of a thermonuclear weapon. It missed by just a few thousand miles. Please forgive the nearly universal artistic license of showing the meteor glowing with heat; that would only happen if it impinged on the atmosphere.

20.1 A differential equation with multiple dimensions

The phrase "the two-body problem" has been applied as a joke to the difficulty that couples sometimes have finding professional careers in the same location, especially when both people are in academia. That joke is now so universal that it may be the most common use of the phrase "two-body problem." Here we are concerned with the phrase only as it applies to Newtonian physics.

The problem of calculating how a collection of N masses move under classical gravitational force has been around for hundreds of years, and the only case for which the shape of the paths can always be expressed with elementary functions is when $N = 2$. The two-body problem is solved when the masses follow paths that are *conic sections*: Ellipse, parabola, hyperbola. Since we *know* the elementary function for the path shape, we can test any numerical method against that exact solution.

However, the reader may be surprised to learn that there is no elementary function for the *trajectory*, that is, where the two masses are as a function of *time*. Like the pendulum problem, there are trajectory formulas expressed as infinite combinations of elementary functions, which leads us to wonder if a numerical method can give a more practical answer, one with numerical values for where the masses are, when.

Also like the pendulum problem, most physics textbooks are riddled with subtle changes to the two-body problem to make the equations look prettier, even if it means making them wrong. The first approximation most physicists reach for is "Assume one mass is much larger than the other and can be regarded as stationary." We will not do that here. It seems reasonable, say, in the case of the Earth orbiting the Sun, where the Sun is about 300 000 times more massive than the Earth. Notice, however, that if that approximation caused the calculation to be off by 1 part in 300 000, it would make the location of the Earth off by almost 300 miles!

One approximation we *will* permit, for now, is that the dynamics are straight Newtonian physics, without the effects of relativistic bending of space-time. In our solar system, the effects of general relativity on planetary orbits are so subtle that they can take centuries to become large enough to observe.

20.1.1 Setting up a traditional simulation

First try the traditional numerical time stepping approach. It is easy to start drowning in variable names and definitions when explaining physics equations, so every effort will be made here to keep the number of variables to a bare minimum. It also is hard to read subscripts, like "v_x" to mean "velocity in the x direction," so the only ones we use will be a zero, like "x_0" to indicate an initial value, or "new," like "y_{new}" to mean a value computed from previous values. Otherwise we try to use single letters for physical quantities. It helps that the orbit always lies in a plane relative to the center of mass, so we only need to use two dimensions for the experiment, x and y. The velocity has x and y components too, which we call u and v. The acceleration components will be called a and b.

Suppose the two masses are $M = 8$ and $m = 1$. The "inverse square" force law says that the masses experience a force of $\frac{GMm}{r^2}$, where r is the distance between the centers of the masses and G is the gravitational constant. We are not using standard units like meters or seconds here, so suppose the gravitational constant is $G = 2.25$ in some system of units. Start the lighter mass m at an initial position $(x_0, y_0) = (4, 0)$ and initial velocity $(u_0, v_0) = (0, 2.25)$. (These numbers were selected to make the graphics clear; it is only a coincidence that both G and the initial value v_0 wound up being the same number, $2.25 = \frac{9}{4}$.) In computer terms:

```
{M, m} = {8, 1};  G = 9 / 4;  {t0, x0, y0, u0, v0} = {0, 4, 0, 0, 9 / 4};
```

The position and velocity of mass *M* are simply those of *m*, scaled by $-\frac{m}{M}$, so that the center of mass is at the origin and the total momentum is zero. This saves us from having to use a lot more variable names, because we only need to follow the behavior of *m*. The behavior of *M* is point-symmetric about the origin, scaled:

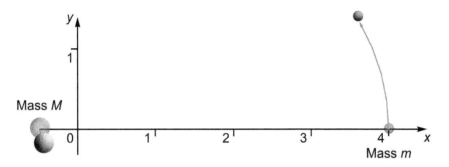

The quantities that change in the simulation are position, velocity, and acceleration. The usual methods employ *time stepping:* Use positions to find the acceleration forces, figure out where the masses will be after a time step, update the positions and velocities for the new time, and repeat. For dynamics problems, this is the simplest traditional numerical technique of them all, the *Euler method.*

Mass *m* is distance $\sqrt{x^2 + y^2}$ from the origin, but mass *M* is closer to the origin by a factor of $\frac{m}{M}$. That means the total distance *r* between the masses is $r = \left(1 + \frac{m}{M}\right)\sqrt{x^2 + y^2}$, which is sometimes easier to write as $\frac{M+m}{M}\sqrt{x^2 + y^2}$. Expressions like $\frac{M+m}{M}$ crop up as the price we pay for tracking only the motion of mass *m*. They can get complicated, but at least expressions with only *m*, *M*, and *G* do not change throughout a classical dynamics calculation, so they need be computed only at the beginning. The expression $\frac{GM^3}{(M+m)^2}$ occurs often enough that we assign it the name **Gm**. The acceleration on mass *m* is computed here by the **acc[{x,y}]** function:

```
Gm = G M³
     ───────── ;
     (M+m)² 

acc[{x_, y_}] :=      -Gm        {x, y}
                 ─────────────
                 (√ x²+y²)³
```

For example, the initial acceleration is to the left (negative) in the *x* direction, and there is no initial acceleration in the *y* direction because the masses start at the same *y* position:

```
{a, b} = N[acc[{x0, y0}]]
```

```
{-0.888889, 0.}
```

Small rounding errors are already showing up in the numerical calculation of the instantaneous acceleration. An **acc** calculation is unlikely to land on exact floats.

Sampling errors happen next (and are much larger than rounding errors), since a time step is an *interval* of time, not an instant of time. There is a raft of literature on the best way to update the position in a time step, but suppose you do not want to do a lot of reading and just try to apply first principles regarding how objects move, like good old $F = ma$. Acceleration is the change of velocity with time, so it is pretty straightforward how to update the velocities after time t:

$$u_{new} = u_0 + a\,t$$
$$v_{new} = v_0 + b\,t$$

Obvious, yes, but think about it a minute: What are the accelerations in x and y *during* the time step? Accelerations vary from the initial values of a and b that were calculated from x_0 and y_0 as shown above, because the masses change positions. That is the sampling error: We are using the acceleration at the *beginning* of the time step as a stand-in for the acceleration *throughout* the time step. The bigger the time step, the bigger the sampling error will be in pretending that a varying quantity like the acceleration can be treated as a constant.

Similarly, update the position with

$$x_{new} = x_0 + u_0\,t + \frac{1}{2}\,a\,t^2 \text{ and}$$

$$y_{new} = y_0 + v_0\,t + \frac{1}{2}\,b\,t^2.$$

Those are exact equations for a trajectory with a constant acceleration, so it seems perfectly reasonable that we can piece together a good approximation to the trajectory using a computer to do this update to the next time step. Try a time step that a computer can represent without rounding error, like $t = 0.5$, and use the formulas above to compute new positions and velocities:

```
t = 0.5;
{unew, vnew} = {u0, v0} + {a, b} t;
{xnew, ynew} = {x0, y0} + {u0, v0} t + 1/2 {a, b} t²;
```

Kinetic energy depends only on the velocities. *Potential energy* depends only on the positions. Since kinetic energy is $\frac{1}{2}$ mass $\times$ velocity^2 for each mass, and potential energy is $-\frac{GMm}{r}$ for the whole system, you can express the total energy as the sum of a constant times $u^2 + v^2$ (the kinetic part) and a constant times $\frac{1}{\sqrt{x^2 + y^2}}$ (the potential part):

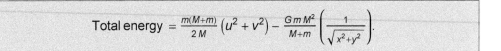

$$\text{Total energy} = \frac{m(M+m)}{2M}\left(u^2 + v^2\right) - \frac{GmM^2}{M+m}\left(\frac{1}{\sqrt{x^2+y^2}}\right).$$

The computer version of this formula will allow us to monitor the energy *e* of the state of the system at any point in the simulation:

$$\texttt{e[\{x\_, y\_, u\_, v\_\}]} := \left(\frac{m\ (M+m)}{2\ M}\right)\left(u^2 + v^2\right) - \left(\frac{G\ m\ M^2}{M+m}\right)\frac{1}{\sqrt{x^2+y^2}};$$

Physics tells us the total energy in the system should be constant. It *should* be, but with float arithmetic and a traditional numerical method, it *won't* be. For these initial conditions, it starts out with this value:

```
e0 = e[{x0, y0, u0, v0}];
N[e0]
```

-1.15234

The negative value for the total energy tells us the trajectory is a bound orbit (ellipse); positive total energy indicates an unbound trajectory (hyperbola). Check the total energy after just *one* time step:

```
N[e[{xnew, ynew, unew, vnew}]]
```

-0.993467

That is not even constant to one decimal place. Another quantity that is supposed to be conserved is the *angular momentum*, which is proportional to the cross product $x\,v - y\,u$. It starts out equal to 9, which we save in **cross0**. Check how close to staying constant it is after the time step:

```
cross0 = x0 v0 - y0 u0
xnew vnew - ynew unew
```

9

9.25

The floating point arithmetic might be accurate to fifteen decimal places, but just with this one small time step, we can see major errors in the energy and the angular momentum. The cause of the sampling error is obvious, because we used the *initial* time to predict what the masses will do for the *entire* time step. That is the hallmark of the Euler method, and there are certainly better approximations. But first, we can plot what the trajectory looks like with this simple approach, by looping on the preceding time step update procedure.

20.1.2 Trying a traditional method out on a single orbit

The theoretical orbital period is 30.476···. If we use 61 time steps of size 0.5 each, the total simulated time will be 30.5, so in principle that should produce a good approximation to one complete elliptical orbit. The following program attempts this, storing each (x, y) and (u, v) time step result in **steps**.

```
Module[{i, period = 30.476, t = 0.5,
   u = u0, unew, v = v0, vnew, x = x0, xnew, y = y0, ynew},
  steps = {{{x, y}, {u, v}}};
  For[i = 1, i < ⌈period / t⌉, i++,
   {a, b} = acc[{x, y}] ;
   {unew, vnew} = {u, v} + t {a, b};
   {xnew, ynew} = {x, y} + t {u, v} + 1/2 {a, b} t²;
   {x, y, u, v} = {xnew, ynew, unew, vnew};
   AppendTo[steps, {x, y, u, v}]]]
```

Here is what that computed trajectory looks like, using blue dots for mass m and red dots for mass M; we also show the correct elliptical orbit shape for m in solid green:

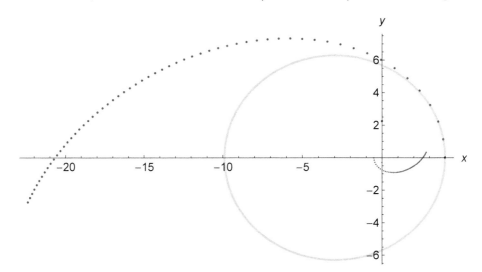

The expression "spiraling out of control" is literally correct here. The simulation fails the "eyeball metric" of correctness: It looks wrong, but only because we have the theoretically correct answer against which to compare. To be more quantitative, we can check the total energy of the last time step, compared to the value of about −1.15234 for the initial conditions. (A subscript of ⟦−1⟧ picks the last item of a list):

```
e[steps⟦-1⟧]
```

```
-0.520659
```

Even without graphing the results for a human to check, a computer can detect and report that the energy is far from the correct value. Also, the cross product $x v - y u$ has risen from 9 to $10.944\cdots$, showing that the angular momentum is drifting upward instead of holding constant. Despite using double-precision throughout and seemingly rock-solid elementary physics, the result does not even have one decimal of correctness left after all the *accumulated sampling errors*. If the energy keeps increasing with more time steps, it will eventually become positive and the masses will fly apart instead of being in a bound orbit!

A wrong-looking simulation usually leads to tinkering with the sampling method. For example, we could try using the *average* of the old and new positions, velocities, and accelerations to update the position, much like the midpoint method of computing the area under a curve. But we do not know the new positions, velocities, and accelerations, so we have to use a *rough* guess to get estimates for the values we need to compute a *better* guess. There are quite a few schemes out there for doing this back-and-forth guessing game. A practicing engineer or scientist might reach for a venerable reference work to look up the "best" method, and receive some folksy advice like the following passage from *Numerical Recipes in Fortran, Second Edition*:

> "For many scientific users fourth-order Runge-Kutta is not just the first word on ODE integrators, but the last word as well. In fact, you can get pretty far on this old workhorse, especially if you combine it with an adaptive step size algorithm. Keep in mind, however, that the old workhorse's last trip may well be to take you to the poorhouse: Bulirsch-Stoer or predictor-corrector methods can be very much more efficient for problems where very high accuracy is a requirement. Those methods are the high-strung racehorses. Runge-Kutta is for plowing the fields."

Correct computing is a science, but advice like the above makes it sound like an art, or even like a teenager comparing sports cars. After reading it, a computer user is likely to conclude, "To heck with all that, **let's just use smaller time steps.**"

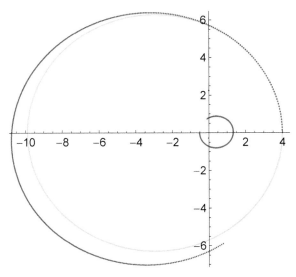

So here is the simulation again with eight times as many time steps, each one-eighth as large as before so that it still simulates what *should* be one complete orbit.

As we plot more points, the plot thickens. The outcome is still obviously wrong; using such simulations to replace theory or experiment is not sound science. It still produced something far from a full orbit, which inspires little confidence that this computed guess is physically accurate.

Just as with the deeply unsatisfying expressions for bounding the error in computing area, not one of the textbook time stepping methods can state bounds on the error in terms of computable numbers; they only can say that "the error decreases with a smaller time step, so please use the smallest time steps you have the patience to wait for." Even if you craft a tenth-order time stepping method with trillions of tiny time steps, its output will still be nothing more than... *a guess.*

20.1.3 What about interval arithmetic methods?

If you have *bounds* on each of the variables in the right-hand side of the equations $x = x_0 + u_0\, t + \frac{1}{2}\, a\, t^2$ and $x = y_0 + v_0\, t + \frac{1}{2}\, b\, t^2$, you can bound x and y. People have tried using interval arithmetic to get rigorous bounds on dynamics calculations like the two-body problem, but it does not work very well. First of all, it is not obvious how to bound the accelerations. If the masses *touch*, their acceleration is unbounded and the velocity goes to infinity. If you take a really short time step, does that guarantee they will not touch? No, because infinite velocity means you can go anywhere or *have come from anywhere* in any time step greater than zero. You could assert "the masses probably won't touch with a time step this small," but the "probably" gives up guaranteed validity, which was the reason for trying to use interval bounds in the first place. It is another chicken-and-egg problem.

Even if you can bound the accelerations, the uncertainty in the resulting position leads to more uncertainty in the acceleration, leading to even more uncertainty in the position, and bounds grow rapidly. As soon as bounds are so large that the ranges of possible positions of the two masses actually *overlap*, the computation has to stop because acceleration goes to infinity as the distance between masses goes to zero. Or the bounds could be kept of constant size and the time step shortened to avoid collisions, but eventually the largest safe time step becomes zero in size, and again progress halts. The rigor of interval arithmetic leads to rigor mortis in the simulation.

Is there a way to stop a bounded computation of two-body dynamics from exploding, and thereby get useful levels of accuracy and a scientific result instead of a guess?

20.2 Ubox approach: The initial space step

20.2.1 Solving the chicken-and-egg puzzle

> Michelangelo was once asked how he would sculpt an elephant. He replied, "I would take a large piece of stone and take away everything that was not the elephant."

That is exactly how we will proceed; start with a block of space-time, and carve away everything that is not the trajectory. The ubox approach to orbital dynamics starts out quite differently from time stepping. Like other ubox methods, it makes no errors. It treats time as just another degree of freedom in the calculation. Like the pendulum problem, it uses space steps to compute the time instead of the other way around.

This solves the chicken-and-egg problem. It works very much like the paint bucket technique, but without keeping track of a "fails" set.

Surprisingly, we only need three degrees of freedom: The time t and the x-y coordinates of one of the two masses. (The position of the other mass is again determined from the fact that the center of mass is at the origin.) The velocities u and v can actually be derived from the position, using conservation of energy and conservation of angular momentum. A *trajectory* is a set of ubounds with dimensions $\{t, x, y\}$. The physically correct trajectory is guaranteed to lie completely within the computed set. We will often work with a subset of these dimensions, depending on what we are doing; for example, to find the acceleration, all we need is x and y.

As usual, we use very low precision to exaggerate the approximate nature of the calculation, yet bound the behavior. An environment of $\{2, 3\}$ has a small dynamic range (about 6×10^{-5} to 5×10^2) and only one to eight bits of fraction. These unums have sizes ranging from just nine to nineteen bits. If we can produce decent bounded results with so few bits, imagine what we could do with the luxury of a $\{2, 5\}$ environment that has up to ten decimals of accuracy! The following sets the environment, and initializes the ubound version of the masses, gravitational constant, and so on. We use the same variable names as before, but append a "**ub**" to each one to remind us that the quantities are ubounds.

```
setenv[{2, 3}]
{Mub, mub} = {{M̂}, {m̂}}; Gub = {Ĝ}; Gmub = {Ĝm};
e0ub = {ê0}; cross0ub = {crôss0};
{t0ub, x0ub, y0ub, u0ub, v0ub} = {{t̂0}, {x̂0}, {ŷ0}, {û0}, {v̂0}};
```

The **trajectory** set initially has just the one ubox, storing the initial time, location, and velocity. We also mark this point as a solution in **new**, meaning that it can be used to look for additional members of the solution set. The approach used here is to store ubounds in **trajectory** but five-dimensional uboxes in **new**. The reason is that there is no wrapping problem or dependency problem in the x, y, or t dimensions. The bound on x and y is always a rectangular one because the coordinates are independent, so it saves storage and effort to use ubounds. The velocities, however, are a complicated shape when computed from x and y, so it is important to track the shape with ubox elements instead of creating a wrapping problem.

```
trajectory = {{t0ub, x0ub, y0ub}}; new = {{t̂0, x̂0, ŷ0, û0, v̂0}};
```

With gravitational dynamics, in any time step there is the possibility that two masses touch, since their velocity approaches infinity as their separation approaches zero. With infinite velocity, such a singularity is possible for any time step greater than zero, so there is no obvious way to get a bound on the behavior. But with a *space step*, we can bound the acceleration and thus the velocity, and thus bound the set of possible locations and times within that space step.

20.2.2 First, bound the position

A space step is a sort of "buffer region" around the *x-y* part of a ubox that is a distance *r* away in both directions. Again to exaggerate accuracy losses, we can attempt a rather big space step: Surround the smaller mass with a square space step bound of radius, say, one-eighth the distance to the origin. That guarantees that the space step does not include the origin, so there is no way that the two masses can come in contact with each other. There is nothing special about squares; any enclosing shape for mass *m* will work that does not touch its scaled reflection about the origin that encloses mass *M*. Squares are easy to work with. With the one-eighth rule, the diameter of the space step will automatically get larger if the mass goes farther from the origin, keeping the relative uncertainty in the acceleration about the same. The following pulls the *x* and *y* values out of the **new** set, and assigns one-eighth the distance of the closest corner of the ubox to the origin, initially 0.5, to **r**.

```
{x0u, y0u} = {new[1,2], new[1,3]};
r = √((Min[Abs[u2g[x0u][1]]])² + Min[Abs[u2g[y0u][1]]]²) / 8.;
```

The following figure is the last one where we will show Mass 1; after this, we focus just on the trajectory of Mass 2. The figure reminds us that any space step bound on Mass 2 automatically puts a corresponding bound on Mass 1, with everything scaled by a ratio of $\frac{m}{M}$.

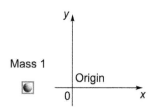

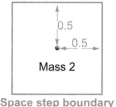

Space step boundary

Given any finite ubox (with any number of dimensions greater than zero) and a radius *r*, the **spacestepbounds**[*ubox*, *r*] function in the reference implementation returns a list of ubounds, one for each dimension, that can be used to test to see if other uboxes intersect the interior. It is always an *open* bound, not a closed bound, in every dimension. The **trials** set is initially just this single square.

```
bound = spacestepbounds[{x0u, y0u}, r];
trials = {bound};
```

The figure on the next page shows the trial region in amber surrounded by the space step boundary. To emphasize that the algorithm knows *nothing* about the trajectory yet, we show several arrows with question marks at the tips.

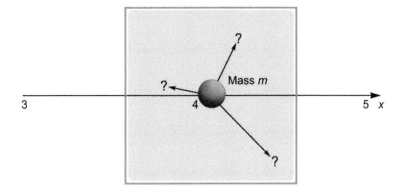

The large space step will make obvious how much the acceleration can vary within the range of motion of the mass. All we need is a *finite bound* on the acceleration, to start an iterative refinement. As always, the unum accuracy, minimum ULP sizes, and space step sizes are things that a computer can choose automatically to keep all of the accuracy losses on par with one another.

The **accub[{**xub,yub**}]** function is the unum arithmetic version of **acc[{**x,y**}]**. It bounds the acceleration of mass **mub** located in the rectangle described by ubounds {xub, yub}. It returns a pair of ubound acceleration values: One for the x direction and one for the y direction. There are two ways to approach the task of creating such a bound. Here is a quick one that works:

```
accub[{xub_, yub_}] := Module[{tempub = squareu[xub] ⊕ squareu[yub]},
   tempub = tempub ⊗ sqrtub[tempub];
   tempub = negateu[Gmub ⊙ tempub];
   {xub ⊗ tempub, yub ⊗ tempub}]
```

This works well for ULP-sized rectangles, but because the formula for the acceleration has x or y in both the numerator and the denominator, there is some accuracy loss caused by the dependency problem. That can be eliminated by a slightly more sophisticated **accub** function which is shown in Appendix D.7. As previously mentioned, whenever a particular function becomes crucial to a calculation, it is often helpful to look for the minima and maxima and create a tighter bound the way we do, say, for the square of a number. Creating such functions is *not* a mandatory skill for users of unums and uboxes the way it is for interval arithmetic, but if you are the sort of person who wants to track down and crush every unnecessary ULP-sized information loss in a sequence of calculations, the option is there.

By evaluating **accub** on the space step interior and finding the maximum and minimum values of the acceleration for the ranges of x and y, we gain quite a bit of information about the trajectory of the mass. Notice that evaluating the acceleration on each x-y ubound pair inside the space step can easily be done *in parallel*, since it is the same operation on different uboxes (data parallelism). The **boundacc[***set***]** function on the next page evaluates the acceleration on every x-y ubound pair in *set*, and returns the minimum and maximum as a pair of ubounds.

```
boundacc[set_] := Module[{aub, bub, aminub = {posinfu},
    bminub = {posinfu}, amaxub = {neginfu}, bmaxub = {neginfu}},
  For[i = 1, i ≤ Length[set], i++,
    {aub, bub} = accub[{set⟦i,-2⟧, set⟦i,-1⟧}];
    {aminub, amaxub} =
      {{(minub[aub, aminub])⟦1⟧}, {(maxub[aub, amaxub])⟦-1⟧}};
    {bminub, bmaxub} = {{(minub[bub, bminub])⟦1⟧},
      {(maxub[bub, bmaxub])⟦-1⟧}}];
  {{aminub⟦1⟧, amaxub⟦-1⟧}, {bminub⟦1⟧, bmaxub⟦-1⟧}}]
```

Here is the initial acceleration bound in the *x* and *y* directions:

```
{aub, bub} = boundacc[trials];
view[aub]
view[bub]
```

(−1.12890625, −0.689453125)

(−0.1611328125, 0.1611328125)

These bounds for the acceleration are loose because the space step is so large relative to the separation of the two masses. However, they *are* bounded, which is enough to get the refinement process started. The next step is to compute the *x-y* part of a trial trajectory, applying the unum version of the usual formulas for position under constant acceleration. (By using the lossless polynomial routine of Chapter 16, we avoid any unnecessary information loss.)

$$x = x_0 + u_0 t + \frac{1}{2} a t^2 \qquad \texttt{xub = polyu[\{x0ub,u0ub,1 \hat{/} 2\},tub\}}$$

$$y = y_0 + v_0 t + \frac{1}{2} b t^2 \qquad \texttt{yub = polyu[\{y0ub,v0ub,1 \hat{/} 2\},tub\}}$$

Apply this for increasing values of *t* until the computed *x* and *y* go completely outside the space bound. Because *a* and *b* are rigorously bounded, the formulas account for *every possible location* the mass can be within the space step.

There is no "sampling" of space *or* time, since unums tile a range of real numbers without leaving any out. Time is just another dimension, so time can advance as a ubox set, alternating between exact values and ULP-wide open ranges.

For example, with a preferred *minpower* for *t* set to −6, the preferred minimum ULP size is $2^{-6} = \frac{1}{64}$. The following unum routine finds a set of uboxes that bounds the trajectory within the space step and stores it in **trials**. It also assigns some *minpower* values that work well for the other trajectory values later on.

```
{minpowert, minpowerx, minpowery, minpoweru, minpowerv} =
  {-6, -5, -5, -7, -7};
trials = {};
Module[{halfaub = 0.5 ⊗ aub, halfbub = 0.5 ⊗ bub, temp,

  tub = t0ub, xub = x0ub, yub = y0ub},
 While[xub ≈ bound₍₁₎ ⋀ yub ≈ bound₍₂₎,
  tub = {nborhi[tub₍₁₎, minpowert]};
  If[exQ[tub₍₁₎], tub = {nborhi[tub₍₁₎, minpowert]}];
  {xub, yub} = {polyu[{x0ub, u0ub, halfaub}, tub],
    polyu[{y0ub, v0ub, halfbub}, tub]};
  {xub, yub} = {intersectu[xub, bound₍₁₎], intersectu[yub, bound₍₂₎]};

  If[xub ≠ {qNaNu} ⋀ yub ≠ {qNaNu}, AppendTo[trials, {xub, yub}]]]]
```

Space step
boundary

The orbit (shown with a blue arrow) is guaranteed to lie within the overlapping amber rectangles, left. This set of possible *x* and *y* positions constitute a *much* smaller set than the original space step. In fact, it already looks like an orbit calculation.

That means we can evaluate the acceleration in this more limited space range and get an improved bound on the acceleration. The bound can never tighten all the way down to an exact value because the acceleration is *not* exact; it varies within the space step. The key is to make that variation on par with the accuracy of the unums used for the calculation.

Here is the new range of acceleration:

```
{aub, bub} = boundacc[trials];
view[aub]
view[bub]
```

(−0.892578125, −0.87109375)

(−0.111328125, 0)

That bound on the acceleration is much tighter. With this improved information, we can repeat the position calculation. An automatic loop to repeat this alternating process converges quickly. Unlike a calculation with floats, there is no need to pick some arbitrary point at which to stop iterating; eventually, the iteration produces *exactly* the same information about the position as before, limited by the precision, so there is no reason to do it again. The calculation can monitor the information (the reciprocal of the total volume of the uboxes). When information does not increase, it is time to stop iterating. End of Stage 1.

The Stage 2 iterations do something subtly different: Time is advanced until the trajectory goes anywhere *outside* the space step boundary. We want to use the *maximum exact unum time* before the trajectory goes outside the boundary. The test is to intersect the computed *x-y* position with the space bound interior.

As soon as the intersection is not the same as the computed *x-y* position, it means something has been clipped off by the space step bound, and we need to back off to the previous time step. That way, we will have a complete set of possible *x-y* positions up to a maximum time. In Stage 2, the time is attached to each set of possible positions; thus, we have the beginnings of a *trajectory* ubound set, not just the location part of the orbit.

```
Module[{halfaub = 1 /̂ 2 ⊗ aub, halfbub = 1 /̂ 2 ⊗ bub,

  newvol = volume[trials], oldvol = ∞, tub, xub, yub},

 While[newvol < oldvol, trials = {};
  tub = {nborhi[tOub⟦1⟧, minpowert]};
  {xub, yub} = {polyu[{xOub, uOub, halfaub}, tub],
    polyu[{yOub, vOub, halfbub}, tub]};
  While[sameuQ[intersectu[xub, bound⟦1⟧], xub] ⋀
    sameuQ[intersectu[yub, bound⟦2⟧], yub],
   AppendTo[trials, {tub, xub, yub}];
   tub = {nborhi[tub⟦1⟧, minpowert]};
   {xub, yub} = {polyu[{xOub, uOub, halfaub}, tub],
     polyu[{yOub, vOub, halfbub}, tub]}];
  {aub, bub} = boundacc[trials];
  {halfaub, halfbub} = {{0.̂5} ⊗ aub, {0.̂5} ⊗ bub};

  oldvol = newvol; newvol = volume[trials]]]
```

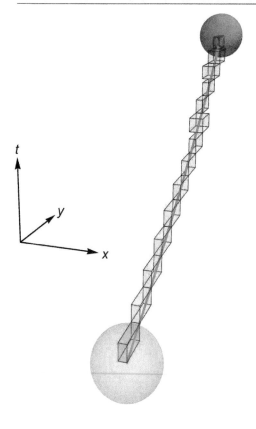

Here is the trajectory bound, with time as the vertical dimension. It is a sort of spiral staircase with steps (equivalent to time steps of a marching method) that are just one ULP high.

Unlike traditional numerical methods, the computed trajectory is a *provable bound*, not a guess. Yet, it requires very few bits of storage. Even if it were possible to calculate rigorously with floats, storing this first space step trajectory in IEEE 64-bit precision would take 10 752 bits. The **trials** set requires *one-sixth* that much storage space. And since the multiplies and adds use only 1 to 4 bits of exponent and 1 to 8 bits of fraction, unums in hardware would probably do the arithmetic about an order of magnitude faster than double-precision floats, even with the higher complexity of unum logic.

Notice how the rectangular blocks touch each other in a way that barely overlaps their corners. The overlaps are the exact time calculations, and show that the calculation is much more precise than it looks. The figure shows the trajectory in blue passing through the tiny rectangles where the stair steps overlap and the time is exact, because the real bound is a set of parabolic arcs stitched together. In effect, the space step produced many exact time steps, not just one.

Here is a closer view that shows the trajectory passing through the very slight overlap of the ubox corners.

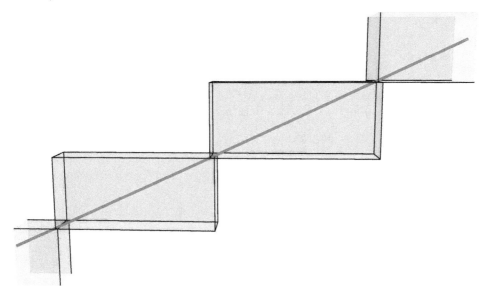

If the *x-y* range is more than a couple of ULPs wide at the last exact time, that indicates the space step radius *r* is too large for the amount of precision being used. We could then reduce the size of the space step radius so there is less variation in the force, making the sampling variation closer to the size of one ULP. Or, because unum precision is completely adjustable down to the bit level, the fix could be to reduce the precision of the environment. If the number of ULPs at the last exact time is always exactly one, that indicates that the space step is too *small* and should be increased, or the precision could be increased. An application program for orbit calculations can make those adjustments automatically; all the user has to do is express the desired accuracy in the trajectory. The program also might automatically take into account the amount of parallelism in the available hardware, and allow more ULPs within each iteration since the uboxes in the *x* and *y* dimensions can be computed in parallel.

In this example, the last exact-time trajectory was a single ubox in the *x* and *y* dimensions: $t = 0.21875$, $x = (3.9765625, 3.984375)$, $y = (0.48828125, 0.4921875)$. That is, there is only one location at the last time, and it happened to be just one ULP wide in both *x* and *y* dimensions.

In the more general case, any ubound results with a pair of endpoints would be split using **uboxlist** so that the new starting points are always just one ULP wide. The last time step, together with the velocities at that time is saved in **laststate**, and the **trials** set is united with the **trajectory** set. In updating the velocities with $u = u_0 + (t - t_0)\, a$ and $v = v_0 + (t - t_0)\, b$, we use a fused multiply-add to protect against any unnecessary expansion in the bound; in this case, it does not affect the result but it is a good habit to get into.

```
laststate = Join[trials[-1], {fmau[trials[-1,1] ⊖ t0ub, aub, u0ub],
    fmau[trials[-1,1] ⊖ t0ub, bub, v0ub]}];
trajectory = trajectory ⋃ trials;
```

The **laststate** now has everything needed to start the next space step. The acceleration values {a, b} and the old t_0 are no longer needed.

20.3 The next starting point, and some state law enforcement

To launch another space step from any *t-x-y*, we need new starting values for the velocities, u_{new} and v_{new}. There are three constraints that can be applied to increase information about the new velocities:

The obvious starting point is this: The change in velocity is acceleration times the change in time. $\{u, v\}_{new} = \{u, v\}_{old} + \{a, b\}\, t$, where t is the difference between the exact start time and the last exact end time, and $\{u, v\}_{old}$ and $\{a, b\}$ have uncertainty ranges in general. That creates a rectangular bound in the *u-v* plane because the bounds are independent. This is the straightforward approach used to calculate new values for *u* and *v* at the end of the previous section.

The second constraint is *the law of conservation of energy*, for which *u* and *v* are definitely *not* independent. Remember that the total energy is

$$e = \frac{m(M+m)}{2\,M}\left(u^2 + v^2\right) - \left(\frac{G\,m\,M^2}{M+m}\right)\frac{1}{\sqrt{x^2+y^2}}$$

The *x* and *y* values have a range, which implies a bound on $u^2 + v^2$ if the energy does not change from its initial value, e_0:

$$C_1 e_0 + \frac{C_2}{\sqrt{\text{Maximum}(x^2+y^2)}} \le u^2 + v^2 \le C_1\, e_0 + \frac{C_2}{\sqrt{\text{Minimum}(x^2+y^2)}}$$

where C_1 and C_2 are constants involving *m*, *M*, and *G*. Since $u^2 + v^2 = r^2$ defines a circle of radius *r* in the *u-v* plane, a bound on $u^2 + v^2$ means the values of *u* and *v* lie inside a ring, as shown in light blue in the figure on the next page:

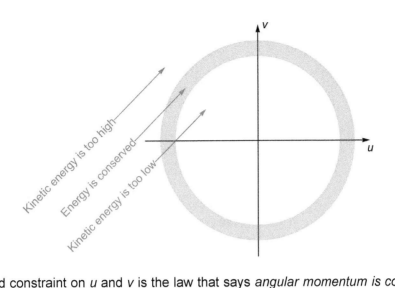

The third constraint on *u* and *v* is the law that says *angular momentum is conserved*. The cross product $x\,v - y\,u$ will be a range because of ULP-wide inexact operations, but it should always contain the original value, $cross_0$. Graphically, this constraint looks like a band in the *u-v* plane that is slightly bent wherever it crosses an axis, shown in magenta:

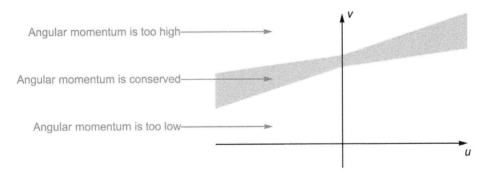

Traditional interval arithmetic would have you create a function to describe the bow tie-like shape above, one that takes into account which quadrant *x* and *y* are in to find the slopes and axis intercepts of the bounding lines, including special cases to avoid dividing by zero. The code for that takes about half a page. Here is what the test looks like with unum arithmetic, in the reference implementation:

```
xub ⊗ vub ⊖ yub ⊗ uub ≈ cross0ub
```

We simply apply the philosophy expressed in Chapter 17: There is no need to "solve" an equation, that is, invert operations to get one value expressed in terms of another. Simply plug in the value of the four-dimensional ubox and see if it is compliant, `True` or `False`. Discarding non-compliant uboxes in a starting set in **new** often eliminates about half of them, which not only saves work and tightens the bound, it does a better job of simulating the actual physical behavior.

Here is what the **compliantQ** tests looks like in the reference implementation:

```
compliantQ[ubox_] :=
  Module[{xu = ubox[[-4]], yu = ubox[[-3]], uu = ubox[[-2]], vu = ubox[[-1]],
```
$$klu = \left(\tfrac{\widehat{M+m}}{M}\right)^2, \quad k2u = 2\,\widehat{G}\,M, \quad k3u = \frac{2\,\widehat{G}\,M}{\sqrt{x0^2+y0^2}}, \quad k4u = \left(\tfrac{M+m}{M}\right)^2\left(\widehat{u0^2 + v0^2}\right)\},$$
```
    k2u ⊙ sqrtu[squareu[xu] ⊕ squareu[yu]] ⊖ k3u ≈
      klu ⊗ (squareu[uu] ⊕ squareu[vu]) ⊖ k4u ∧
    cross0ub ≈ fdotu[{{xu, negateu[yu]}, {vu, uu}}]]
```

Remember that **laststate** has the same form as **new** did: The ubounds for t, x, y, u, v. Why not simply use **laststate** as we used **new**, to start the next space step? This is another critical difference between interval arithmetic and the ubox approach. Always split the **laststate** ubounds into ULP-sized uboxes. That way, the loss of accuracy will be approximately the same with each space step, and the trajectory bound will grow linearly. With interval arithmetic, the bound grows at least as fast as the *square* of the number of space steps, and it does not take long before the bound includes the origin and creates a division by zero.

Use the end time as the new t_0.
Create ubox lists for x, y, u, and v bounds with preferred minimum ULP sizes.
Form a ubox *set* consisting of tuples of all possible combinations of the linear lists.
Initialize *new* to the empty set, {}.
For each entry in *set*,
　　If the entry is compliant with conservation laws, append it to *new*.

The entries in *new* will thus always be no more than one ULP in any dimension. To show how sensitive the conservation test is, here is what the u-v dimensions of the starting set look like after compliance testing:

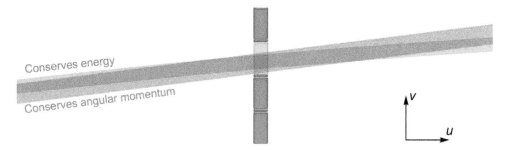

The ubox list for v resulted in seven elements. Only one was compliant, shown in cyan; the others failed and are shown in red. The light blue arc is the ring for which energy is conserved. The light magenta band is the region for which angular momentum is conserved. Only the compliant cyan ubox intersects *both* regions and thus passes, giving a fourfold refinement in the information about the trajectory.

With floats, we could only *hope* that the computed results were consistent with conservation of energy and angular momentum. In general, they are not. In a conventional numerical simulation, you might be tempted to "correct" deviations in the energy and angular momentum by adjusting *x*, *y*, *u*, or *v* to keep those quantities constant. The problem is, which of the four variables do you adjust? And with what mathematical or physical justification? This type of correction technique even has its own time-honored slang term: "Fudging" the answer.

With uboxes, there is no fudging. We can rigorously enforce the conservation laws simply by *eliminating any uboxes that do not obey the laws*. Checking which of the *x*-*y*-*u*-*v* combinations violate conservation of energy and angular momentum is another task that is easy to perform in parallel. The **enforcelaw[**state**]** function takes any set of *t*-*x*-*y*-*u*-*v* ubounds, splits it into uboxes, and returns the set that follow the laws. Using this on **laststate** provides the set of starting points for the next space steps:

```
enforcelaw[state_] :=
  Module[{set, compliant = {}, tub, xub, yub, uub, vub},
    {tub, xub, yub, uub, vub} = state;
    set =
      Tuples[{tub, uboxlist[xub, minpowerx], uboxlist[yub, minpowery],
        uboxlist[uub, minpoweru], uboxlist[vub, minpowerv]}];
    For[i = 1, i ≤ Length[set], i++,
      If[compliantQ[set⟦i⟧], AppendTo[compliant, set⟦i⟧]]];
    compliant]

new = enforcelaw[laststate];
```

In general, **enforcelaw** will trim away a considerable amount of the "fat" in the computed trajectory ubox set, since it recognizes that the *u* and *v* pairs lie inside a complicated shape and not a simple rectangular bound.

20.4 The general space step

The first space step started with a single exact ubox. In general, we start with a set of uboxes that may not be exact in position and velocity; we selected the time to be exact from the computed trajectory uboxes. Each starting ubox leads to its own trajectory within its own space box, refined for acceleration bound the way we did the first step. It is then merged with the known set of trajectory boxes, but with a catch: When uniting two trajectory sets, we find the maximum time attained by each trajectory set, and discard any uboxes greater than the minimum of those two maximum times. In that way, we can again select a single exact time to use as the starting point for the next space step.

With traditional intervals, the uncertainty in the starting values of each time step gets rapidly larger, leading to wider and wider ranges for the accelerations, so that the bound on the position soon includes the origin; the simulation blows up at that point.

> *With ubox trajectories, the initial conditions are always*
> *minimum ULP sized, so uncertainty per trajectory does **not** grow.*

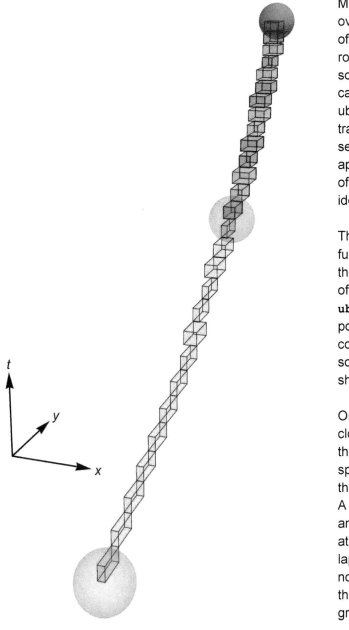

Most of the trajectories overlap, so the expansion of the uncertainty is roughly linear, not something that grows catastrophically fast. The uboxes are displayed with transparency, and the second space step appears darker because of overlap of the nearly identical trajectories.

The `findtraj[ubox]` function does exactly what the preceding computation of a space step did, using `ubox` as a starting point in position and velocity. The code is simply the consolidation of each stage shown previously.

On the following page is a closeup of the trajectory at the end of a third time space step, about 2% of the way around a full orbit. A loss of accuracy in *x*, *y*, and *t* dimensions is visible at this scale as overlapping uboxes. There is no exponential growth in the error, only linear growth.

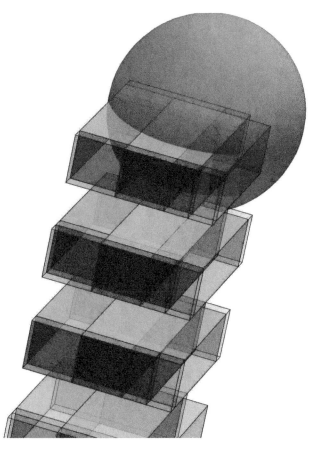

Notice that the union of all the *x-y* ranges is not always a rectangle; the conservation laws clipped off a corner on the top step of the staircase. There is little point in going further with such low precision. Kahan has noticed that the "corners" seem to cause the rapid expansion of interval bounds in dynamics problems, and has created an orbit simulator that uses *hyper-ellipsoids* to represent the states of each mass. Kahan's method also appears to exhibit only linear error increases in error with each time step. The ubox method may show empirical validation of Kahan's observation.

> The example shows that tight bounds can be achieved not with high precision or high-order sampling methods, but by simply using the physics and math that are taught in high school, and letting the computer use sound arithmetic.

Suppose the method were applied to the Sun-Earth system, and imagine that we somehow had exact knowledge of the masses of the Sun and Earth as well as the gravitational constant. How accurately could the ubox method predict where the Earth will be after a full year, one orbit around the Sun? If an environment of {3, 6} is used (with up to $2^6 = 64$ bits of precision in the fraction), a space step of about 100 meters would produce about one smallest-possible ULP of range in the acceleration. About nine billion space steps would suffice to do a full orbit, with an information loss of about nine billion ULPs. That works out to an error of about *two kilometers*.

To put it another way, the ubox method could predict the duration of a year to *0.07 second* at that precision, if we ignore all the other masses in the solar system and the tiny effects of general relativity.

20.5 The three-body problem

The picture above is a frame from the motion picture "Game of Shadows." When Sir Arthur Conan Doyle created the arch-enemy of Sherlock Holmes, he drew inspiration from mathematicians and astronomers of that era who were rivaling to be the first to solve gravitational dynamics problems that had defied solution for centuries. Doyle published his short story *The Valley of Fear* describing Moriarty's rarefied achievements in gravitational dynamics the same year floats were proposed by Torres y Quevado for automatic computation, 1914. The story was set in 1888, one year after publication of a proof that the general three-body problem is unsolvable.

Isaac Newton was the first to look for an analytical solution to the motion of the Sun-Earth-Moon system, in 1687. He had, after all, managed to find an analytical proof for Kepler's observation that the planets move in elliptical orbits around the Sun, regarded as two-body problems. How much harder could it be to find the mathematics governing the behavior of *three* bodies acting under gravitational attraction?

Much harder. After centuries of frustrated attempts by many brilliant thinkers, Henri Poincaré and Ernst Bruns showed in 1887 that there *is* no analytical solution for the general problem, even when you allow the answer to be expressed using integrals that no one knows how to evaluate. The trajectories in three-body systems are generally non-repeating, that is, chaotic. It is also very easy to set up an unstable problem, meaning that a small variation in the initial speeds and positions creates a large change in the trajectories. Even now, over 327 years since Newton took a shot at the problem, mathematical physicists are chipping away at special case after special case of the three-body problem that happens to be solvable in the conventional sense of "solvable."

The preceding method for two-body gravitational dynamics extends easily to multiple bodies. The author has bounded the behavior of the three-body problem in three dimensions in a simulated low-precision environment using space stepping and bounds that grow by only about one ULP per step. Which raises the question: *What does it mean to "solve" the three-body problem?*

The original posers of the question probably were envisioning something like the algebraic formula for the two-body orbit, with sines and cosines and square roots covering the blackboard. But what if "solve" means to *bound the answer to arbitrary accuracy using a finite number of operations*? If a procedure produces as much information as demanded, with a finite amount of work, does that not constitute a solution to the problem? If you allow that, the three-body problem *can* be solved with the ubox method. Without caveats or restrictions of any kind. Stable or unstable, repeating or chaotic. If the behavior is unstable or chaotic, it is because the *true physical behavior is unstable or chaotic*, not because the numerical method is!

The *details* of extending the two-body method to three bodies do not make for good reading, but are summarized here as follows:

- Each body is tracked for position and velocity, since there is no symmetry about the center of mass.
- The motion is not necessarily planar, so the uboxes involve *x, y,* and *z* positions and velocities for each mass.
- Angular momentum has three dimensions, not one, when filtering "compliant" uboxes.
- Conservation of linear momentum and the motion of the center of mass are part of compliance testing, in addition to energy and angular momentum.
- The uboxes have 19 dimensions: Each of the three bodies has three dimensions in position and velocity, for 18 dimensions. Tracking the time makes 19.

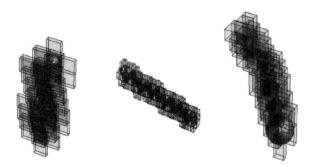

The figure on the left shows the first few space steps of a three-body problem in three dimensions with rigorous ubox bounds at very low precision: 2 bits of exponent, 4 bits of fraction. Each trajectory is surrounded by complementary-colored uboxes that provably cannot touch the trajectory.

20.6 The *n*-body problem and the galaxy colliders

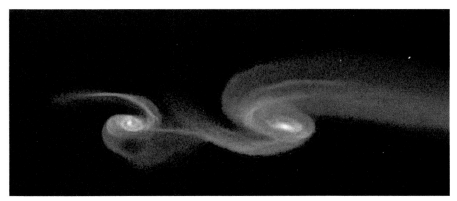

The Milky Way is heading for a collision with the Andromeda galaxy in about four billion years. Many supercomputer users have set up simulations involving billions of individual masses interacting under gravitational force, and have produced lots of gorgeous graphics like the picture above. But there's a catch.

Some computational scientists have pushed their time-stepping *n*-body simulations to handle over a trillion interacting bodies, allowing them to simulate the collision of entire galaxies. While the pictures look amazing, the more amazing thing is how little scientific validity there is in the simulations. Remember that the gravitational force between two bodies is

$$F = \frac{G\,m\,M}{r^2}$$

where r is the distance between the bodies. Computer simulations showed a tendency to blow up with a large number of masses having pairwise interaction, because it is very likely that r gets close to zero for at least one pairwise interaction. So the appalling workaround, a dirty little secret of the *n*-body research community, is to use *this* force equation instead:

$$F = \frac{G\,m\,M}{r^2 + q}$$

where the q in the denominator is a fudge number selected by the programmer that seems to prevent the answer from looking wrong. It has no scientific basis. If a galaxy collider is asked to solve the two-body problem with the same software used for galaxies, the results are likely to be embarrassingly wrong. Yet, users are simulating millions or billions of interacting masses.

Many computer simulations are very pretty, but as science they become a lot more interesting when they are actually *correct*.

21 Calculus considered evil: Discrete physics

21.1 Continuum versus discrete physics

The Preface mentioned that the reader does not need to know calculus for the material presented here. Calculus is being avoided for a reason other than ease of reading: Calculus and computers make for strange bedfellows, and their combination can destroy the validity of results. Calculus deals with infinitesimal quantities; computers do not calculate with infinitesimals. Some calculus appears in this chapter, but only to contrast it with approaches that are much easier to understand, and which offer the promise of provably correct, bounded calculations for physical behavior instead of estimates.

We saw a little of this conflict in the chapter on pendulum physics. In the limit where the oscillation of the pendulum is infinitesimal, the restoring force on the pendulum is proportional to its displacement (and not the sine of its displacement). Calculus then dictates the incorrect answer to the differential equation describing the motion of the pendulum when oscillations are *not* infinitesimal.

A more serious kind of calculus error occurs when a computational physicist tries to program computer arithmetic to mimic an equation that has *derivatives*.

The mathematical operations plus, minus, times, divide, square root, and so on have computer versions that physicists and engineers use to imitate mathematics numerically; but there is no **numerical** function that takes a function $u(t, x)$ as input and produces, say, the partial first derivative $\frac{\partial u}{\partial t}$ or second derivative $\frac{\partial^2 u}{\partial t^2}$ as output.

Derivatives are approximated with *difference* equations, like

$$\frac{\partial u}{\partial t} \cong \frac{u(x,t+h)-u(x,t)}{h} \quad \text{and} \quad \frac{\partial^2 u}{\partial t^2} \cong \frac{u(x,t+h)-2\,u(x,t)+u(x,t-h)}{h^2}.$$

The derivatives are defined as the limit as h becomes infinitesimal; so it is tempting to think about computing them with some small nonzero value for h. That is, use *finite differences* instead of calculus. Then the two expressions above look like something you can actually compute, but it is just about the worst thing you can ask a computer to do, numerically. If u is computed with floats, the numerator will be similar rounded numbers that cancel out their significant digits when subtracted; the smaller h is, the larger the rounding error relative to the result. If you try using a larger value of h to reduce the rounding error, it increases the *sampling* error, since you are sampling the function u over a range instead of a single point, so you can't win. If unums are used for the calculation, at least the loss of information from using difference approximations becomes visible to the programmer; with floats, the programmer generally has no idea just how inaccurate the finite difference is. There is a different approach, one that bypasses calculus and that can produce rigorous bounds on the behavior of physical systems.

To pick just a simple example, the *wave equation* in one dimension describes how something like a string under tension will vibrate, where the string is modeled as a continuum. The function u describes the displacement of the string at position x and time t, and c is a constant that depends on the tension, the length of the string, and the mass of the string. (Don't worry if the following looks completely cryptic to you:)

$$\frac{\partial^2 u}{\partial t^2} = c^2 \frac{\partial^2 u}{\partial x^2}.$$

Where did that equation come from in the first place? In most derivations, it is the result of reasoning about a *discrete* problem for which calculus is not needed! The string is broken up into mass elements under spring tension, and the acceleration is derived using approximations that are correct if all the quantities are infinitesimal.

> Instead of turning a discrete problem into a continuum problem with calculus, only to then approximate the calculus poorly on a computer, simply **model the discrete problem directly, and with rigorous bounds.**

The more powerful the computer, the more discrete elements can be correctly modeled. In *that* sense, one can "take the limit" of making the model more and more like a continuum problem, without rounding error or sampling error. The first example we examine is the one above, the physics of a vibrating string.

21.2 The discrete version of a vibrating string

A string is, of course, not really a continuum. It is made of atoms held together by chemical bonds. Instead of treating it as a continuum, think of it as being made of *really big* atoms, with well-behaved springs between atoms instead of chemical bonds. Both the masses and the spring constants can be scaled to duplicate the mass and tension of the string. Solve that problem by bounded methods like those shown in the previous chapters, and then increase the number of "big atoms" until the bound on the behavior is so tight that it exceeds the accuracy of the input values.

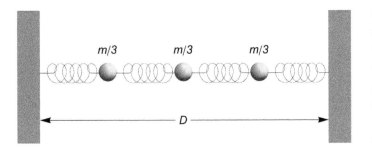

Consider three bodies of mass $m/3$ under tension T, equally spaced. This discrete problem approximates a continuous string of mass m under the same tension.

The calculus derivation estimates the force on each mass for a "small vertical displacement," then takes the limit to create the differential equation. The displacement has to be small and only in the vertical direction; that allows the replacement of complicated expressions with simple ones for which solutions are known, just like the first-year physics way to model pendulum motion. If we have just three discrete masses instead, is it possible to solve the problem with a rigorous bound on the Newtonian forces, and no approximations?

Imagine pulling down the leftmost mass and letting go. The masses would jangle up and down (and slightly left and right) and form a pattern approximating that of a stretched string with the same mass and tension that had been plucked there. With ubox methods, the trajectories of the three masses can be computed without error of any kind; this is, after all, just an *n*-body problem where the forces come from springs instead of gravity. It is easier than a free space *n*-body gravity problem since the force on each mass depends only on its two springs, not on all $n - 1$ other masses.

Besides the two spring forces, we can add the vertical gravity force on each mass as well, to account for *all* forces. Classical derivations discard the effect of gravity, assuming it negligible compared to the acceleration due to the springs, but there is no need to make that approximation in a computer simulation. Classical derivations also assume motion is *planar*, but the masses can actually move in three dimensions.

Once the behavior can be rigorously bounded for three bodies, why not a thousand bodies each of mass $m/1000$, or more? The bounds on the behavior can be driven down to any desired size, *without* making the errors of expressing the problem with derivatives and then approximating the derivatives.

The methods detailed in the previous chapter on pendulum and two-body motion apply here as well; begin by selecting a box-shaped space step, a range of motion about each mass, and then bound acceleration and velocity to derive the amount of time before the first mass reaches the boundary of its space step. Use that traversal time as the time step, tighten the bounds on positions, use that to tighten bounds on the forces (accelerations), and iterate until there is no change in information about the solution. It only takes two or three iterations.

While we only present that sketch of the approach here and not an actual calculation like that done in the previous two chapters, it is worth pointing out the value of using a complete and correct model of all the Newtonian forces on each mass, instead of the approximations used in most models of the vibrating string. For example, most approximations assume the tension on the string is T, unaffected by the additional tension of pulling it out of line. Imagine pulling the center mass so far down that its springs form a 90° angle; that would stretch each spring to $\sqrt{2}$ longer than shown in the figure, which certainly increases the tension.

The idealized model of a spring is that the force it exerts is linearly proportional to the amount it is stretched from some unstressed length. Unlike the pendulum model, it is possible for actual springs to behave close to the ideal, not just for infinitesimal displacements, but for a considerable range of motion. The constant of proportionality is the stiffness of the spring, k. Suppose L_0 is the length of a spring with no force on it. Then the force it exerts when stretched to length L is

$$T = -k(L - L_0).$$

If T is the measured tension of the masses being in a line, and we can measure L_0 as the length of a spring under no tension and L is simply $\frac{D}{4}$ here, then that determines the spring constant k:

$$k = -\frac{T}{D/4 - L_0}.$$

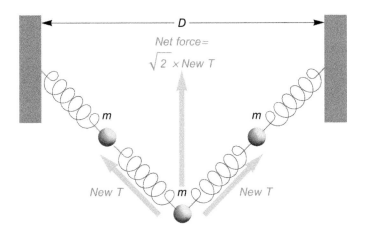

Compared to the masses being in a straight line, imagine pulling the center one down: Not by a small displacement, but a large one.

The forces add, precisely, as vectors. No terms are treated as "negligible."

Each spring lengthens from $\frac{D}{4}$ to $\frac{\sqrt{2}\,D}{4}$, so the tension on each spring increases to

$$\text{new } T = -k\left(\sqrt{2}\,L - L_0\right) = T\,\frac{\sqrt{2}\,D/4 - L_0}{D/4 - L_0}.$$

The force on the center mass is in the vertical direction, the sum of the two force vectors, which geometrically form a square so the force on the center mass is $\sqrt{2}$ times the new tension T, plus mg, where g is acceleration due to gravity. For the kinds of strings that are under enough tension to make musical notes, mg will be much smaller than the tension force, but there is no reason to discard the gravity term in the general case.

Just as we used a big 60° angle to show that the pendulum approximation is quite far off, we here use such a large displacement that it is clear the tension cannot simply be treated as constant. Pick up any physics textbook on the behavior of a vibrating string, however, and you will find a derivation littered with approximations. Why?

First, it makes the equations of motion look simpler, and scientists that have to wrestle with ugly-looking equations are always looking for ways to replace the terms in the equations with simpler ones that are "close" under certain conditions. Second, with enough simplifications, the limiting case of an infinite number of masses and an infinite number of tiny springs with only infinitesimal deflection is one for which calculus can supply nice-looking solutions in terms of elementary functions.

With the advent of high-speed computing, we need to rein in our instinct for simplification because it is no longer necessary, and it starts us down the road to accepting wrong answers as guidance about the actual physical behavior. Instead, we could have provable bounds, that is, real *science*.

When an experiment is done with an actual stretched string, any difference between the computed result and the experimental result *proves* something is incomplete in the physical model. For example, if a steel wire is put under tension and plucked while observed with a high-speed camera, the first assumption that breaks down is that the "string" is perfectly flexible, with no stiffness. A sharp kink at the point where the string is plucked is actually rounded out slightly by the stiffness. In the stiff-string case, the computational model then has to use a three-dimensional mesh of masses on springs to match the material properties. Perhaps the string vibrates as predicted, but not for very long; that indicates the need to account for energy lost to sound energy in the air and heating of the string. **Every physical effect can be modeled without rounding error or sampling error if the model is discrete.** The computer can be used to detect incompleteness in the physical model, if making the discrete model finer and finer does not converge to match experimental results.

21.3 The single-atom gas

The following thought experiment may be easier for the reader to conjure up if we shamelessly use old-fashioned English units.

Imagine six hard squares forming a box in the vacuum of space (no gravity), measuring a foot on each side. Inside it is a one-ounce steel ball 2/3 inch in diameter moving at a thousand miles per hour. Like a trapped bullet, it hits each face hundreds of times per second. Imagine the bounces are perfect elastic collisions and that the ball does not spin but simply bounces off the walls. It is a pretty good puzzle to figure out what path the ball can take that hits each face once, at equal-spaced times. The picture shows a solution that treats each of the three dimensions of the cube equally, seen from two different angles:

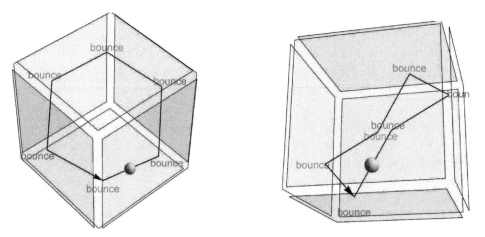

If the cube is a unit cube where *x*, *y*, and *z* range from 0 to 1, then the coordinates {*x*, *y*, *z*} of the bounce locations make a cyclic pattern:

$$\left\{0, \frac{1}{3}, \frac{2}{3}\right\}$$

$$\left\{\frac{1}{3}, 0, \frac{1}{3}\right\}$$

$$\left\{\frac{2}{3}, \frac{1}{3}, 0\right\}$$

$$\left\{1, \frac{2}{3}, \frac{1}{3}\right\}$$

$$\left\{\frac{2}{3}, 1, \frac{2}{3}\right\}$$

$$\left\{\frac{1}{3}, \frac{2}{3}, 1\right\}$$

Every bounce imparts momentum, since the velocity v is reversed in one dimension. The velocity in each dimension is v times $\dfrac{1}{\sqrt{3}}$. If L is the length of an edge, then the distance between bounces is

$$L \times \sqrt{\left(\frac{1}{3}\right)^2 + \left(\frac{1}{3}\right)^2 + \left(\frac{1}{3}\right)^2} = \frac{L}{\sqrt{3}}.$$

Every six of those is a circuit, so the rate at which the ball hits any particular wall is velocity divided by that distance:

$$\text{collision rate} = \frac{v}{6\frac{L}{\sqrt{3}}} = \frac{v}{2\sqrt{3}\,L}$$

Each collision transfers momentum from the glancing angle:

$$\text{momentum change} = m\frac{v}{\sqrt{3}} - m\left(\frac{-v}{\sqrt{3}}\right) = \frac{2\,m\,v}{\sqrt{3}}$$

The rate of momentum change is the force on the wall; things start to look simpler:

$$\text{Force} = \text{collision rate} \times \text{momentum change} = \frac{m\,v^2}{3\,L}$$

Pressure is force per unit area, so if we divide by the area of a cube face L^2, it turns out the denominator is three times the volume of the cube, V:

$$\text{Pressure} = \frac{m\,v^2}{3\,L^3} = \frac{m\,v^2}{3\,V}$$

The average force on the walls works out to about fifteen pounds per square inch.

Atmospheric pressure!

At sea level and 78 °F (25 °C), the weight of a cubic foot of air is about one ounce. The speed of air molecules at room temperature is about a thousand miles per hour. The number of molecules of oxygen and nitrogen in a cubic foot of air is close to Avogadro's number, the value 6×10^{23} that was used as an example at the very beginning of this book. But it appears possible to scale the problem of simulating 6×10^{23} molecules all the way down to *one* particle, a sort of super-atom (the steel ball), and derive the properties of pressure using the elementary physics of elastic collisions. No calculus, and no continuum approximations.

And no *statistics*. This thought experiment is a form of gas *kinetic theory*, a very old idea, so old that the original developers (Maxwell, Boltzmann, and others) predated the computer era by many decades. The difference between conventional kinetic theory and the thought experiment here is that kinetic theory uses statistical averages of **random** velocities.

By the Pythagorean theorem in 3D,

$$\overline{v^2} = \overline{v_x^2} + \overline{v_y^2} + \overline{v_z^2}$$

where the $\overline{\text{over - bar}}$ means "average value" of everything underneath it. But with random motion, there is no difference between the average value of the square of the velocity in the x, y, or z direction. Picking just the x direction, that means

$$\overline{v^2} = 3\,\overline{v_x^2}.$$

In the thought experiment where the steel ball bounces around in a path the shape of a bent hexagon, each edge of the hexagon is in a direction of the line connecting opposite corners of the cube. In other words, the velocity v goes in the direction of a vector $\{\pm 1, \pm 1, \pm 1\}$. So $v_x = \pm \dfrac{1}{\sqrt{3}}\, v$.

Which means, remarkably, that the single-atom gas model *exactly* matches reasoning based on statistical averages:

$$v^2 = 3\,v_x^2.$$

Wikipedia has a good explanation of the statistical model in its *Kinetic theory* entry, and an effort has been made here to match the variable names if anyone wants to compare the derivations side by side.

Incidentally, the word "gas" comes from the Greek word we pronounce as *chaos*. Our ballistic one-atom model takes all the chaos out of the system, yet it winds up with the same result.

If the steel ball was replaced by two steel balls half the weight, moving along the same path at the same speed, the pressure on the walls would be the same. If you replaced the steel ball by 6×10^{23} tiny masses at the same speed, each weighing $1/(6 \times 10^{23})$ as much, the pressure on the walls would be the same, and each mass would be almost exactly that of a nitrogen or oxygen molecule! It's amusing to think of all of them forming a convoy whizzing around that bent-hexagon shape, the world's most organized gas. The point is that when scaling mass, the *velocity* stays the same.

Now let's do the computation again, using international units and more accuracy. The standard unit of pressure is the Pascal (Newtons per square meter), and the U.S. Standard Atmosphere is 101 325 Pascals. Solve for the velocity of the big atom, because there is one more feature of the atmosphere this model can derive:

$$\text{Pressure} = 101\,325\,\text{Pascals} = \frac{m\,v^2}{3\,V}, \text{ where}$$

$$m = 0.028964 \text{ kg (mean molar mass of air)}$$
$$V = 0.023644 \text{ cubic meters at sea level}$$

$$\Rightarrow v = 498.14 \text{ meters per second.}$$

Imagine the cube is airtight and set down on the surface of the earth at sea level, with gravity at 9.80 meters per second squared. The ball will hit the floor of the cube with slightly more force than it hits the ceiling, because of the acceleration of gravity. Instead of straight-line paths, the mass makes a ballistic path that is ever so slightly curved into a parabola. The location where it hits the side wall only changes by about 3 microns $(3 \times 10^{-6}$ meter), so it does not require much alteration to the geometry to get the ball banging around at the same steady frequency. If you work out the pressure difference, it is 0.344 Pascals less at the ceiling of the cube than the bottom. The cube is 0.287 meters on a side, so pressure drops with altitude at a rate of $0.344 \div 0.287 = 1.20$ Pascals per meter, near sea level.

That agrees, to three decimals, with actual measurements. The model predicts not just the pressure, but also how atmospheric pressure changes with altitude.

Exercise for the reader: Calculate the velocity drop and the pressure difference, using the ballistic model of a one-atom gas. Remember that the change in kinetic energy, $\frac{1}{2}m\,v^2$, equals the change in potential energy, $m\,g\,h$ where h is the height of the cube, and that only the vertical component of the velocity is affected.

So it is possible to model pressure by such a simple setup, at least. What about effects like viscosity and turbulence? Obviously we need more "big atoms" to start to see such effects. Precise, bounded computations are also possible with more inter-esting and realistic random distributions of speed for collections of particles. The first deviation from the perfect gas model is to allow the gas particles to collide not just with the walls of the container, but with each other. This happens at rates that can be scaled by adjusting the sizes of the elastic spheres in the simulation; the smaller the sphere diameter, the lower the collision rate.

They also are not necessarily *spheres* in a real gas, but molecules that have rotational energy and vibrational energy. A computer is certainly capable of modeling a very large ensemble of balls-on-springs that do a crude job of simulating diatomic oxygen and nitrogen molecules. By adjustment of parameters on the "big molecules," we should either be able to match experimental properties of gases or discover the physics missing from the model, and iterate until they *do* match. At some point, with a match to measured properties of gases, the model should be able to provide bounded answers to questions about the lift and drag on an airfoil, or the drag coefficient of a car.

The calculus alternative is the Navier–Stokes equations, some of the nastiest partial differential equations you will ever see. (Also some of the most commonly mispro-nounced. Navier was French, so that part sounds like NAH-vee-ay.) There are five of them; one for conservation of mass, three for conservation of momentum (one for each space dimension), and one for conservation of energy. They are nonlinear and coupled, and look very different when working with an incompressible fluid like water as compared to a compressible fluid like air. When you consider that they have to describe how heat diffuses in a gas and how sound waves travel, the Navier–Stokes equations seem like all the equations of computational physics rolled together.

The usual approach to the Navier–Stokes equations is to simplify them as much as possible, artistically deleting aspects of the physics in the hope that they have little effect on the result. Then a programmer starts figuring out how to turn all the funny little $\frac{\partial}{\partial t}$ and other derivatives into something discrete, like difference approximations, and use a marching method to step through time. As the simulation drifts ever further from the truth, there is absolutely no way to know how far off it is other than to compare with a real experiment. Which can be *really* expensive, say, if the simulation is for the design of a hypersonic aircraft.

The programs that do computational fluid dynamics (CFD) are amazing works of craftsmanship developed over many decades, and somehow their authors have managed to wrestle the rounding and sampling errors to the point where the results can give practical guidance to engineers. Armed with unum arithmetic, it might be interesting to develop programs that bound the behavior of discrete particles to see where the classic CFD approximations based on continuum approximations and calculus diverge from a rigorous model based on kinetic theory, one that needs no calculus whatsoever.

21.4 Structural analysis

On the left, the billion-dollar North Sea oil platform Sleipner A, as it appeared August 23, 1991. On the right, the same location on August 24, 1991. Oops. It collapsed to the ocean floor, causing an earthquake of 3.0 magnitude. Amazingly, no one was killed or injured, not even the engineer in charge of the project when his manager found out the disaster was caused by a computing error, overestimating the *shear strength* of parts of the concrete pillars by 47%. This happened despite the use of 64-bit floats and a well-respected software package for finite element analysis (NASTRAN).

Structural analysis is an engineering field that is almost as deeply invested in difficult calculus for continuum physics as is CFD. Envision a large rubber eraser that can be squeezed, pulled, twisted, or shear-stressed in any of the three space dimensions. It is like a spring that does not just have a spring *constant*, it has a spring *matrix* to describe all the interacting ways it will strain when placed under stress.

Rubber is well-behaved and predictable, but stress and strain do *not* have a linear relationship for many materials, which messes up modeling efforts that usually cast the structural analysis problem as a system of linear equations. (The shear strength of concrete is a good example of nonlinear stress-strain response. Or if you owned stock in the Sleipner A oil platform above, a very *bad* example.)

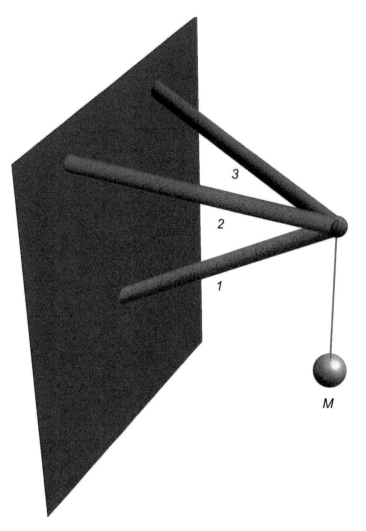

Here is a problem in structural analysis that is about as simple as they get, at least in three dimensions. It requires no calculus nor any knowledge of how things bend under shearing forces or twisting forces.

Find the structural forces necessary to hold up a mass *M* with three members: A strut (shown as a blue cylinder, labeled *1*) and two cables (shown as red cylinders, labeled *2* and *3*) attached to a vertical wall. The cables experience tension force, and the strut experiences compression force.

The three members join at a point, but that joint can exert no torque. It is a "pin connector," to be thought of as allowing the members to swivel at the point of attachment. That means the only thing that can happen to the members is compressive forces (the strut) and tensile force (the cable).

For now, ignore the weights of the members, cables, and joint where they connect; **what is the force on each part of the structure?** This can be answered using just two common-sense facts:

- The forces must balance, since nothing is accelerating.
- Forces *add*, separately in each dimension. In other words, they are vectors.

Assume that the wall supplies the counter-force to whatever force is put on it by the cables and struts. **Since these are simply cables and struts, the only direction they can exert force is along their length.**

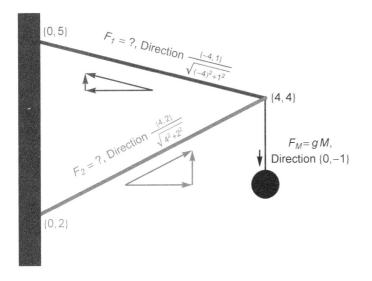

In a two-dimensional version of the problem, force balance is a little easier to visualize.

In the diagram on the left, the red cable has slope $-1/4$ and the blue strut has slope $1/2$; the force directions are made into unit vectors by dividing each one by the length of its member.

In the vertical direction z, the mass M exerts a force of magnitude gM where g is gravity. Therefore, the strut must exert a counter-force in the z direction of $+gM$, to cancel. But the fact that it can only exert force in the direction of its length means there must also be a horizontal force on the strut four times as large, or $+4gM$. That in turn means the cable must pull horizontally (the x direction) with force $-4gM$. Then all the forces in each dimension add to zero.

Balance *horizontal* forces at the joint where the two members meet and also support the mass M:

$$\frac{4}{\sqrt{4^2+(-1)^2}}\, F_1 + \frac{4}{\sqrt{4^2+2^2}}\, F_2 + \frac{0}{\sqrt{0^2+(-1)^2}}\, F_M = 0$$

Balance *vertical* forces at that same point:

$$\frac{-1}{\sqrt{4^2+(-1)^2}}\, F_1 + \frac{2}{\sqrt{4^2+2^2}}\, F_2 + \frac{-1}{\sqrt{0^2+(-1)^2}}\, F_M = 0$$

Since we know $F_M = gM$, the two equations above form a pair of linear equations in two unknown quantities F_1 and F_2. In matrix form, it looks like this:

$$\begin{pmatrix} \dfrac{4}{\sqrt{17}} & \dfrac{4}{\sqrt{20}} \\ \dfrac{-1}{\sqrt{17}} & \dfrac{2}{\sqrt{20}} \end{pmatrix} \begin{pmatrix} F_1 \\ F_2 \end{pmatrix} = \begin{pmatrix} 0 \\ gM \end{pmatrix}$$

The exact solution to that system is $F_1 = -\frac{\sqrt{17}}{3} g M$, $F_2 = \frac{\sqrt{20}}{3} g M$.

To get a better visceral understanding of this result, suppose M is one metric ton, which is about the same as a ton in English units. Then the upper cable has to provide tension (negative direction force) of $\frac{\sqrt{17}}{3} = 1.37 \cdots$ times the weight of a ton, and the strut has to provide compressive strength to hold $\frac{\sqrt{20}}{3} = 1.49 \cdots$ times the weight of a ton.

With unum math, all the inputs can be bounded, and resulting forces can be computed with a c-solution that contains all representable solutions and no F_1, F_2 pairs that fail to solve the problem. If the structure is "stiff," that is, very sensitive to changes in inputs, then the c-solution will be a very skinny quadrilateral and **an engineer would instantly see that the design is dubious**. That would happen, say, if the problem instead looked like this:

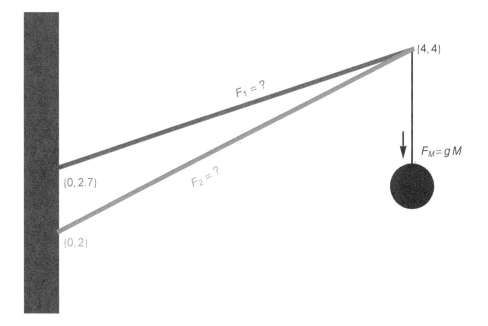

An approach that is presently used is called "sensitivity analysis," where the input parameters are varied to see how the answer is affected. This tends to require some artistic choices since there are many degrees of freedom in actual engineering structural problems. With uboxes, all the possible parameter sensitivity is already in the c-solution, for all to see.

To apply the technique in three dimensions and a more complicated structure with multiple joints, simply write equations describing balanced forces in each of x, y, and z directions at each joint.

> **One last exercise for the reader**: If the coordinates of the attachment points of the three-dimensional structure shown at the beginning of this section are {0, −1, 4.5}, {0, 1, 4.5}, {0, 0, 2} and the members meet at {4, 0, 3}, set up the three force equations.

The forces from the weights of the members themselves can be taken into account by using the fact that the weight of each member is equally borne by both ends of the member. For example, if the red cable shown above has mass m, the equations would need to add $g\,m/2$ to the weight of the suspended mass M.

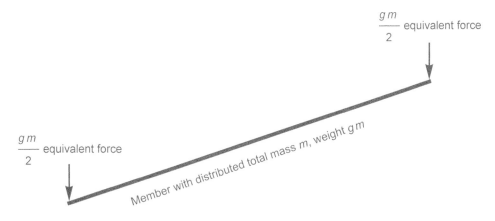

$\dfrac{g\,m}{2}$ equivalent force

$\dfrac{g\,m}{2}$ equivalent force

Member with distributed total mass m, weight $g\,m$

This is when the engineering starts getting interesting, because if building a steel bridge, say, you need to look up the tensile and compressional strength of steel and figure out the minimum weight that stays within safety limits. The trial-and-error process can be automated. The design process is made both easier and safer if the uncertainty of all the inputs is expressed with ubounds, and if the solution of the system of equations is not smeared by the dependency problem as it is with naive use of interval arithmetic. Engineers have plenty of uncertainty to cope with already, without also having to wonder if the underlying arithmetic in their computer is sound.

Now imagine a large three-dimensional mesh of members connected by joints, something like this:

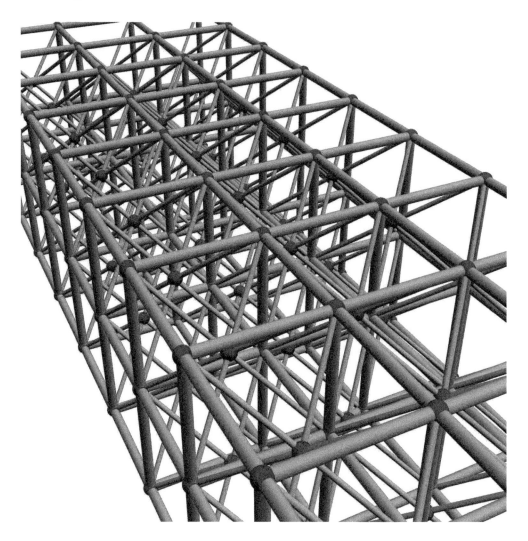

Such a mesh can model torque and shear, despite being made of simple straight members connected at joints that cannot exert torque or shear forces.

So far, the assumption has been that there is no *strain* from the stress placed on the structure, so the coordinates of each joint are fixed. Once you allow members to act as linear springs and the joints to move, and can adjust each spring stiffness separately, then all kinds of properties can be simulated that mimic the behavior of continuous materials. The *finite element method* is much like what is described here, but notice that unlike any book on that method you will find, there is *no calculus in the derivation*. Just discrete objects governed by Newtonian mechanics. The large system of equations that results would be hopeless to solve by hand but is possible to solve by computer with *provable bounds on the results*.

22 The end of error

The title of this text is provocative to some, who are anxious to point out that the "end of error" is still nowhere in sight. That depends on what is meant by *error*. The term, as used here, has nothing to do with "soft errors" caused by cosmic rays striking memory chips and flipping their bits. It does not refer to the human errors of incorrect data entry, or programming mistakes. It does not mean experimental error, the inevitable variability of measurement. Here, "error" means mathematical incorrectness, something stated as numerically true when it is not. *That* is the kind of error that we unfortunately have learned to tolerate, but that now have the ability to **end** through advances in computer technology.

The early computer designers were well aware that their designs incorporated compromises to mathematical correctness, but costs prohibited the building of correct numerics. When the first electronic, automatic computers were built starting around 1940, a logic gate cost tens of dollars and a single bit of storage cost over a dollar, in 2014 dollars. As long as the programmer was someone as bright as John von Neumann, the errors were manageable; the early computers were used by a "priesthood" and it was never envisioned that someday billions of people would rely on automatic computers, with over 99.99% of those users completely clueless about the hazards of high-speed numerical errors. They also never envisioned that an hour of human talent would eventually cost hundreds of times *more* than an hour of time on a very powerful computer, so it was reasonable to ask humans to bear all the burden of creating correct methods for performing calculations.

As hardware got cheaper and faster, the work expanded to fill the budget and time available. An example is what happened to computer printouts. In 1970, a printer might produce something that looks like this, and take about 30 seconds to do so:

```
C       PRIME NUMBERS
        DO 100 I=1,1000
        J=2
        K=2
2       L=J*K
        IF (L-I) 10,100,10
10      M=2+3
        IF (K-I) 20,3,3
20      K=K+1
        GO TO 2
3       K=2
        IF (J-I) 5,4,4
5       J=J+1
        GO TO 2
4       WRITE (3,6) I
6       FORMAT (I10)
100     CONTINUE
        STOP
        END
```

Over forty years later, a laser printer still might take 30 seconds to put out a single page, but technology has improved to allow full-color, high-resolution output.

Initially, IBM used laser printing to generate many pages per second, with each page consisting of mono-spaced, all-capital, letters. Eventually, the work done per page rose to match the speed available. The repeated mistake of thinking the need is for *time reduction* when it really is for *quality improvement* is such a widespread and recurring fallacy, it is surprising that it is not more widely known.

There are still those who assert that parallel processing should be used to do existing sequential workloads in *less time*, but parallel computation is almost invariably used to increase the *work done in the time people are willing to wait.*

Why did technology speed improvement not turn into quality improvement for numerical computing as well? While processing speeds increased by a factor of a trillion, floats got better by maybe a factor of two in their *quality*, with the emphasis on doing more of them per second. One possible explanation is that intellectual inertia varies depending on the technology. It is easy, for example, to get people to change hardware if it can be made to run the same software. It is tough to change software. Tougher still is to change the lowest level meanings of bits, like going from EBCDIC to ASCII standard characters to Unicode. Eventually it becomes so obvious that change is necessary that a wrenching shift finally takes place. Such a shift is overdue for numerical computing, where we are stuck with antediluvian tools.

Once we make the shift, it will become possible to solve problems that have baffled us for decades. The evidence supporting that claim is here in this book, which contains quite a few results that appear to be new. Many were enabled by the unum-ubox approach, with its ability to express real numbers properly. Consider this partial list of the breakthroughs that have been demonstrated here for the first time:

- Solution of time stepping dynamics equations made massively parallel even if they only involve one variable

- Addition operations that fully obey the associative law, even when a summation is done by different computers in parallel with arbitrary ordering

- Provable bounds on n-body simulations that expand only linearly in the number of time steps

- A fixed-size, fixed-time method of computing x^y to maximum accuracy where x and y are both floats, and that can identify when the result is exact

- A compact way to express intervals with open or closed endpoints, eliminating "edge condition" errors

- A solution to the paradox of slow arithmetic for complex arithmetic, via fused operation definition

- A system to guarantee bitwise identical answers across computing systems, not by dumbing down the arithmetic but by *raising* its quality through a precisely defined standard scratchpad

- A way to evaluate the polynomial of an interval without information loss

- A practical solution to "The Table-Maker's Dilemma"

- Calculations with 8-bit unums that get the right answers where 128-bit floats fail

- A way to represent the entire real number line as well as exception values that uses only four bits

- A root finder that works on any continuous function, including high-order polynomials that have no closed-form expression for their roots

- Trig functions that produce exact answers when the inputs are special angles

- A Fast Fourier Transform that produces tightly bounded results but uses fewer bits per number than a single-precision float

- An information-based, precision-independent definition of computing speed

- A polynomial-complexity method of finding extrema of continuous functions

- A way to eliminate time wasted checking for n-bit exception values by maintaining pre-decoded single-bit exception states that are testable with a single logic gate

- A solver for systems of equations, linear or nonlinear, that finds *all* the expressible solutions and automatically makes obvious when the answer is sensitive to small changes in input values

- An arbitrarily precise solution method for nonlinear ordinary differential equations that uses *no calculus*, just elementary algebra, geometry and Newtonian physics

- Derivation of gas pressure and pressure gradient using ballistic physics but absolutely no statistical mechanics

- A technique for finding fixed points of functions that works on both stable and unstable points

There will almost certainly be many attackers who think unums are like interval arithmetic and try to denigrate the approach, based on that mischaracterization. There is an interesting analogy between the intervals-versus-floats debate and the issues of *medical testing*.

In medical testing, "false positives" refers to the reporting of a problem when none exists. "False negatives" are when a problem exists but the test fails to detect it. Both are forms of error. The debate between advocates of floats and advocates of interval arithmetic is really a debate about *which type of error is worse*. Floats commit false negatives by silently producing incorrect results, but traditional intervals commit false positives by declaring many legitimate calculations as wildly unstable and having huge bounds. Which side is right?

Neither one. We have finally reached the technology level where we can eliminate both false positives *and* false negatives in numerical calculations. The concept of information as the reciprocal of the size of the uncertainty makes clear what is wrong with both float and interval arithmetic approaches. A float cannot convey its error range, so its information is zero. An interval that expands to infinite width also has zero information, since the information is the reciprocal of infinity. Unums, ubounds, and uboxes are designed to monitor and maximize information produced by a calculation, which is the actual purpose of computing.

Because unums are a superset of floats, unum arithmetic can do everything that floats can do, but it opens up the ability to communicate numerical concessions and demands to the computer system for the first time. Unums can finally distinguish between exact and inexact values as easily we currently distinguish between positive and negative numbers. When a real number has no exact representation, it is honest to mark a result as inexact and provide a tight bound for its value. The burden is on the defenders of floats: Why should anyone continue to use floats when unum arithmetic can mimic floats but also has so many advantages? And why should we tolerate numerical methods that are hard to write and treacherous to use, when bounded methods like ubox sets can produce results that are complete solutions of maximum expressible accuracy?

It is also not dishonest to produce a computational result without rigor if it is *labeled* as a guess, instead of quietly guessing about every result the way rounded floats do. As numerical recipes are rewritten for unums, they could begin by making plenty of calls to the "guess" function to make them duplicate what floats do, and then gradually reduce the use of those calls to the point where there are none, and the calculation can claim to be "guess-free." Because unums are also a superset of traditional interval arithmetic, they can provide the rigorous bounds offered by that technique, but their ability to control unnecessary range expansion (information loss) goes well beyond what traditional intervals can deliver. Unums provide a rich vocabulary for describing *sets* of real numbers, not just exact points or closed sets.

Suppose the situation was reversed: Imagine that for the last hundred years, all calculations were labeled as exact or inexact, number storage was always allowed to grow and shrink automatically for optimal use of memory and energy, and physical simulations were held to the same standards of scientific validity as the correctness of the equations on which they are based. Imagine that someone now proposes that instead we make every number *look* exact even when it is not, always use a fixed, over-sized storage independent of the precision and dynamic range of any particular datum, that we let scratchpad calculations vary from computer to computer, and that we replace rigorous computational physics with approximations that create attractive visualizations of dubious validity.

How would the technical community react? Probably with something sarcastic, like "Great idea. It wastes memory, slows everything down by using more bandwidth, produces different answers on different systems, prevents the use of parallel processing, makes arithmetic with small bit strings take just as long as those with big ones, and can make huge calculation errors without any kind of warning of having done so. What are you, crazy? Get out of here."

Remember this figure from the Preface?

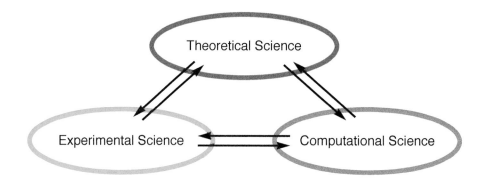

When Kenneth Wilson proposed this elevation of Computational Science to the same level as the other two ovals, he had recently won the Nobel Prize in Physics, the first one ever granted for computational accomplishments as opposed to theoretical or experimental results. He had used some array processors from Floating Point Systems to model the behavior of quarks ("Quantum Chromodynamics," or QCD) using something called Lattice Gauge Theory. There was no way to solve the equations by hand, and there certainly was no way to do a comparable experiment with sub-nuclear particles. But with enough calculations based on the theory, Wilson was able to achieve the same kind of guidance a physicist would get from an experimental setup, showing approximate agreement between the calculation and the measured mass of a neutron. The awarding of his Nobel Prize was controversial; some thought it a complete violation of the rules of science to use a computer, because a computer could offer guidance but no *proof* of correctness.

We have come a long way since the 1980s, both in how much we respect the value of computational science and in the technology we have at our disposal with which to perform it. With unum-based simulations that provably bound the behavior of physical systems, we can at last start putting examples into the blue Computational Science oval above that really *are* science, not merely guidance. ∎

Glossary

↓: The "↓" symbol, when placed at the end of a set of displayed fraction digits, emphasizes that there are no more nonzero digits to the right; the fraction shown is *exact*. The opposite of "···", which means there are nonzero digits after the last one shown. Section 3.1, page 26.

c-solution: In computing, a *c-solution* to a problem involving real numbers is the smallest complete set of uboxes that satisfy the problem statement. Section 15.3, page 199.

computational speed: In the context of a particular computing task, *computational speed* is information achieved per second. Note that this is *not* the conventional meaning of the phrase, which is often used to mean activity performed per second (like operations per second, or instructions per second) whether that activity produces information about the answer or not. Section 15.6, page 211.

dependency problem: The *dependency problem* is excessively large bounds in interval calculations caused by treating multiple occurrences of the same variable as if they were independent, when they are not. Section 16.3, page 220.

esizesize: The *esizesize* is the number of bits allocated to store the maximum number of bits in the exponent field of a unum. The number of exponent bits, *es*, ranges from 1 to $2^{esizesize}$. Section 4.3, page 39.

fsizesize: The *fsizesize* is the number of bits allocated to store the maximum number of bits in the fraction field of a unum. The number of fraction bits, *fs*, ranges from 1 to $2^{fsizesize}$. Section 4.3, page 39.

g-layer: The *g-layer* is the scratchpad part of a computing system where results are computed such that they are always correct to the smallest representable uncertainty when returned to the *u*-layer. Section 5.3.1, page 66.

gbound: A *gbound* is the data structure used for temporary calculations performed at higher precision than in the unum environment; that is, the scratchpad. The higher precision is whatever suffices to determine the correct unum or ubound result in the current environment. Pronounced "jee-bound." Section 5.3.1, page 65.

guess: A *guess* is an exact representable value selected from (or adjacent to) a range of real values. A rounded number is a form of guess. In the context of unums, a guess is an explicit command to replace a unum or ubound with a float-like rounding error instead of following mathematical rules. Section 18.1, page 257.

***h*-layer**: The *h-layer* is where numbers (and exception quantities) are represented in a form understandable to humans. The *h*-layer is not limited to character strings that represent numbers; *h*-layer quantities can be graphic information, mechanical motion, or anything else that is physically perceivable by a human or can be supplied by a human as input to a computer system. Section 5.4.1, page 74.

information: *Information*, in the context of a particular computing question, is the reciprocal of the total volume of a set that bounds the answer. The maximum information possible is the reciprocal of the volume of the *c*-solution. The information when there is no known bound for a result (a float calculation) or when the bound is infinite (common with casual use of interval arithmetic) is, by this definition, zero. Section 15.6, page 210.

maxreal: The value *maxreal* is the largest finite value that a format can express. It is one ULP less than the representation of infinity. The opposite end of the dynamic range from *smallsubnormal*. Section 3.2, page 28.

maxubits: *maxubits* is the maximum number of bits a unum can have in a given environment. When the unum environment is set to {*esizesize, fsizesize*}, the value of *maxubits* is $2 + esizesize + fsizesize + 2^{esizesize} + 2^{fsizesize}$. (The minimum number of bits is simply $3 + utagsize$.) Section 4.4, page 40.

smallsubnormal: The value *smallsubnormal* is the smallest value greater than zero that a format can express. It is one ULP larger than zero, using the maximum fraction length allowed by the format. The opposite end of the dynamic range from *maxreal*. Section 3.2, page 29.

summary bit: A *summary bit* indicates a crucial feature of a unum that can otherwise be derived only by examining combinations of bits; it can be used to save computing time and energy by storing "decoded" attributes of a unum. For instance, a single bit can indicate that a unum has infinite magnitude, saving the arithmetic logic the trouble of checking every bit of a unum other than the sign bit. Section 7.3.2, page 100.

***u*-layer**: The *u-layer* is the level of computer arithmetic where all the operands are unums (as well as data structures made from unums, like ubounds and uboxes). Section 5.2.1, page 63.

ubit: The *ubit* is a bit in the unum format that is 0 if a unum is exact and 1 if the unum represents the open interval between two exact unums. That is, a 1 indicates there are more bits in the fraction but no space to store them in the particular format settings of the unum. Rhymes with "cubit." Section 3.1, page 27.

ubound: A *ubound* is either a single unum or a pair of unums that represent a mathematical interval of the real line. Closed endpoints are represented by exact unums, and open endpoints are represented by inexact unums. They are more powerful than traditional "interval arithmetic" intervals in that they can represent both open and closed endpoints, which has wide-ranging ramifications for their arithmetic behavior. Section 5.2.1, page 62.

ubox: A *ubox* is an *n*-dimensional shape represented by a list of *n* unums. They are the fundamental building blocks for expressing solution sets as well as intermediate calculations. Note that any particular dimension can be zero (exact) or an open interval of width a power of two (that is, one ULP, at some precision). In three dimensions, for example, a ubox shape could be a point, a line segment that does not include its endpoints, the interior of a rectangle, or the interior of a box. Section 15.3, page 199.

ULP: An *ULP* (rhymes with gulp) is the difference between exact values represented by bit strings that differ by one Unit in the Last Place, the last bit of the fraction. Some texts use "Unit of Least Precision" as the abbreviated phrase. Section 3.1, page 27.

unum: A *unum* is a bit string of variable length that has six sub-fields: sign bit, fraction, uncertainty bit, exponent size, and fraction size. It represents either a floating point number or the open interval between two floating point numbers that differ only in the last bit of their fraction. Pronounced "you-num," since it is short for "universal number." Introduced in Section 1.1, page 4; complete discussion in Chapter 4, pages 35 – 54.

utag: The *utag* is the set of three self-descriptive fields that distinguish a unum from a float. The fields are the ubit, the exponent size bits, and the fraction size bits, located to the right of the three fields of an IEEE-style float (sign bit, exponent bits, and fraction bits). Section 4.3, page 40.

utagsize: The *utagsize* is the length (in number of bits) of the utag bit string. Its value is $1 + esizesize + fsizesize$. Section 4.4, page 40.

Warlpiri unums: The *Warlpiri unums* are the smallest unum possible, a {0, 0} environment. They have a sign bit, an exponent bit, a fraction bit, and a ubit. They are thus four bits long and the sixteen possible bit strings represent fifteen distinct values (exact numbers, open number ranges, positive and negative infinity, and both signaling NaN and quiet Nan) since zero is represented twice. Section 7.1, page 94.

wrapping problem: The *wrapping problem* is the occurrence of excessively large bounds in interval calculations that are caused by bounding a multidimensional result set with a single axis-aligned box. That is, a simple box-shaped bound is used for what is really a complicated shape of smaller volume, resulting in loss of information. Section 16.2, page 216.

Appendix A Glossary of unum functions

A.1 Environment and bit-extraction functions

`setenv[{es,fs}]`	Sets the environment for maximum exponent field size 2^{es} bits and maximum fraction field size 2^{fs} bits.
`numbits[u]`	Finds the total number of bits in a unum u.
`nbits[ub]`	Finds the total number of bits in a ubound ub.
`inexQ[u]`	True if u is inexact; else False, meaning it is exact.
`exQ[u]`	True if u is exact; else False, meaning it is inexact.
`expovalue[u]`	The numerical meaning of exponent bits of a unum u.
`exact[u]`	Sets ubit of unum u to 0, making it the nearest exact unum.

A.2 Constructors

`utag[e,f]`	Constructs the utag for exponent size e and fraction size f.
`bias[u]`	Finds the exponent bias for a unum u.
`bigu[u]`	Constructs unum representing the biggest exact real expressible using the utag of u.
`big[u]`	Constructs the biggest exact real expressible with utag of u.

A.3 Visualization functions

`utagview[ut]`	Displays a utag ut with color-coded, annotated fields.
`colorcode{u]`	Displays the six fields of a unum u with color coding.
`autoN{x]`	Displays a float rational x exactly, using decimals.
`gview[x]`	Displays x as a general interval, converting it first if necessary.
`uview[x]`	Displays x as a color-coded, annotated unum or ubound, converting it first if necessary.
`view[x]`	Displays a u- or g-layer x, in conventional math notation.
`reportdatamotion [nn, nu]`	Produces a table showing nn numbers moved, nu unum bits moved, and average bits per number.

A.4 Conversion functions

`u2f`[*u*]	Converts an exact unum *u* to its float value.
`f2g`[*x*]	Converts a real number *x* to general interval format.
`x2u`[*x*] or $\hat{x}$	Converts *x* to a float or an open interval containing *x*.
`unum2g`[*ub*]	Converts a unum *u* to a general interval.
`ubound2g`[*ub*]	Converts a ubound *ub* to a general interval.
`u2g`[*u*]	Converts a unum or ubound *u* to a general interval.
`g2u`[*g*]	Converts a general interval *g* to a ubound.
`unify`[*ub*]	Looks for a single unum that contains a unum bound *ub*.
`smartunify`[*ub,r*]	Unifies *ub* if it improves information-per-bit by at least *r*.

A.5 Argument validity tests

`unumQ`[*x*]	True if *x* is a valid unum, else False.
`floatQ`[*x*]	True if *x* can convert to a float (real, ±∞, or NaN), else False.
`gQ`[*x*]	True if *x* is a valid general interval, else False.
`uboundQ`[*x*]	True if *x* is a valid unum bound (single or paired), else False.
`uboundpairQ`[*x*]	True if *x* is a valid unum bound pair, else False.
`uQ`[*x*]	True if *x* is a valid unum or ubound, else False.

A.6 Helper functions for functions defined above

`esizeminus1`[*u*]	Finds the exponent size minus 1 of a unum *u*.
`esize`[*u*]	Finds the exponent size of a unum *u*.
`fsizeminus1`[*u*]	Finds the fraction size minus 1 of a unum *u*.
`fsize`[*u*]	Finds the fraction size of a unum *u*.
`signmask`[*u*]	Finds the binary that has a **1** where the sign bit is of a unum *u*.
`hiddenmask`[*u*]	Finds binary that has a **1** left of the first fraction bit of a unum *u*.
`fracmask`[*u*]	Finds the binary with **1**s in all the fraction field bits of a unum *u*.
`expomask`[*u*]	Finds the binary with **1**s in all exponent field bits of a unum *u*.
`floatmask`[*u*]	Finds binary with all **1**s for the float bits of a unum *u*.
`sign`[*u*]	Extracts the sign bit of a unum *u*; **0** if positive, **1** if negative.
`expo`[*u*]	Extracts the exponent of a unum *u*, before bias adjustment.
`hidden`[*u*]	Value (**0** or **1**) of the hidden bit of a unum *u*.
`frac`[*u*]	Extracts fraction part of unum *u*, without hidden bit or scaling.
`scale`[*x*]	Finds power-of-2 exponent for converting a real *x* to a unum.
`ne`[*x*]	Finds number of bits needed to express scale of a real *x*.
`padu`[*u*]	Pads fraction bits of unum *u* with **0**s. Alters meaning if inexact.

A.7 Comparison operations

`ltgQ`[*g,h*]	True if a *g*-layer *g* is less than a *g*-layer *h*, else False.
`ltuQ`[*u,v*] or *u<v*	True if a *u*-layer *u* is less than a *u*-layer *v*, else False.
`gtgQ`[*g,h*]	True if a *g*-layer *g* is greater than a *g*-layer *h*, else False.
`gtuQ`[*u,v*] or *u>v*	True if a *u*-layer *u* is greater than a *u*-layer *v*, else False.
`neqgQ`[*g,h*]	True if *g*-layer values *g* and *h* are disjoint, else False.
`nequQ`[*u,v*] or *u≠v*	True if *u*-layer values *u* and *v* are disjoint, else False.
`nneqgQ`[*g,h*]	True if *g*-layer values *g* and *h* overlap, else False.
`nnequQ`[*u,v*] or *u≈v*	True if *u*-layer values *u* and *v* overlap, else False.
`samegQ`[*g,h*]	True if *g*-layer values *g* and *h* are identical, else False.
`sameuQ`[*u,v*] or *u≡v*	True if the *u*-layer values *u* and *v* represent the same general interval, else False.
`intersectg`[*g,h*]	Returns the general interval for the intersection of *g*-layer *g* and *h*; returns a *g*-layer NaN if the intersection is empty.
`intersectu`[*u,v*]	Returns the ubound for the intersection of *u*-layer *u* and *v*; returns a quiet NaN ubound if the intersection is empty.

A.8 Arithmetic operations

Basic plus-minus-times-divide operations, and some algebraic functions.

`plusg`[*g, h*]	Adds general intervals *g* and *h*.
`plusu`[*u, v*] or *u⊕v*	Adds *u*-layer values *u*, *v*; tracks data motion.
`negateg`[*g*]	Negates a general interval *g*.
`minusg`[*g, h*]	Subtracts general interval *h* from *g*.
`minusu`[*u, v*] or *u⊖v*	Subtracts *u*-layer values *u* and *v*; tracks data motion.
`timesg`[*g, h*]	Multiplies general intervals *g* and *h*.
`timesu`[*u, v*] or *u⊗v*	Multiplies *u*-layer values *u* and *v*; tracks data motion.
`divideg`[*g, h*]	Divides general interval *g* by *h*.
`divideu`[*u, v*] or *u⊘v*	Divides *u*-layer *u* by *u*-layer *v*; tracks data motion.
`spanszeroQ`[*ub*]	True if a ubound *ub* contains zero, else False.
`squareg`[*g*]	Squares a general interval *g*.
`squareu`[*u*]	Squares a *u*-layer *u*; tracks data motion.
`sqrtg`[*g*]	Finds the square root of a general interval *g*.
`sqrtu`[*u*]	Finds the square root of a *u*-layer *u*; tracks data motion.
`powg`[*g, h*]	Takes general interval *g* to the power of a general interval *h*.
`powu`[*u, v*]	Takes a *u*-layer *u* to the power of a *u*-layer *v*; tracks data motion.
`absg`[*g*]	Finds the absolute value of a general interval *g*.
`absu`[*u*]	Finds absolute value of a *u*-layer *u*; tracks data motion.

A basic set of transcendental functions: exponential, logarithm, and trigonometric.

`expg[g]`	Finds exp(g) for a general interval g.
`expu[u]`	Finds exp(u) for a u-layer u; tracks data motion.
`logg[g]`	Finds log(g) for a general interval g.
`logu[u]`	Finds log(u) for a u-layer u; tracks data motion.
`cosg[g]`	Finds cos(g) for a general interval g, where the argument is in degrees.
`cosu[u]`	Finds cos(u) for a u-layer u, where the argument is in degrees; tracks data motion.
`sing[g]`	Finds sin(g) for a general interval g, where the argument is in degrees.
`sinu[u]`	Finds sin(u) for a u-layer u, where argument is in degrees; tracks data motion.
`tang[g]`	Finds tan(g) for a general interval g, where the argument is in degrees.
`tanu[u]`	Finds tan(u) for a u-layer u, where the argument is in degrees; tracks data motion.
`cotg[g]`	Finds cot(g) for a general interval g, where the argument is in degrees.
`cotu[u]`	Finds cot(u) for a u-layer u, where the argument is in degrees; tracks data motion.

A.9 Fused operations (single-use expressions)

`fmag[ag, bg, cg]`	Fused g-layer multiply-add: $(ag \times bg) + cg$.
`fmau[au, bu, cu]`	Fused u-layer multiply-add: $(au \times bu) + cu$; tracks data motion.
`fdotg[glist, hlist]`	Finds the exact dot product of g-layer lists *glist* and *hlist*.
`fdotu[ulist, vlist]`	Finds closest u-layer value to the exact dot product of u-layer lists *ulist* and *vlist*; tracks data motion.
`fsumg[glist]`	Finds the exact sum of a g-layer list *glist*.
`fsumu[ulist]`	Finds closest u-layer value to the exact sum of the u-layer list *ulist*; tracks data motion.
`fprodg[glist]`	Finds the exact product of a g-layer list *glist*.
`fprodu[ulist]`	Finds closest u-layer value to the exact product of the u-layer list *ulist*; tracks data motion.
`famg[ag, bg, cg]`	Fused g-layer add-multiply: $(ag + bg) \times cg$.
`famu[au, bu, cu]`	Fused u-layer add-multiply: $(au + bu) \times cu$; tracks data motion.
`fprodratiog` [*numg, deng*]	Fused g-layer ratio of the exact product of a list *numg* to the exact product of a list *deng*, to u-layer accuracy
`fprodratiou` [*numg, deng*]	Closest u-layer value to the ratio of exact product of a list *deng* to the exact product of a list *denu*.

A.10 Some defined data values

Some data values are defined in the prototype, including the color palette if anyone is interested in using similar colors in describing unums and uboxes. The levels are set by **RGBColor**$[r, g, b]$, where r, g, and b range from 0 (black) to 1 (saturated). The text is shown in the actual color, to make it easier to spot. Quantities NaN, open, and closed are defined in terms of their *Mathematica* equivalents and should not be changed. The default values of **esizesize** and **fsizesize** are adjustable. The tally counters **ubitsmoved** and **numbersmoved** are initialized to 0, and need to be set back to 0 at the beginning of any measurement sequence. The **relwidthtolerance** defaults to 0.001 for at least three decimals answer accuracy, adjustable by user.

gogreen = RGBColor[0,0.75,0.625]	Traffic light color. For solution uboxes.
cautionamber = RGBColor[0.96,0.72,0]	Traffic light color. For trial uboxes.
stopred = RGBColor[1,0.25,0]	Traffic light color. For failure uboxes.
brightblue = RGBColor[0.25,0.5,1]	Exponent bits and many other places. Easier to distinguish from black.
sanegreen = RGBColor[0,0.75,0]	*es* bits in utag and elsewhere. Easier to distinguish from white.
paleblue = RGBColor[{0.75,1,1}]	Background color for the *g*-bound data structure.
brightpurple = RGBColor[0.75,0.5,1]	Easier to distinguish from black. Used in various graphics.
brightmagenta = RGBColor[1,0.25,1]	Easier to distinguish from red. Used in various graphics.
textamber = RGBColor[0.75,0.5,0]	Amber is too light for text; a legible darker version.
chartreuse = RGBColor[0.875,1,0.5]	Background color for 2×2 general intervals.
NaN = Indeterminate	*Mathematica* name for NaN.
open = True	Boolean for an open endpoint; ubit = 1.
closed = False	Boolean for a closed endpoint; ubit = 0.
esizesize = 3	Maximum unum exponent size default: $2^3 = 8$ bits.
fsizesize = 4	Maximum unum fraction size default: $2^4 = 16$ bits.
ubitsmoved = 0 **numbersmoved = 0**	Initializes the data motion tallies. Reset for a new measurement.
relwidthtolerance = 0.001	Default relative width tolerance, for auto-accuracy control. Set as needed.

A.11 Auto-precision functions

needmoreexpQ [*ub*]	True if a *u*-layer value *ub* is using the limits of the current dynamic range, indicating a need for more exponent bits, else False.
relwidth [*ub*]	Computes the relative width of a *u*-layer value *ub* as a measure of accuracy.
needmorefracQ [*ub*]	True if the relative width of a *u*-layer value is too big, indicating a need for more fraction bits, else False.

Appendix B: Glossary of ubox functions

B.1 ULP-related manipulations

promotef [u]	Appends a **0** to the fraction bits of an exact unum u, if possible.
promotee [u]	Increases the length of the exponent field of an exact unum u, if possible (preserving its numerical meaning).
promote [u, v]	Finds a pair of unums that are equal to exact unums u and v, but have identical utags.
demotef [u]	Demote fraction length of a unum u, even if it makes it inexact.
demotee [u]	Demote exponent length of a unum u, even if it makes it inexact.

B.2 Neighbor-finding functions

nborhi [$u, minpower$]	Finds the neighbor on the right of a unum u (closer to $+\infty$), and if u is exact, prefer a neighbor with ULP width $2^{minpower}$, where $minpower$ is an integer.
nborlo [$u, minpower$]	Finds the neighbor on the left of a unum u (closer to $-\infty$), and if u is exact, prefer a neighbor with ULP width $2^{minpower}$ if possible, where $minpower$ is an integer.
findneighbors [$ubx, minpowers$]	Find all neighbors in all dimensions of a multidimensional ubox ubx using the set of integers $minpowers$ as the preferred minimum ULP sizes for inexact neighbor widths.

B.3 Ubox creation by splitting

uboxlist [$ub, minpower$]	Splits a ubound ub into unums such that inexact unums have preferred width $2^{minpower}$, where $minpower$ is an integer.
uboxlistinexact [$ub, minpower$]	Splits a ubound ub into unums such that inexact unums have preferred width $2^{minpower}$, where $minpower$ is an integer, but returning only the *inexact* unums.
splitub [ub]	Split a ubound ub if possible. Prioritizes splitting off exact endpoints first.
bisect [g]	Bisect a general interval g along an ULP boundary.

B.4 Coalescing ubox sets

coalescepass[*set*]	Makes a single pass over a two-dimensional ubox set *set* to find trios of inexact-exact-inexact unums that can be replaced by a single unum of twice the ULP width. *Note:* Prototyped quickly, and very inefficient!
coalesce[*set*]	Calls **coalescepass** repeatedly, until no further ubox consolidation is possible.
coalesce1D[*set*,*ub*]	Incorporate a new ubound *ub* into a set of ubounds *set*, simplifying any overlap or adjacency.
ubinsert[*ub*]	Split a ubound *ub* if possible. Prioritizes splitting off exact endpoints first.
hollowout[*set*]	Remove interior uboxes from a collection of uboxes *set*.
guessu[*ub*]	Replace a ubound or unum *u* with an exact value, either a coarse-boundary ULP interior point near the mean, or an endpoint imitating float rounding.

B.5 Ubox bounds, width, volume

minub[*ub*,*vb*]	Find the minimum left endpoint of two ubounds *ub* and *vb*, taking into account the closed-open endpoint qualities.
maxub[*ub*,*vb*]	Find the maximum right endpoint of two ubounds *ub* and *vb*, taking into account the closed-open endpoint qualities.
width[*u*]	Find the (real-valued) width of a unum or ubound *u*.
volume[*set*]	Find the multidimensional volume of a set of uboxes *set* (a simplistic measure of solution uncertainty).

B.6 Test operations on sets

setintersectQ[*ubox*,*set*]	True if *ubox* intersects any members of *set*, else False.
gsetintersectQ[*g*,*set*]	True if a general interval *g* intersects any member of a set of general intervals *set*, else False.
glistQ[*set*]	True if every member of *set* is a general interval, else False.
glistNaNQ[*set*]	True if *any* member of a set of general intervals *set* is a NaN, else False.
ulistQ[*set*]	True if every member of *set* is a unum or ubound, else False.

B.7 Fused polynomial evaluation and acceleration formula

polyg[*coeffsg,g*]	Fused polynomial evaluation using general interval coefficients *coeffsg* for a general interval argument *g*, with no dependency problem.
polyu[*coeffsu,u*]	Fused polynomial evaluation using *u*-layer coefficients *coeffsu* for a unum or ubound argument *u*, with no information loss in the *u*-layer.
accub[*xub,yub*]	Example of a library function for bounding gravitational acceleration on a ubox {*xub, yub*} that uses knowledge of extrema to eliminate dependency error.

B.8 The try-everything solver

solveforub[*domain*]	Exhaustive search of a domain specified by a set of ubounds *domain* to find the *c*-solution to a user-supplied test **ConditionQ**.

B.9 The guessing function

guessu[*range*]	Finds the closest exact unum to the midpoint of a unum, ubound, or general interval *range*. Equivalent to rounding a float, but an explicit operation.

Appendix C Algorithm listings for Part 1

Since in-line code makes for difficult reading, almost all of the code listings for important functions are here in the appendices. For those unfamiliar with *Mathematica* language conventions, here are a few things to know.

To define a function $f(x) = \sqrt{x}$, the notation is

```
f[x_] := √x
```

Square brackets [ ], not parentheses, enclose function arguments; parentheses ( ) are used to group expressions. Curly braces { } contain lists. Double square brackets ⟦ ⟧ indicate subscripts of a list, and the convention here is to actually typeset them as subscripts, like $list_{⟦2⟧}$ to indicate the second element of `list`.

The /; characters are followed by a condition that guards whether to evaluate a function. For example,

```
f[x_ /; x ≥ 0] := √x
```

only evaluates if x is greater than or equal to zero. This means a function defined with

```
f[x_ /; unumQ] := ...
```

will not execute unless `unumQ[u]` is True (that is, u meets all the requirements for being a legitimate unum).

If statements are in the three-part form If[condition, action to take if True, action to take if False]. The third part, the "else," is optional.

When a digit string is to be interpreted as binary, it is preceded with "2^^"; for example, "2^^1101" represents the decimal number 13.

Comments are in **gray** in the code. Where possible, standard mathematical notation is used to make the code easier to read. For example, a logical AND is written "∧" instead of "&&". Where a general interval is used, like $\begin{array}{cc} -1 & \infty \\ \textbf{open} & \textbf{closed} \end{array}$, the light yellow-green background is used to make it distinct as a two-by-two table.

Whereas Appendices A and B mainly show the routines that an application programmer would invoke to solve a problem, Appendices C and D also show some of the auxiliary functions that go into making the application-level functions work. They are critical to understanding how unum arithmetic might be supported more directly, in compiled libraries or actual hardware.

C.1 The set-the-environment function

Here is the code for **setenv**, showing exactly how the following environmental variables are set based on the values of **esizesize** and **fsizesize**:

utagsize	smallsubnormalu	qNaNu
maxubits	smallnormalu	sNaNu
ubitmask	signbigu	negopeninfu
fsizemask	posinfu	posopeninfu
esizemask	maxrealu	negopenzerou
efsizemask	neginfu	posbig
utagmask	negbigu	smallsubnormal
ulpu		

The environment is initially set, as a default, to {3, 4}, which has a slightly larger dynamic range than single-precision IEEE floats and slightly less precision (about 5 decimal digits). Such unums range in size from 11 to 33 bits. The default **setenv[{3,4}]** call assures that the above variables are defined. In the prototype, **esizesize** is limited to 4 and **fsizesize** is limited to 11.

```
(* Set the environment variables based on esizesize and fsizesize.
   Here, maximum esizesize is 4 and maximum fsizesize is 11. *)
setenv[{e_Integer /; 0 ≤ e ≤ 4, f_Integer /; 0 ≤ f ≤ 11}] :=
```
$$\left(\right.$$
```
{esizesize, fsizesize} = {e, f};
```
```
{esizemax, fsizemax} = 2^{e, f};
utagsize = 1 + f + e;
maxubits = 1 + esizemax + fsizemax + utagsize;
ubitmask = BitShiftLeft[1, utagsize - 1];
fsizemask = (BitShiftLeft[1, f] - 1);
esizemask = (ubitmask - 1) - fsizemask;
efsizemask = BitOr[esizemask, fsizemask];
utagmask = BitOr[ubitmask, efsizemask];
ulpu = BitShiftLeft[1, utagsize];
smallsubnormalu = efsizemask + ulpu;
smallnormalu = efsizemask + BitShiftLeft[1, maxubits - 1 - esizemax];
signbigu = BitShiftLeft[1, maxubits - 1];
posinfu = signbigu - 1 - ubitmask;
maxrealu = posinfu - ulpu;
minrealu = maxrealu + signbigu;
neginfu = posinfu + signbigu;
negbigu = neginfu - ulpu;
qNaNu = posinfu + ubitmask;
sNaNu = neginfu + ubitmask;
negopeninfu =
  If[utagsize == 1, 2^^1101, BitShiftLeft[2^^1111, utagsize - 1]];
posopeninfu = If[utagsize == 1, 2^^0101,
  BitShiftLeft[2^^0111, utagsize - 1]];
negopenzerou = BitShiftLeft[2^^1001, utagsize - 1];
```
$$\texttt{maxreal} = 2^{2^{esizemax-1}} \frac{(2^{fsizemax}-1)}{2^{fsizemax-1}};$$
$$\texttt{smallsubnormal} = 2^{2-2^{esizemax-1}-fsizemax}; \left.\right)$$

C.2 Type-checking functions

Most of the functions in the prototype are careful to check that its arguments are of the proper data type before proceeding. If they are not, the command is simply echoed back; for example, the **setenv** function expects a *list* of two integers, the first between 0 and 4 and the second between 0 and 11. If you accidentally forget the curly braces indicating a list, or if you ask for integers outside those ranges, the function will not execute and will only echo back the command:

```
setenv[2, 2]
```

```
setenv[2, 2]
```

Since the type-check functions return `True` or `False`, their names end in **Q** to indicate a question with a Boolean result.

The **floatQ**[*x*] function tests if *x* is representable as a float, which includes NaN and ±∞ representations. If it is a numeric value, it cannot be a complex type (of the form $a + bi$). Notice that to test if *x* is NaN, we have to write "x === NaN" where "===" means "is identical to". It is not possible to test non-numeric values with the usual double-equal sign, "==".

```
(* Test if x is representable as a float. *)
floatQ[x_] := If[NumericQ[x], If[Head[x] =!= Complex, True, False],
    If[x === ∞ ∨ x === -∞ ∨ x === NaN, True, False]]
```

The **unumQ**[*x*] function tests if *x* is a legitimate unum in the current environment, by treating it as a bit string representing an integer and making sure that integer is between zero and the largest possible unum (all **1** bits at the maximum length of the string, meaning signaling NaN):

```
(* Test if a value is a legitimate unum. *)
unumQ[x_] := If[IntegerQ[x], If[x ≥ 0 ∧ x ≤ sNaNu, True, False], False]
```

The test for whether *x* is a ubound, **uboundQ**[*x*], is more complicated. A ubound is a list containing one or two unums, and if there are two, then the interval it describes must make mathematical sense. For example, [3, 2] is not allowed but [2, 2] is. More subtly, (2, 2] is not allowed. This shows the first function where a negative subscript is used; negative subscripts count backwards from the end of a list, so writing $x_{[-1]}$ means to take the *last* item in the list x. This is a concise way to get the rightmost endpoint of a ubound, whether the ubound has one unum or two unums. The built-in **Length**[*x*] function returns the length of a list *x*.

```
(* Test if a value is a ubound, with either one or two unums. *)
uboundQ[x_] :=
 If[Head[x] === List,
  If[Length[x] == 1 ⋁ Length[x] == 2,
   Module[{xL = x⟦1⟧, xR = x⟦-1⟧, gL, gR},
    If[unumQ[xL] ⋀ unumQ[xR],
     {gL, gR} = {unum2g[xL], unum2g[xR]};
     If[Length[x] == 1 ⋁
        (xL == qNaNu ⋁ xL == sNaNu ⋁ xR == qNaNu ⋁ xR == sNaNu) ⋁
        ((gL⟦1,1⟧ < gR⟦1,2⟧) ⋁ (gL⟦1,1⟧ == gR⟦1,2⟧ ⋀ (exQ[xL] ⋀ exQ[xR]))),
      True, False], False]], False], False]
```

At times, such as for functions that display ubounds, it is important to know whether the ubound is a pair. The **uboundpairQ**[*x*] returns True if *x* is a ubound and has a list length of 2.

```
(* Test for a value being in the form of a ubound with two unums. *)
uboundpairQ[x_] := uboundQ[x] ⋀ Length[x] == 2
```

For most *u*-layer operations, it is easiest if functions tolerate arguments being unums or ubounds and are not strict about the presence of curly braces to indicate a list. The **uQ**[*x*] test returns True if *x* is either type.

```
(* Test for a value being in the u-layer: unum or ubound. *)
uQ[x_] := unumQ[x] ⋁ uboundQ[x]
```

A test specific to unum data types is whether the unum is an exact or inexact value. The **inexQ**[*x*] test returns True if *x* is an inexact unum (the ubit is set), and **exQ**[*x*] returns True if *x* is an exact unum. Both are included for readability, since "not inexact" becomes "¬inexQ" and that looks a little cryptic.

```
inexQ[u_ /; unumQ[u]] := BitAnd[ ubitmask, u] > 0
exQ[u_ /; unumQ[u]] := BitAnd[ubitmask, u] == 0
```

The test for an argument *x* being in the form of a general bound is **gQ**[*x*]. It takes the form of nested **If** statements because it makes no sense to test if a list has length 2 until you have tested if the item is even a list at all. In English: The code on the following page returns True if the argument is a 2-by-2 list with two values that can be expressed as floats on the top row, and two True-False values on the bottom row; if the floats are numeric, they must be in order, and can be identical only if they are *exact* (that is, both bottom row entries are False).

```
(* Test for a value being in the form of a general bound. *)
gQ[x_] :=
 If[Head[x] === List,
  If[Length[x] == 2,
   If[Head[x[[1]]] === List,
    If[Length[x[[1]]] == 2,
     If[Head[x[[2]]] === List,
      If[Length[x[[2]]] == 2,
       If[x[[2,1]] ∈ Booleans ∧
          x[[2,2]] ∈ Booleans ∧ floatQ[x[[1,1]]] ∧ floatQ[x[[1,1]]],
        If[x[[1,1]] === NaN ∨ x[[1,2]] === NaN, True,
         If[(x[[1,1]] == x[[1,2]] ∧ ¬ x[[2,1]] ∧ ¬ x[[2,2]]) ∨ x[[1,1]] < x[[1,2]], True,
          False], False], False], False], False], False], False],
   False], False]
```

C.3 The unum-to-float converter and supporting functions

The supporting functions extract the bits from each field of a unum. Those are used to compute a floating point value via u2f. First, some functions that depend only on the environment settings and not the bits stored in the utag of a unum:

```
(* Bit masks and values used for taking apart a unum.
   These are independent of the utag contents. *)
esizeminus1[u_ /; unumQ[u]] :=
 BitShiftRight[BitAnd[u, esizemask], fsizesize]
esize[u_ /; unumQ[u]] := 1 + esizeminus1[u]
fsizeminus1[u_ /; unumQ[u]] := BitAnd[u, fsizemask]
fsize[u_ /; unumQ[u]] := 1 + fsizeminus1[u]
utag[esize_Integer /; 1 ≤ esize ≤ esizemax,
    fsize_Integer /; 1 ≤ fsize ≤ fsizemax] :=
 BitOr[fsize - 1, BitShiftLeft[esize - 1, fsizesize]]
numbits[u_ /; unumQ[u]] := 1 + esize[u] + fsize[u] + utagsize
signmask[u_ /; unumQ[u]] := BitShiftLeft[1, numbits[u] - 1]
hiddenmask[u_ /; unumQ[u]] := BitShiftLeft[1, fsize[u] + utagsize]
fracmask[u_ /; unumQ[u]] :=
 BitShiftLeft[BitShiftLeft[1, fsize[u]] - 1, utagsize]
expomask[u_ /; unumQ[u]] :=
 BitShiftLeft[BitShiftLeft[1, esize[u]] - 1, fsize[u] + utagsize]
floatmask[u_ /; unumQ[u]] := signmask[u] + expomask[u] + fracmask[u]
```

Those functions describe where the data is located in the unum. The functions on the next page use them to extract unum data. (Note: The **Boole** function returns 1 if the argument is True and 0 if the argument is False.)

```
(* Values and bit masks that depend on what is stored in the utag. *)
bias[u_ /; unumQ[u]] := 2^esizeminus1[u] - 1
sign[u_ /; unumQ[u]] := Boole[BitAnd[u, signmask[u]] > 0]
expo[u_ /; unumQ[u]] :=
  BitShiftRight[BitAnd[u, expomask[u]], utagsize + fsize[u]]
hidden[u_ /; unumQ[u]] := Boole[expo[u] > 0]
frac[u_ /; unumQ[u]] :=
  BitShiftRight[BitAnd[u, fracmask[u]], utagsize]
inexQ[u_ /; unumQ[u]] := BitAnd[ ubitmask, u] > 0
exQ[u_ /; unumQ[u]] := BitAnd[ubitmask, u] == 0
exact[u_ /; unumQ[u]] := If[inexQ[u], BitXor[u, ubitmask], u]
```

The bit string in the exponent field needs two adjustments to find the actual power of 2 that it represents: First, the bias is subtracted. Second, the value (1 − hidden bit) is added to adjust for subnormal values (that is, values for which the hidden bit is zero.) The function **expovalue**[u] performs this function for a unum u. That makes it easy to express **u2f**[u], the function that converts an exact unum u to its float value.

```
(* Numerical value meant by exponent bits;
helper function for u2f: *)
expovalue[u_ /; unumQ[u]] := expo[u] - bias[u] + 1 - hidden[u]

(* Convert an exact unum to its float value. *)
u2f[u_ /; unumQ[u] ⋀ exQ[u]] := Which[
  u == posinfu, +∞,
  u == neginfu, -∞,
  True, (-1)^sign[u] 2^expovalue[u] (hidden[u] + frac[u]/2^fsize[u])]
```

C.4 The u-layer to general interval conversions

Mainly for completeness, the trivial **f2g**[x] function is shown below; it converts an exact float x into a two-by-two general interval representation, where both endpoints are closed and the same value.

```
(* Expression of a float in the form of a general interval. *)
                                    NaN   NaN      x        x
f2g[x_ /; floatQ[x]] := If[x === NaN, open  open , closed  closed]
```

Functions **bigu**[u] and **big**[u] return the unum and real-valued representations of the largest number that can be expressed by a particular utag bit string. They are used by the conversion functions.

```
(* Biggest unum possible with identical utag contents. *)
bigu[u_ /; unumQ[u]] := expomask[u] + fracmask[u] +
  BitAnd[efsizemask, u] - ulpu Boole[BitAnd[u, efsizemask] == efsizemask]

(* Biggest real value representable with identical utag contents. *)
big[u_ /; unumQ[u]] := u2f[bigu[u]]
```

The `unum2g[u]` function converts a unum bit string into a general interval:

```
(* Conversion of a unum to a general interval. *)
```

$$\texttt{unum2g[u\_ /; unumQ[u]] := If}\left[\texttt{u == qNaNu} \bigvee \texttt{u == sNaNu,} \begin{array}{cc} \texttt{NaN} & \texttt{NaN} \\ \texttt{open} & \texttt{open} \end{array},\right.$$

$$\texttt{Module}\left[\texttt{\{x = u2f[exact[u]], y = u2f[exact[u] + ulpu]\}, Which}\left[\right.\right.$$

$$\texttt{exQ[u],} \begin{array}{cc} \texttt{x} & \texttt{x} \\ \texttt{closed} & \texttt{closed} \end{array},$$

$$\texttt{u == bigu[u] + ubitmask,} \begin{array}{cc} \texttt{big[u]} & \infty \\ \texttt{open} & \texttt{open} \end{array},$$

$$\texttt{u == signmask[u] + bigu[u] + ubitmask,} \begin{array}{cc} -\infty & \texttt{-big[u]} \\ \texttt{open} & \texttt{open} \end{array},$$

$$\texttt{sign[u] == 1, ,} \begin{array}{cc} \texttt{y} & \texttt{x} \\ \texttt{open} & \texttt{open} \end{array},$$

```
(* If negative, left endpoint is farther from zero. *)
```

$$\texttt{True,} \begin{array}{cc} \texttt{x} & \texttt{y} \\ \texttt{open} & \texttt{open} \end{array}\left.\left.\left.\right]\right]\right]$$

The `ubound2g[u]` converts a ubound u to a general interval, using the `unum2g` function.

```
(* Conversion of a ubound to a general interval. *)
```

$$\texttt{ubound2g[ub\_ /; uboundQ[ub]] := Module}\left[\texttt{\{gL, gR, uL = ub}_{[\![1]\!]}\texttt{, uR = ub}_{[\![-1]\!]}\texttt{\},}\right.$$

$$\texttt{If}\left[\texttt{uL == qNaNu} \bigvee \texttt{uL == sNaNu} \bigvee \texttt{uR == qNaNu} \bigvee \texttt{uR == sNaNu,} \begin{array}{cc} \texttt{NaN} & \texttt{NaN} \\ \texttt{open} & \texttt{open} \end{array},\right.$$

$$\texttt{\{gL, gR\} = \{unum2g[uL], unum2g[uR]\};} \begin{array}{cc} \texttt{gL}_{[\![1,1]\!]} & \texttt{gR}_{[\![1,2]\!]} \\ \texttt{gL}_{[\![2,1]\!]} & \texttt{gR}_{[\![2,2]\!]} \end{array}\left.\left.\right]\right]$$

The `u2g[u]` function converts a u-layer value to a general interval. This shows how u-layer functions accept either unum or ubound data types, a simple way to make the prototype a little more object-oriented.

```
(* Conversion of a unum or ubound to a general interval. *)
u2g[u_ /; uQ[u]] := If[unumQ[u], unum2g[u], ubound2g[u]]
```

C.5 The real-to-unum or $\hat{x}$ conversion function

The `x2u[x]` or $\hat{x}$ function converts an arbitrary real number or exception value x into a unum. Think of it as the routine that moves values from the human-understandable h-layer into the u-layer where real numbers (or ULP-wide ranges that contain them) are finite-precision bit strings.

The helper function `scale[x]` function determines the amount to scale a real number x the way we would do it to convert a number to floating point; it is essentially the integer part of the logarithm base 2 of the magnitude. Dividing x by $2^{scale[x]}$ then yields a number with magnitude in the general interval $[1, 2)$. It does not attempt to work with infinities or NaNs, but if x is zero, it returns zero as an exception.

```
(* Helper function; find the scale factor, with exceptions. *)
scale[x_ /; floatQ[x] ∧ x ≠ ∞ ∧ x =!= NaN] := If[x == 0, 0, ⌊Log[2, Abs[x]]⌋]
```

The `ne[x]` function determines the *minimum* number of exponent bits to use in the unum representation of a real number x. There are many subtleties in this very short bit of code, so **beware of simplifying it in ways that seem harmless**! It is careful to avoid a bug that would occur in `x2u` when the minimum exponent size requires the use of a subnormal number.

```
(* Find best number of scale bits, accounting for subnormals. *)
ne[x_ /; floatQ[x] ∧ x ≠ ∞ ∧ x =!= NaN] :=
  If[x == 0 ∨ scale[x] == 1, 1, ⌈Log[2, 1 + Abs[scale[x] - 1]]⌉ + 1]
```

With those helper functions, here is the conversion function `x2u`, with plenty of comments (continues on the next page):

```
(* Conversion of a floatable real to a unum. Same
  as the "^" annotation. Most of the complexity stems
  from seeking the shortest possible bit string. *)
x2u[x_ /; floatQ[x]] := Which[
  (* Exceptional nonnumeric values: *)
  x === NaN, qNaNu, x == +∞, posinfu, x == -∞, neginfu,
  (* Magnitudes too large to represent: *)
  Abs[x] > maxreal, maxrealu + ubitmask + If[x < 0, signbigu, 0],
  (* Zero is a special case. The smallest unum for it is just 0: *)
  x == 0, 0,
  (* Magnitudes too small to represent become
    "inexact zero" with the
    maximum exponent and fraction field sizes: *)
  Abs[x] < smallsubnormal, utagmask + If[x < 0, signbigu, 0],
  (* For subnormal numbers, divide by the ULP value to get the
    fractional part. The While loop strips off trailing bits. *)
  Abs[x] < u2f[smallnormalu],
```

```
Module[{y}, y = Abs[x] / smallsubnormal;
  y = If[x < 0, signbigu, 0] + efsizemask +
    If[y ≠ ⌊y⌋, ubitmask, 0] + BitShiftLeft[⌊y⌋, utagsize];
  While[BitAnd[BitShiftLeft[3, utagsize - 1], y] == 0,
    y = (y - BitAnd[efsizemask, y]) / 2 + BitAnd[efsizemask, y] - 1]; y],
(* All remaining cases are in the normalized range. *)
True, Module[{n = 0, y, z},
  y = Abs[x] / 2^scale[x]; n = 0; While[⌊y⌋ ≠ y ⋀ n < fsizemax, {n++, y *= 2}];
  If[y == ⌊y⌋, (* then the value is representable
    exactly. Fill in fields from right to left: *)
    (* Size of fraction field,
    fits in the rightmost fsizesize bits... *)
    y = n - Boole[n > 0]
      (* Size of exponent field minus 1,
      fits in the esizesize bits... *)
      + BitShiftLeft[ne[x] - 1, fsizesize]
      (* Significant bits after hidden bit,
      fits left of the unum tag bits... *)
      + If[n == 0, 0, BitShiftLeft[⌊y⌋ - 2^scale[y], utagsize]]
      (* Value of exponent bits, adjusted for bias... *)
      + BitShiftLeft[scale[x] + 2^(ne[x]-1) - 1, utagsize + n + Boole[n == 0]]
      (* If negative, add the sign bit *)
      + If[x < 0, BitShiftLeft[1, utagsize + n + Boole[n == 0] + ne[x]], 0];
    (* If a number is more concise as a subnormal, make it one. *)
    z = Log[2, 1 - Log[2, Abs[x]]];
    If[IntegerQ[z] ⋀ z ≥ 0, BitShiftLeft[z, fsizesize] + ulpu +
      Boole[x < 0] signmask[BitShiftLeft[z, fsizesize]], y],
    (* else inexact. Use all available fraction bits. *) (
    z = ⌈(Abs[x] / 2^(scale[x]-fsizemax))⌉ 2^(scale[x]-fsizemax); n = Max[ne[x], ne[z]];
    (* All bits on for the fraction size,
    since we're using the maximum *)
    y = fsizemask
      (* Store the exponent
      size minus 1 in the exponent size field *)
      + BitShiftLeft[n - 1, fsizesize]
      (* Back off by one ULP and make it inexact *)
      + ubitmask - ulpu
      (* Fraction bits are the ones to the left of the
      binary point after removing hidden bit and scaling *)
      + BitShiftLeft[⌊(z / 2^scale[z] - 1) 2^fsizemax⌋, utagsize]
      (* Exponent value goes in the exponent field *)
      + BitShiftLeft[scale[z] + 2^(n-1) - 1, utagsize + fsizemax];
    (* If x is negative, set the sign bit in the unum. *)
    If[x < 0, y += signmask[y], y])]]]]
```

Defining the `OverHat[x]` notation to mean `x2u[x]` is what allows us to write $\hat{x}$ to convert *x* into a unum:

```
(* Assign the x2u function to the "^" notation. *)
OverHat[x_] := x2u[x];
```

The `autoN[x]` function displays a float x with enough decimals that it is exact; for example, $\frac{1}{16}$ displays as `0.0625`, stopping exactly where it should as a decimal without trailing zeros. It also displays NaN values as `Indeterminate`. It has no protection against improper inputs like "0.5" or $\frac{1}{3}$; it expects an input that is a special value, an integer, or a fraction with a power of 2 in the denominator.

```
(* View a float rational value as an exact decimal. *)
autoN[x_] := Module[{y, z}, Which[
    x === NaN ⋁ x == 0 ⋁ x == ∞, x,
    x < 0, Row[{"−", autoN[-x]}],
    y = Log[2, Denominator[x]];
    y == 0, IntegerString[x, 10, 1 + ⌊Log[10, x]⌋],
    Head[x] == Rational ⋀ y == ⌊y⌋,
    y = x - ⌊x⌋;
    z = ⌊Log[2, Denominator[y]]⌋;
    Row[{{⌊x⌋, ".", IntegerString[y * 10ᶻ, 10, z]}}],
    True, "?"]]
```

C.6 Unification functions and supporting functions

First, some support functions. The `padu[u]` function pads the fraction field of a unum u with as many trailing zeros as possible, effectively reversing the compression of the fraction field and altering the meaning of the ULP width if the unum is inexact.

```
(* Pad the fraction out to its maximum size
   by shifting in zero bits. *)
padu[u_ /; unumQ[u]] :=
  BitShiftLeft[BitAnd[floatmask[u], u], fsizemax - fsize[u]] +
    BitAnd[utagmask, u] + fsizemax - fsize[u]
```

We will also need the helper function `negateu[u]`, that finds the unum representing the negative of the value of u, taking into account hazards like "negative zero."

```
(* Negate a unum or ubound. *)
negateu[u_ /; uQ[u]] := Which[
    uboundQ[u],
    If[Length[u] == 1,
        {If[u2g[u⟦1⟧] === u2g[0], 0̂, BitXor[signmask[u⟦1⟧], u⟦1⟧]]},
        {If[u2g[u⟦2⟧] === u2g[0], 0̂, BitXor[signmask[u⟦2⟧], u⟦2⟧]],
         If[u2g[u⟦1⟧] === u2g[0], 0̂, BitXor[signmask[u⟦1⟧], u⟦1⟧]]}],
    True, If[u2g[u] === u2g[0], 0̂, BitXor[signmask[u], u]]]
```

The `unifypos` function takes a ubound argument where both endpoints are nonnegative, and attempts to find a single unum that contains the ubound. The code for `unifypos` is made complicated by the fact that we cannot simply test for equality or inequality of endpoint values. We also need to check whether the endpoints are closed or open. Constructs like "¬blo" mean "the Boolean indicating that the low end of the bound is closed," since "not inexact" means exact. In developing the prototype, `unifypos` generated more bugs than any other routine; be careful if attempting to make improvements.

```
(* Seek a single-ULP enclosure for a ubound ≥ zero. *)
unifypos[ub_ /; uboundQ[ub]] := Module[{u = ub[[1]], v = ub[[-1]], n},

  Which[(* Trivial case where endpoints express the same value *)
    u2g[u] === u2g[v], g2u[u2g[u]],
    (* Cannot unify if the interval includes exact 0,
    1,  2,  or  3 *)
    ub ≈ {0̂} ⋁ ub ≈ {1̂} ⋁ ub ≈ {2̂} ⋁ ub ≈ {3̂}, ub,
    (* Refine the endpoints for the tightest possible unification. *)
    u = promote[{u2g[û][[1,1]], efsizemask}][[1]] +
      If[inexQ[u], ubitmask, -ubitmask];
    v = promote[{u2g[v̂][[1,2]], efsizemask}][[1]] -
      If[inexQ[v], ubitmask, -ubitmask];
    u == v, {u},
    (* Upper bound is open ∞ and lower bound > maxreal: *)
    u2g[v][[1,2]] == ∞ ⋀ u2g[v][[2,2]],
    If[ltuQ[{maxrealu}, {u}], {maxrealu + ubitmask},
      (* Demote the left bound until the upper bound is open ∞ *)
      While[u2g[u][[1,2]] < ∞,
        u = If[esize[u] > 1, demotee[u], demotef[u]]]; {u}],
    True,
    (* While demoting exponents is possible and still
    leaves unumswithin the ubound, demote both exponents *)
    While[u ≠ v ⋀ (((u2g[demotee[u]][[1,1]] < u2g[demotee[v]][[1,1]] ⋀
          u2g[demotee[u]][[1,2]] < u2g[demotee[v]][[1,2]] < ∞))) ⋀
      esize[u] > 1, u = demotee[u]; v = demotee[v]];
    While[u ≠ v ⋀ frac[v] ≠ frac[u] ⋀ fsize[u] > 1,
      u = demotef[u]; v = demotef[v]];
    (* u is inexact zero and v < 1: *)
    If[u ≠ v ⋀ BitAnd[floatmask[u] + ubitmask, u] == ubitmask ⋀
      ltuQ[{v}, {1̂}], n = Min[esizemax,
        ⌊Log[2, 1 - Log[2, (u2g[v + If[exQ[v], ubitmask, 0]])[[1,2]]]]⌋]];
      {2^(-2̂^n+1) - ubitmask},
      If[u == v, {u}, ub]]]]]
```

With `negateu` and `unifypos` defined, the `unify` function merely needs to look out for NaN exceptions and flip the sign of the argument if it is negative (and then flip it back after unification):

```
(* Seek a single-ULP enclosure for a ubound. *)
unify[ub_ /; uboundQ[ub]] :=
  Module[{ulo, uhi, lo = u2g[ub]⟦1,1⟧, hi = u2g[ub]⟦1,-1⟧}, Which[
    lo === NaN ⋁ hi === NaN, {qNaNu}, (* Exception case *)
    (* No need to check for intervals spanning 0; unifypos does it. *)
    lo ≥ 0, unifypos[ub],
    (* Negate both operands, call unifypos, and negate them again. *)
    True, negateu[unifypos[negateu[ub]]]]]
```

There are many ways to make `unify` more intelligent, and one of these ways is discussed in Section 6.2. First, we need the function `nbits[u]`; it measures the number of bits required to store a *u*-layer value as a ubound, including the bit that says whether the ubound has one or two unums.

```
(* Find bits needed to store a ubound.
   If argument is a unum, make it a ubound, first. *)
nbits[u_ /; uQ[u]] := Module[{ub = If[unumQ[u], {u}, u]},
  1 + If[Length[ub] == 2, numbits[ub⟦1⟧] + numbits[ub⟦2⟧], numbits[ub⟦1⟧]]]
```

The `smartunify[ub, ratio]` function applies unify to a ubound *ub* only if doing so increases information-per-bit by a factor of *ratio*.

```
smartunify[ub_, ratio_] :=
  Module[{g = u2g[ub], v = unify[ub], widthbefore,
    widthafter, nbitsbefore, nbitsafter},
    widthbefore = g⟦1,2⟧ - g⟦1,1⟧;
    widthafter = u2g[v]⟦1,2⟧ - u2g[v]⟦1,1⟧;
    nbitsbefore = nbits[ub];
    nbitsafter = nbits[v];
    Which[widthafter == ∞, If[ratio ≥ 1, v, ub],
      widthafter == 0, v,
      (widthbefore/widthafter) (nbitsbefore/nbitsafter) ≥ ratio, v,
      True, ub]]
```

C.7 The general interval to unum converter

The overall plan for `g2u[g]`, which converts a general interval *g* into a ubound, is described in Section 7.3. The helper functions are `ubleft` and `ubright`, each of which take as an argument a set consisting of a value and a Boolean that is `True` if the endpoint is open and `False` if it is closed. The helper functions assume their arguments are not NaN values, since those are screened out by the calling function.

First, the one for `ubleft`:

```
(* Find the left half of a ubound
   (numerical value and open-closed bit). *)
ubleft[xleft_List] :=
  Module[{open = xleft[[2]], u = x2u[xleft[[1]]], x = xleft[[1]]}, Which[
    x == -∞, If[open, negopeninfu, neginfu],
    u2f[u] === x, If[open, BitOr[u - ulpu Boole[x < 0], ubitmask], u],
    True, BitOr[u, Boole[open] ubitmask]]]
```

Notice the subtle asymmetry in the routines for `ubleft` and `ubright`, caused by special handling of "negative zero":

```
(* Find the right half of a ubound
   (numerical value and open-closed bit). Not exactly
   the reverse of ubleft, because of "negative zero". *)
ubright[xright_List] :=
  Module[{open = xright[[2]], u = x2u[xright[[1]]], x = xright[[1]]}, Which[
    x == ∞, If[open, posopeninfu, posinfu],
    x == 0 ∧ open, negopenzerou,
    u2f[u] === x, If[open, BitOr[u - ulpu Boole[x ≥ 0], ubitmask], u],
    True, BitOr[u, Boole[open] ubitmask]]]
```

The code for `g2u` makes one of the rare uses of the "⊻" symbol, which means a logical exclusive OR. The code relies on the `unify` function as a final step, since it is quite possible that the general interval is expressible as a single unum without loss. Whenever there is more than one way to express the same ubound, the one with fewer bits is preferred.

```
(* Convert a general interval to the closest possible ubound. *)
g2u[g_ /; gQ[g]] :=
  Module[{ulo = g[[1,1]], uhi = g[[1,2]], blo = g[[2,1]], bhi = g[[2,2]]},
    (* Get rid of the NaN cases first *) Which[
    ulo == qNaNu ∨ uhi === qNaNu ∨
      g[[1,1]] > g[[1,2]] ∨ (g[[1,1]] == g[[1,2]] ∧ (blo ∨ bhi)), {qNaNu},
    (* If both unums are identical, and both open or both closed,
    we have a single-unum bound. *)
    ulo == uhi ∧ ¬ (blo ⊻ bhi), {x2u[g[[1,1]]]},
    True, Module[{u1 = ubleft[{g[[1,1]], blo}], u2 = ubright[{g[[1,2]], bhi}]},
      If[ u2g[unify[{u1, u2}]] === u2g[{u1, u2}],
        unify[{u1, u2}], {u1, u2}]]]]
```

C.8 Comparison tests and ubound intersection

Comparisons of *u*-layer values are performed by converting to general intervals and comparing those, using open-closed endpoint qualities. Comparison tests return `True` if the comparison is strictly true; if it is not possible to determine or is clearly not true, the test returns `False`.

```
(* Test if interval g is strictly less than interval h. *)
ltgQ[g_ /; gQ[g], h_ /; gQ[h]] := Which[
   g[[1,1]] === NaN ⋁ g[[1,2]] === NaN ⋁ h[[1,1]] === NaN ⋁ h[[1,2]] === NaN, False,
   True, g[[1,2]] < h[[1,1]] ⋁ (g[[1,2]] == h[[1,1]] ⋀ (g[[2,2]] ⋁ h[[2,1]]))]

(* Test if ubound or unum u is
 strictly less than ubound or unum v. *)
ltuQ[u_ /; uQ[u], v_ /; uQ[v]] := ltgQ[u2g[u], u2g[v]]

(* Assign the "less than" test the "<" infix symbol. *)
Precedes[u_, v_] := ltuQ[u, v]
```

```
(* Test if interval g is strictly greater than interval h. *)
gtgQ[g_ /; gQ[g], h_ /; gQ[h]] := Which[
   g[[1,1]] === NaN ⋁ g[[1,2]] === NaN ⋁ h[[1,1]] === NaN ⋁ h[[1,2]] === NaN, False,
   True, g[[1,1]] > h[[1,2]] ⋁ (g[[1,1]] == h[[1,2]] ⋀ (g[[2,1]] ⋁ h[[2,2]]))]

(* Test if ubound or unum u is
 strictly greater than ubound or unum v. *)
gtuQ[u_ /; uQ[u], v_ /; uQ[v]] := gtgQ[u2g[u], u2g[v]]

(* Assign the "greater than" test the ">" infix symbol. *)
Succeeds[u_, v_] := gtuQ[u, v]
```

```
(* Test if interval g is nowhere equal to interval h. *)
neqgQ[g_ /; gQ[g], h_ /; gQ[h]] := Which[
   g[[1,1]] === NaN ⋁ g[[1,2]] === NaN ⋁ h[[1,1]] === NaN ⋁ h[[1,2]] === NaN, False,
   True, ltgQ[g, h] ⋁ gtgQ[g, h]]

(* Test if ubound or unum u is nowhere equal to ubound or unum v. *)
nequQ[u_ /; uQ[u], v_ /; uQ[v]] := neqgQ[u2g[u], u2g[v]]

(* Assign the "nowhere equal" test the "≠" infix symbol. *)
NotTildeTilde[u_, v_] := nequQ[u, v]
```

```
(* Test if interval g is not nowhere equal to interval h. *)
nneqgQ[g_ /; gQ[g], h_ /; gQ[h]] := Which[
   g[[1,1]] === NaN ⋁ g[[1,2]] === NaN ⋁ h[[1,1]] === NaN ⋁ h[[1,2]] === NaN, False,
   True, ¬ (ltgQ[g, h] ⋁ gtgQ[g, h])]

(* Test if ubound or unum u is
 not nowhere equal to ubound or unum v. *)
nnequQ[u_ /; uQ[u], v_ /; uQ[v]] := nneqgQ[u2g[u], u2g[v]]

(* Assign not nowhere equal test the "≠" infix symbol. *)
TildeTilde[u_, v_] := nnequQ[u, v]
```

```
(* Test if interval g is identical to interval h. *)
samegQ[g_ /; gQ[g], h_ /; gQ[h]] := g === h

(* Test if ubound or unum u value
   is identical to ubound or unum v value. *)
sameuQ[u_ /; uQ[u], v_ /; uQ[v]] := samegQ[u2g[u], u2g[v]]

(* Assign the "identical" test the "≡" infix symbol. *)
Congruent[u_, v_] := sameuQ[u, v]
```

The routine to intersect u-layer quantities u and v, `intersectu[u,v]`, is based on a general interval equivalent for two general intervals g and h, `intersectg[g, h]`, that is a surprisingly complicated piece of code. The main reason it is so complicated is that "less than" is very different from "less than or equal" when figuring out intersections, and the intervals can have open or closed endpoints. The code is a good example of how much more careful unum computation is with mathematical sets than is traditional interval arithmetic.

```
(* Find the intersection of two general intervals in the g-layer. *)
intersectg[g_ /; gQ[g], h_ /; gQ[h]] :=
  Module[{glo = g[[1,1]], ghi = g[[1,2]], glob = g[[2,1]], ghib = g[[2,2]],

    hlo = h[[1,1]], hhi = h[[1,2]], hlob = h[[2,1]], hhib = h[[2,2]]}, Which[

                                                        NaN   NaN
    glo === NaN ⋁ ghi === NaN ⋁ hlo === NaN ⋁ hhi === NaN,    open  open  ,

    glo < hlo ⋁ (glo == hlo ⋀ hlob),
    Which[ (* left end of g is left of left end of h.

        Three sub-cases to test. *)
      ghi < hlo ⋁ (ghi == hlo ⋀ (ghib ⋁ hlob)),
      NaN   NaN
      open  open  , (* g is completely left of h. *)

                                              hlo   ghi
      ghi < hhi ⋁ (ghi == hhi ⋀ (ghib ⋁ ¬ hhib)),  hlob  ghib  ,
      (* right part of g overlaps left part of h. *)

             hlo   hhi
      True,  hlob  hhib  ], (* h is entirely inside g. *)
    glo < hhi ⋁ (glo == hhi ⋀ (¬ glob ⋀ ¬ hhib)),
    Which[ (* left end of g is inside h. Two sub-cases to test. *)
      ghi < hhi ⋁ (ghi == hhi ⋀ (ghib ⋁ ¬ hhib)),
      glo   ghi
      glob  ghib  , (* g is entirely inside h. *)

             glo   ghi
      True,  glob  ghib  ], (* left end of g overlaps right end of h. *)

           NaN   NaN
    True,  open  open  ] (* g is entirely to the right of h. *)]


(* Intersect two ubounds or unums. *)
intersectu[u_ /; uQ[u], v_ /; uQ[v]] := g2u[intersectg[u2g[u], u2g[v]]]
```

C.9 Addition and subtraction functions

The `plusg` code implements the *g*-layer addition tables for left and right endpoints shown in Chapter 9.

```
plusg[ xlo_   xhi_  ,  ylo_   yhi_ ] :=
       xlob_  xhib_    ylob_  yhib_

  Module[{sumleft, sumright, openleft, openright},

    If[(* If any value is NaN, the result is also NaN. *)

      xlo === NaN ∨ xhi === NaN ∨ ylo === NaN ∨ yhi === NaN,  NaN  NaN , (
                                                              open open

        (* Compute left endpoint: *)
        {sumleft, openleft} = Which[
          (* Cases involving exact ∞ or -∞: *)
          xlo == -∞ ∧ ¬ xlob,
          If[ylo == ∞ ∧ ¬ ylob, {NaN, open}, {-∞, closed}],
          ylo == -∞ ∧ ¬ ylob, If[xlo == ∞ ∧ ¬ xlob, {NaN, open}, {-∞, closed}],
          (xlo == ∞ ∧ ¬ xlob) ∨ (ylo == ∞ ∧ ¬ ylob), {∞, closed},
          xlo == -∞, If[ylo == ∞ ∧ ¬ ylob, {∞, closed}, {-∞, open}],
          ylo == -∞, If[xlo == ∞ ∧ ¬ xlob, {∞, closed}, {-∞, open}],
          (* Finally, what is left is the simple arithmetic case. But
            watch out for result of sum being out of range. *)
          xlo + ylo < -posbig, {-∞, open},
          xlo + ylo > posbig, {posbig, open},
          True, {xlo + ylo, xlob ∨ ylob}];
        (* Compute right endpoint, using similar logic: *)
        {sumright, openright} = Which[
          (* Cases involving exact ∞ or -∞: *)
          xhi == -∞ ∧ ¬ xhib,
          If[yhi == ∞ ∧ ¬ yhib, {NaN, open}, {-∞, closed}],
          yhi == -∞ ∧ ¬ yhib, If[xhi == ∞ ∧ ¬ xhib, {NaN, open}, {-∞, closed}],
          (xhi == ∞ ∧ ¬ xhib) ∨ yhi == ∞ ∧ ¬ yhib, {∞, closed},
          xhi == ∞, If[yhi == -∞ ∧ ¬ yhib, {-∞, closed}, {∞, open}],
          yhi == ∞, If[xhi == -∞ ∧ ¬ xhib, {-∞, closed}, {∞, open}],
          xhi + yhi < -posbig, {-posbig, open},
          xhi + yhi > posbig, {∞, open},
          True, {xhi + yhi, xhib ∨ yhib}];

        {{sumleft, sumright}, {openleft, openright}})]]
```

Below is the very simple code for `negateg`, which is then used to form the code for general interval subtraction, `minusg`.

```
negateg[ x_   y_  ] :=  -y   -x
         xb_  yb_        yb_  xb_


minusg[ xlo_   xhi_  ,  ylo_   yhi_ ] :=
        xlob_  xhib_    ylob_  yhib_

  plusg[ xlo   xhi  , negateg[ ylo   yhi  ]]
         xlob  xhib           ylob  yhib
```

With the *g*-layer functions defined, the `plusu[`*u*, *v*`]` and `minusu[`*u*, *v*`]` equivalents simply call the *g*-layer versions but also track the number of bits and numbers moved for purposes of comparing storage efficiency. The `CirclePlus` function defines the infix operator ⊕ to mean ubound addition, and the `CircleMinus` function defines the infix operator ⊖ to mean ubound subtraction, which helps make the unum mathematics in this book look more like traditional mathematics.

```
plusu[u_ /; uQ[u], v_ /; uboundQ[vb]] :=
  Module[{wb = g2u[plusg[u2g[ub], u2g[vb]]]}, ubitsmoved +=
     nbits[ub] + nbits[vb] + nbits[wb]; numbersmoved += 3; wb];

CirclePlus[ub_ /; uboundQ[ub], vb_ /; uboundQ[vb]] := plusu[ub, vb]
```

```
minusu[ub_, vb_] := Module[{wb = g2u[minusg[u2g[ub], u2g[vb]]]},
   ubitsmoved += nbits[ub] + nbits[vb] + nbits[wb]; numbersmoved += 3; wb]

CircleMinus[ub_, vb_] := minusu[ub, vb]
```

C.10 Multiplication functions

As usual, we start by stripping off the NaN cases, then the zero-infinity combinations that require special care, and then treat the case of endpoints for we can actually do arithmetic. We take the smallest of the candidates for the minimum and the largest of the candidates for the maximum, and if there are any ties between an open and a closed extremum, we use the more inclusive closed case. The `timesposleft` and `timesposright` helper functions assume $x \geq 0$ and $y \geq 0$, and reproduce the multiplication tables in Section 10.1. They assume `NaN` cases have already been dealt with. We have to exclude `NaN` because we want to take maxima and minima of the results.

```
(* The "left" multiplication table for general intervals. *)
timesposleft[{x_, xb_}, {y_, yb_}] := Which[
  {x, xb} == {0, closed},
  If[{y, yb} == {∞, closed}, {NaN, open}, {0, closed}],
  {y, yb} == {0, closed},
  If[{x, xb} == {∞, closed}, {NaN, open}, {0, closed}],
  {x, xb} == {0, open}, If[{y, yb} == {∞, closed}, {∞, closed}, {0, open}],
  {y, yb} == {0, open}, If[{x, xb} == {∞, closed}, {∞, closed}, {0, open}],
  {x, xb} == {∞, closed} ⋁ {y, yb} == {∞, closed}, {∞, closed},
  True, {x × y, xb ⋁ yb}]

(* The "right" multiplication table for general intervals. *)
timesposright[{x_, xb_}, {y_, yb_}] := Which[
  {x, xb} == {∞, closed},
  If[{y, yb} == {0, closed}, {NaN, open}, {∞, closed}],
  {y, yb} == {∞, closed},
  If[{x, xb} == {0, closed}, {NaN, open}, {∞, closed}],
  {x, xb} == {∞, open}, If[{y, yb} == {0, closed}, {0, closed}, {∞, open}],
  {y, yb} == {∞, open}, If[{x, xb} == {0, closed}, {0, closed}, {∞, open}],
  {x, xb} == {0, closed} ⋁ {y, yb} == {0, closed}, {0, closed},
  True, {x × y, xb ⋁ yb}]
```

The tiny function **neg** helps make the code that follows a little more readable. Notice, however, that **neg** turns a "left" result into a "right" result and vice versa.

```
(*Helper function for timesg; negates numerical part of endpoint.*)
neg[{x_, xb_}] := {-x, xb}
```

The following fixes a bug in *Mathematica*, which orders $-\infty$ as "greater" than any numeric value. Customer Service assures us this is a feature, not a bug.

```
unionfix[end1_, end2_] := Module[{temp = Union[end1, end2], e1, e2},
  If[Length[temp] == 2,
    {e1, e2} = temp;
    If[temp[[2,1]] === -∞, {e2, e1}, temp],
    temp]]
```

Here is the code to rigorously multiply two general intervals:

$$timesg\left[\begin{array}{cc} xlo\_ & xhi\_ \\ xlob\_ & xhib\_ \end{array}, \begin{array}{cc} ylo\_ & yhi\_ \\ ylob\_ & yhib\_ \end{array}\right] := Module\Big[$$

```
  {timesleft, timesright, openleft, openright, lcan = {}, rcan = {}},
  If[(* If any value is NaN, the result is also NaN. *)
    xlo === NaN \/ xhi === NaN \/ ylo === NaN \/ yhi === NaN, NaN  NaN
                                                              open open ,
    (* Lower left corner in upper right quadrant, facing uphill: *)
    If[xlo ≥ 0 /\ ylo ≥ 0,
      lcan = unionfix[lcan, {timesposleft[{xlo, xlob}, {ylo, ylob}]}]];
    (* Upper right corner in lower left quadrant, facing uphill: *)
    If[(xhi < 0 \/ (xhi == 0 /\ xhib)) /\ (yhi < 0 \/ (yhi == 0 /\ yhib)),
      lcan = unionfix[lcan, {timesposleft[{-xhi, xhib}, {-yhi, yhib}]}]];
    (* Upper left corner in upper left quadrant, facing uphill: *)
    If[(xlo < 0 \/ (xlo == 0 /\ ¬ xlob)) /\ (yhi > 0 \/ (yhi == 0 /\ ¬ yhib)), lcan =
      unionfix[lcan, {neg[timesposright[{-xlo, xlob}, {yhi, yhib}]]}]];
    (* Lower right corner in lower right quadrant, facing uphill: *)
    If[(xhi > 0 \/ (xhi == 0 /\ ¬ xhib)) /\ (ylo < 0 \/ (ylo == 0 /\ ¬ ylob)), lcan =
      unionfix[lcan, {neg[timesposright[{xhi, xhib}, {-ylo, ylob}]]}]];
    (* Upper right corner in upper right quadrant, facing downhill: *)
    If[(xhi > 0 \/ (xhi == 0 /\ ¬ xhib)) /\ (yhi > 0 \/ (yhi == 0 /\ ¬ yhib)),
      rcan = unionfix[rcan, {timesposright[{xhi, xhib}, {yhi, yhib}]}]];
    (* Lower left corner in lower left quadrant, facing downhill: *)
    If[(xlo < 0 \/ (xlo == 0 /\ ¬ xlob)) /\ (ylo < 0 \/ (ylo == 0 /\ ¬ ylob)), rcan =
      unionfix[rcan, {timesposright[{-xlo, xlob}, {-ylo, ylob}]}]];
    (* Lower right corner in upper left quadrant, facing downhill: *)
    If[(xhi < 0 \/ (xhi == 0 /\ xhib)) /\ ylo ≥ 0, rcan =
      unionfix[rcan, {neg[timesposleft[{-xhi, xhib}, {ylo, ylob}]]}]];
    (* Upper left corner in lower right quadrant, facing downhill: *)
    If[xlo ≥ 0 /\ (yhi < 0 \/ (yhi == 0 /\ yhib)), rcan =
      unionfix[rcan, {neg[timesposleft[{xlo, xlob}, {-yhi, yhib}]]}]];
```

```
If[MemberQ[Flatten[lcan], NaN] ⋁ MemberQ[Flatten[rcan], NaN],

   timesleft timesright   NaN  NaN ];
   openleft  openright  = open open

   timesleft timesright   lcan⟦1,1⟧ rcan⟦-1,1⟧
   openleft  openright  = lcan⟦1,2⟧ rcan⟦-1,2⟧ ;

   If[Length[lcan] > 1, If[
      lcan⟦1,1⟧ ⩵ lcan⟦2,1⟧ ⋀ (¬ lcan⟦1,2⟧ ⋁ ¬ lcan⟦2,2⟧), openleft = closed]];
   If[Length[rcan] > 1, If[rcan⟦-1,1⟧ ⩵ rcan⟦-2,1⟧ ⋀
      (¬ rcan⟦-1,2⟧ ⋁ ¬ rcan⟦-2,2⟧), openright = closed]];

   timesleft timesright ]]
   openleft  openright
```

The *u*-layer version is `timesu[u, v]`, which is assigned to the infix operator `CircleTimes`, ⊗. As with most other *u*-layer arithmetic functions, it tracks the number of data items moved and total bits moved.

```
(* Multiplication in the u-layer, with data motion tracking. *)
timesu[u_ /; uQ[u], v_ /; uQ[v]] :=
 Module[{w = g2u[timesg[u2g[u], u2g[v]]]},
  ubitsmoved += nbits[u] + nbits[v] + nbits[w]; numbersmoved += 3; w]

(* Assignment of u-layer multiplication to the "⊗" symbol. *)
CircleTimes[ub_, vb_] := timesu[ub, vb]
```

C.11 Division routines

The `divideposleft` and `divideposright` implement the tables in Section 10.2.

```
(* The "left" division table for general intervals. *)
divideposleft[{x_, xb_}, {y_, yb_}] := Which[
  {y, yb} ⩵ {0, closed}, {NaN, open},
  {x, xb} ⩵ {∞, closed},
  If[{y, yb} ⩵ {∞, closed}, {NaN, open}, {∞, closed}],
  {x, xb} ⩵ {0, closed} ⋁ {y, yb} ⩵ {∞, closed}, {0, closed},
  {x, xb} ⩵ {0, open} ⋁ {y, yb} ⩵ {∞, open}, {0, open},
  True, {x / y, xb ⋁ yb}]

(* The "right" division table for general intervals. *)
divideposright[{x_, xb_}, {y_, yb_}] := Which[
  {y, yb} ⩵ {0, closed}, {NaN, open},
  {x, xb} ⩵ {∞, closed},
  If[{y, yb} ⩵ {∞, closed}, {NaN, open}, {∞, closed}],
  {x, xb} ⩵ {0, closed} ⋁ {y, yb} ⩵ {∞, closed}, {0, closed},
  {x, xb} ⩵ {∞, open} ⋁ {y, yb} ⩵ {0, open}, {∞, open},
  True, {x / y, xb ⋁ yb}]
```

The `divideg` function uses the `divideposleft` and `divideposright` helper functions to cover the entire *x-y* plane, and also checks for NaN values in the inputs (next page):

```
(* Division in the g-layer. *)
divideg[ xlo_   xhi_  ,  ylo_   yhi_  ] :=
        [ xlob_  xhib_    ylob_  yhib_ ]

Module[{divleft, divright, openleft, openright, lcan = {}, rcan = {}},

  If[(* If any value is NaN,
    or denominator contains 0, the result is a NaN. *)
    xlo === NaN ∨ xhi === NaN ∨ ylo === NaN ∨
    yhi === NaN ∨ (ylo < 0 ∨ (ylo == 0 ∧ ¬ ylob)) ∧
                                              NaN   NaN
      (yhi > 0 ∨ (yhi == 0 ∧ ¬ yhib)),     open  open ,
    (* Upper left corner in upper right quadrant, facing uphill. *)
    If[xlo ≥  0 ∧ (yhi > 0 ∨ (yhi == 0 ∧ ¬ yhib)),
      lcan = unionfix[lcan, {divideposleft[{xlo, xlob}, {yhi, yhib}]}]];
    (* Lower right corner in lower left quadrant, facing uphill: *)
    If[(xhi < 0 ∨ (xhi == 0 ∧ xhib)) ∧ (ylo < 0 ∨ (ylo == 0 ∧ ¬ ylob)), lcan =
      unionfix[lcan, {divideposleft[{-xhi, xhib}, {-ylo, ylob}]}]];
    (* Lower left corner in upper left quadrant, facing uphill: *)
    If[(xlo < 0 ∨ (xlo == 0 ∧ ¬ xlob)) ∧ ylo ≥ 0, lcan = unionfix[lcan,
        {neg[divideposright[{-xlo, xlob}, {ylo, ylob}]]}]];
    (* Upper right corner in lower right quadrant, facing uphill: *)
    If[(xhi > 0 ∨ (xhi == 0 ∧ ¬ xhib)) ∧ (yhi < 0 ∨ (yhi == 0 ∧ yhib)),
      lcan = unionfix[lcan,
        {neg[divideposright[{xhi, xhib}, {-yhi, yhib}]]}]];
    (* Lower right corner in upper right quadrant,
    facing downhill: *)
    If[(xhi > 0 ∨ (xhi == 0 ∧ ¬ xhib)) ∧ ylo ≥ 0,
      rcan = unionfix[rcan, {divideposright[{xhi, xhib}, {ylo, ylob}]}]];
  (* Upper left corner in lower left quadrant, facing downhill: *)
    If[(xlo < 0 ∨ (xlo == 0 ∧ ¬ xlob)) ∧ (yhi < 0 ∨ (yhi == 0 ∧ yhib)), rcan =
      unionfix[rcan, {divideposright[{-xlo, xlob}, {-yhi, yhib}]}]];
    (* Upper right corner in upper left quadrant,
    facing downhill: *)
    If[(xhi < 0 ∨ (xhi == 0 ∧ xhib)) ∧ (yhi > 0 ∨ (yhi == 0 ∧ ¬ yhib)), rcan =
      unionfix[rcan, {neg[divideposleft[{-xhi, xhib}, {yhi, yhib}]]}]];
    (* Lower left corner in lower right quadrant,
    facing downhill: *)
    If[xlo ≥ 0 ∧ (ylo < 0 ∨ (ylo == 0 ∧ ¬ ylob)), rcan =
      unionfix[rcan, {neg[divideposleft[{xlo, xlob}, {-ylo, ylob}]]}]];
  If[MemberQ[Flatten[lcan], NaN] ∨ MemberQ[Flatten[rcan], NaN],

    divleft  divright  _  NaN   NaN  ];
  openleft openright  =  open  open

      divleft  divright  _  lcan[[1,1]]   rcan[[-1,1]] ;
    openleft openright  =  lcan[[1,2]]   rcan[[-1,2]]
    If[Length[lcan] > 1, If[lcan[[1,1]] == lcan[[2,1]] ∧ (¬ lcan[[1,2]] ∨ ¬ lcan[[2,2]]),
      openleft = closed]];
    If[Length[rcan] > 1, If[rcan[[-1,1]] == rcan[[-2,1]] ∧
        (¬ rcan[[-1,2]] ∨ ¬ rcan[[-2,2]]),
      openright = closed]];  divleft  divright  ]]
                            openleft openright
```

The `divideu[`*u*, *v*`]` function returns the ubound that best represents *u* divided by *v*, where *u* and *v* are unums or ubounds. It is assigned to the `CircleDot` infix operator, ⊙. The reason it was not assigned to something like a circle containing a "/" is that no such operator exists in *Mathematica*, for some reason.

```
divideu[u_ /; uQ[u], v_ /; uQ[v]] :=
  Module[{i = nbits[u], j = nbits[v], w = g2u[divideg[u2g[u], u2g[v]]]},
    ubitsmoved += nbits[u] + nbits[v] + nbits[w]; numbersmoved += 3; w]

CircleDot[u_, v_] := divideu[u, v]
```

C.12 Automatic precision adjustment functions

The values ±*smallsubnormal* and ±*maxreal* are like the "third rail" of unum arithmetic; touch them, and it means something is almost certainly in trouble. That is how `needmoreexpQ` detects the need for a larger dynamic range, so that we do not need the values meaning "almost nothing" and "almost infinite."

```
(* Check if we have hit the dynamic range limits. *)
needmoreexpQ[u_ /; uQ[u]] := Module[{g = u2g[u]},
    If[((g⟦1,2⟧ == -maxreal ⋁ g⟦1,1⟧ == -smallsubnormal) ⋀ g⟦2,1⟧) ⋁
      ((g⟦1,1⟧ == maxreal ⋁ g⟦1,2⟧ == smallsubnormal) ⋀ g⟦2,2⟧), True, False]];
```

One definition of the *relative width* of the range of a value is the one used here. There are others.

```
(* Find the relative width in a unum or ubound. *)
relwidth[u_ /; uQ[u]] :=
  Module[{lo = (u2g[u])⟦1,1⟧, hi = (u2g[u])⟦1,2⟧}, Which[
    lo === NaN, ∞,
    Abs[lo] == ∞ ⋁ Abs[hi] == ∞, 1,
    lo == 0 ⋀ hi == 0, 0,
    True, Abs[hi - lo] / (Abs[lo] + Abs[hi])]];
```

The relative width function is used to detect if more fraction bits are needed.

```
(* Test if a larger fraction field is needed. *)
needmorefracQ[u_ /; uQ[u]] :=
  If[relwidth[u] > relwidthtolerance, True, False]
```

C.13 Fused operations (single-use expressions)

There are six fused operations in this prototype: multiply-add, dot product, sum, product, add-multiply, and multiply-divide. One variant that seems worth adding would be one for operations of the form $(a \times b) - (c \times d)$, a "fused multiply-multiply subtract." It comes up in complex arithmetic and the evaluation of cross products, and would save a step over negating *c* or *d* and doing a fused dot product.

The fused multiply-add is already in the IEEE Standard:

```
(* Fused multiply-add in the g-layer. *)
fmag[ag_ /; gQ[ag], bg_ /; gQ[bg], cg_ /; gQ[cg]] :=
  plusg[timesg[ag, bg], cg]


(* Fused multiply-add in the u-layer, with data motion tracking. *)
fmau[au_ /; uQ[au], bu_ /; uQ[bu], cu_ /; uQ[cu]] :=
  Module[{w = g2u[fmag[u2g[au], u2g[bu], u2g[cu]]]}, ubitsmoved +=
    nbits[au] + nbits[bu] + nbits[cu] + nbits[w]; numbersmoved += 4; w]
```

The fused dot product is extremely helpful in controlling information loss when doing linear algebra, allowing dot products to be hundreds of times more accurate than floats when doing things like matrix multiply, linear system factorization, convolution, etc. This routine is unusual in the collection in that it actually prints out an error message if the input arguments are not of the correct type.

```
(* Fused dot product, entirely in the g-layer. *)
fdotg[ag_List, bg_List] :=
  Module[{i, legitQ = True, sumg =    0      0    }, Which[
                                   closed closed
    Length[ag] ≠ Length[bg], Style["Vectors must be of equal length.",
      "Text", Background → RGBColor[{1, .9, .7}]],
    For[i = 1, i ≤ Length[ag], i++,
      legitQ = legitQ ∧ (gQ[ag[[i]]] ∧ gQ[bg[[i]]])];
    ¬ legitQ, Style["Vector entries must be general intervals.",
      "Text", Background → RGBColor[{1, .9, .7}]],
    True, For[i = 1, i ≤ Length[ag], i++,
      sumg = plusg[sumg, timesg[ag[[i]], bg[[i]]]]]; sumg]]


(* Fused dot product in g-layer, with inputs and output in the
u-layer and with data motion tracking. *)
fdotu[au_List, bu_List] :=
  Module[{i, legitQ = True, sumg =    0      0   , w}, Which[
                                   closed closed
    Length[au] ≠ Length[bu], Style["Vectors must be of equal length.",
      "Text", Background → RGBColor[{1, .9, .7}]],
    For[i = 1, i ≤ Length[au], i++,
      legitQ = legitQ ∧ (uQ[au[[i]]] ∧ uQ[bu[[i]]])];
    ¬ legitQ, Style["Vector entries must be unums or ubounds.",
      "Text", Background → RGBColor[{1, .9, .7}]],
    True, For[i = 1, i ≤ Length[au], i++,
      ubitsmoved += (nbits[au[[i]]] + nbits[bu[[i]]]);
      sumg = plusg[sumg, timesg[u2g[au[[i]]], u2g[bu[[i]]]]]];
    w = g2u[sumg];
    ubitsmoved += nbits[w];
    numbersmoved += 2 * Length[au] + 1;
    w]]
```

Fused sums are a special case of the fused dot product where one vector consists of the number 1 for every entry. However, fused sums are important enough to justify their own routine.

In hardware, the fused sum would probably share much circuitry with the fused dot product, simply bypassing the multiplier.

```
(* Fused sum of a list, entirely in the g-layer. *)
fsumg[ag_List] :=
  Module[{i, legitQ = True, sumg =    0        0      }, Which[
                                   closed  closed
    For[i = 1, i ≤ Length[ag], i++,
      legitQ = legitQ ⋀ gQ[ag[[i]]]];
    ¬ legitQ, Style["List entries must be general intervals.",
      "Text", Background → RGBColor[{1, .9, .7}]],
    True, For[i = 1, i ≤ Length[ag], i++,
      sumg = plusg[sumg, ag[[i]]]]; sumg]]
```

```
(* Fused sum of a list, with inputs and output in the u-layer
   and with data motion tracking. *)
fsumu[au_List] :=
  Module[{i, legitQ = True, sumg =    0        0     , w}, Which[
                                   closed  closed
    For[i = 1, i ≤ Length[au], i++,
      legitQ = legitQ ⋀ uQ[au[[i]]]];
    ¬ legitQ, Style["List entries must be unums or ubounds.",
      "Text", Background → RGBColor[{1, .9, .7}]],
    True, For[i = 1, i ≤ Length[au], i++,
      ubitsmoved += nbits[au[[i]]];
      sumg = plusg[sumg, u2g[au[[i]]]]];
    w = g2u[sumg];
    ubitsmoved += nbits[w];
    numbersmoved += Length[au] + 1;
    w]]
```

The fused product is helpful even if doing only a pair of multiplications in a row, like $a \times b \times c$, not so much because it reduces ULP expansion but because it can often remove spurious intermediate calculations that go beyond *maxreal* or below *smallsubnormal* in magnitude. Often, the only reason for needing a lot of exponent bits with floats is because of the likelihood of underflow or overflow when doing a series of multiplies or divides, even when the final result is well within range.

```
(* Fused product of a list, entirely in the g-layer. *)
fprodg[ag_List] :=
  Module[{i, legitQ = True, prodg =  1      1   }, Which[
                                    closed closed
    For[i = 1, i ≤ Length[ag], i++,
      legitQ = legitQ ⋀ gQ[ag〚i〛]];
    ¬ legitQ, Style["List entries must be general intervals.",
     "Text", Background → RGBColor[{1, .9, .7}]],
    True, For[i = 1, i ≤ Length[ag], i++,

    prodg = timesg[prodg, ag〚i〛]]; prodg]]
```

```
(* Fused product of a list, with inputs and output in the u-
 layer and with tallying of bits and numbers moved. *)
fprodu[au_List] :=
  Module[{i, legitQ = True, prodg =  1      1  , w}, Which[
                                    closed closed
    For[i = 1, i ≤ Length[au], i++,
      legitQ = legitQ ⋀ uQ[au〚i〛]];
    ¬ legitQ, Style["List entries must be unums or ubounds.",
     "Text", Background → RGBColor[{1, .9, .7}]],
    True, For[i = 1, i ≤ Length[au], i++,
        ubitsmoved += nbits[au〚i〛];
        prodg = timesg[prodg, u2g[au〚i〛]]]
      w = g2u[prodg]];
  ubitsmoved += nbits[w];
  numbersmoved += Length[au] + 1;
  w]
```

Having a fused add-multiply is like having a guarantee that the distributive law works: $(a + b) \times c = (a \times c) + (b \times c)$. If the left-hand side is done with a fused add-multiply and the right-hand side is done as a fused dot product, the results will be identical in both the *g*-layer and the *u*-layer.

```
(* Fused add-multiply in the g-layer. *)
famg[ag_ /; gQ[ag], bg_ /; gQ[bg], cg_ /; gQ[cg]] :=
  timesg[plusg[ag, bg], cg]

(* Fused add-multiply in the u-layer,
with tallying of bits and numbers moved. *)
famu[au_ /; uQ[au], bu_ /; uQ[bu], cu_ /; uQ[cu]] :=
  Module[{w = g2u[famg[u2g[au], u2g[bu], u2g[cu]]]}, ubitsmoved +=
    nbits[au] + nbits[bu] + nbits[cu] + nbits[w]; numbersmoved += 4; w]
```

The last fused operation is a fused product ratio. The fixed-point example in Section 18.2 demonstrated the value of this, where a value near 77 was multiplied by something near 77 only to be divided by 77 as the next step. The intermediate result would have required an *esizesize* of three or more, but a fused product ratio allows tight computation with an *esizesize* of only two, thus saving many bits per value.

```
(* Fused product ratio, entirely in the g-layer. *)
fprodratiog[numeratorsg_List, denominatorsg_List] :=
 Module[{i, legitQ = True},
  For[i = 1, i ≤ Length[numeratorsg], i++,
   legitQ = legitQ ∧ gQ[numeratorsg[[i]]]];
  For[i = 1, i ≤ Length[denominatorsg], i++,
   legitQ = legitQ ∧ gQ[denominatorsg[[i]]]];
  If[¬ legitQ, Style["Vector entries must be general intervals.",
    "Text", Background → RGBColor[{1, .9, .7}]],
   divideg[fprodg[numeratorsg], fprodg[denominatorsg]]]]
```

```
(* Fused product ratio in the u-layer, with data motion tracking. *)
fprodratiou[numeratorsu_List, denominatorsu_List] :=
 Module[{i, legitQ = True,
   lengthden = Length[denominatorsu], lengthnum = Length[numeratorsu],
   proddeng =  1       1    , prodnumg =  1       1    , w},
           closed closed             closed closed
  Which[
   For[i = 1, i ≤ lengthden, i++,
    legitQ = legitQ ∧ uQ[denominatorsu[[i]]]];
   For[i = 1, i ≤ lengthnum, i++,
    legitQ = legitQ ∧ uQ[numeratorsu[[i]]]];
   ¬ legitQ ∨ lengthden < 1 ∨ lengthnum < 1,
   Style["Vector entries must be unums or ubounds",
    "Text", Background → RGBColor[{1, .9, .7}]],
   True,
   For[i = 1, i ≤ lengthden, i++,
    ubitsmoved += nbits[denominatorsu[[i]]];
    proddeng = timesg[proddeng, u2g[denominatorsu[[i]]]]];
   For[i = 1, i ≤ lengthnum, i++,
    ubitsmoved += nbits[numeratorsu[[i]]];
    prodnumg = timesg[prodnumg, u2g[numeratorsu[[i]]]]];
   w = g2u[divideg[prodnumg, proddeng]];
   ubitsmoved += nbits[w];
   numbersmoved += lengthden + lengthnum + 1;
   w]]
```

C.14 Square and square root

The **squareu[*ub*]** operation relies on a **squareg[*g*]** function that is careful to note if the value being squared spans zero, resulting in a closed left endpoint at zero.

```
(* Square in the g-layer. *)
squareg[g_ /; gQ[g]] :=
  Module[{g1 = g[[1,1]], g2 = g[[1,2]], b1 = g[[2,1]], b2 = g[[2,2]]}, Which[
    g1 === NaN ⋁ g2 === NaN, f2g[NaN],
    True, Module[{t1 = g1², t2 = g2², tset},
      tset = Sort[{{t1, b1}, {t2, b2}}];
      Which[(* See if 0 is in the range *)
        (g1 < 0 ⋀ g2 > 0) ⋁ (g1 > 0 ⋀ g2 < 0) ⋁ (g1 == 0 ⋀ ¬ b1) ⋁ (g2 == 0 ⋀ ¬ b2),
        If[t1 == t2, {{0, t1}, {closed, b1 ⋀ b2}},
          {{0, tset[[2,1]]}, {closed, tset[[2,2]]}}],
        True, Transpose[tset]]]]]

(* Square in the u-layer, with data motion tracking. *)
squareu[u_ /; uQ[u]] := Module[{i = nbits[u], v = g2u[squareg[u2g[u]]]},
  ubitsmoved += nbits[u] + nbits[v]; numbersmoved += 2; v]
```

The square root functions for general intervals, unums, and ubounds return NaN if the input contains NaN or is less than zero:

```
(* Square root in the g-layer. *)
sqrtg[g_ /; gQ[g]] := If[g[[1,1]] === NaN ⋁ g[[1,2]] === NaN ⋁ g[[1,1]] < 0,
  f2g[NaN], {{√g[[1,1]], √g[[1,2]]}, {g[[2,1]], g[[2,2]]}}]

(* Square root in the u-layer, with data motion tracking. *)
sqrtu[u_ /; uQ[u]] := Module[{i = nbits[u], v = g2u[sqrtg[u2g[u]]]},
  ubitsmoved += nbits[u] + nbits[v]; numbersmoved += 2; v]
```

C.15 The power function x^y and $\exp(x)$

Like multiplication, there are tables for left-facing and right-facing interval bounds, and a saddle-shaped surface to account for; unlike multiplication, the level point of the saddle occurs at $x = 1$, $y = 0$ instead of $x = 0$, $y = 0$, and the function does not have real-valued output for the three quadrants where x or y is negative, unless y happens to be an integer.

```
(* The "left" power function table. *)
powposleft[{x_ /; x ≥ 1, xb_}, {y_ /; y ≥ 0, yb_}] := Which[
   {x, xb} == {1, closed},
   If[{y, yb} == {∞, closed}, {NaN, open}, {1, closed}],
   {y, yb} == {0, closed},
   If[{x, xb} == {∞, closed}, {NaN, open}, {1, closed}],
   {x, xb} == {1, open}, If[{y, yb} == {∞, closed}, {∞, closed}, {1, open}],
   {y, yb} == {0, open}, If[{x, xb} == {∞, closed}, {∞, closed}, {1, open}],
   {x, xb} == {∞, closed} ∨ {y, yb} == {∞, closed}, {∞, closed},
   True, {x^y, xb ∨ yb}]
```

```
(* The "right" power function table. *)
powposright[{x_ /; x ≥ 1, xb_}, {y_ /; y ≥ 0, yb_}] := Which[
   {x, xb} == {∞, closed},
   If[{y, yb} == {0, closed}, {NaN, open}, {∞, closed}],
   {y, yb} == {∞, closed},
   If[{x, xb} == {1, closed}, {NaN, open}, {∞, closed}],
   {x, xb} == {∞, open}, If[{y, yb} == {0, closed}, {1, closed}, {∞, open}],
   {y, yb} == {∞, open}, If[{x, xb} == {1, closed}, {1, closed}, {∞, open}],
   {x, xb} == {1, closed} ∨ {y, yb} == {0, closed}, {1, closed},
   True, {x^y, xb ∨ yb}]
```

The **rec** helper function is like the **neg** function for multiplication and division; it reciprocates the *numerical* part of an endpoint but leaves the open-closed bit alone. In this case, we can define $\frac{1}{0} = \infty$ because the function excludes $-\infty$ as a possibility.

```
(* Reciprocal helper function for the power function. *)
rec[{x_, xb_}] := {Which[x === NaN, NaN, x == 0, ∞, True, 1 / x], xb}
```

The general interval version of the power function follows. This is probably the record-holder for exception case handling needed for correct results, and for the length of code needed to compute an algebraic function in the prototype. It requires more than two pages to list.

```
(powg[ xlo_  xhi_    ylo_  yhi_ ] :=
      xlob_ xhib_ ' ylob_ yhib_

  Module[{NaNg =  NaN  NaN , lcan = {}, openleft,
                  open open

    openright, powleft, powright, rcan = {}},

  Which[(* If any value is NaN, the result is also NaN. *)

    xlo === NaN \/ xhi === NaN \/ ylo === NaN \/ yhi === NaN, NaNg,
    (* Do not allow exact zero to a negative or zero power,
    unless the negative power is an exact even integer. *)
    (xlo < 0 \/ (xlo == 0 /\ ¬ xlob)) /\ (xhi > 0 \/ (xhi == 0 /\ ¬ xhib)) /\
     (ylo < 0 \/ (ylo == 0 /\ ¬ ylob)) /\
     ¬ (ylo == yhi /\ ylo < 0 /\ EvenQ[ylo]), NaNg,
    (* Weird case: complex number of zero
       magnitude is real. Zero. *)
    yhi == -∞ /\ ¬ yhib /\ ((xlo > 1 \/ (xlo == 1 /\ xlob)) \/
        (xhi < 1 \/ (xhi == 1 /\ xhib))),  0     0   ,
                                         closed closed '
    (* If y is an exact integer, loads of special cases. *)
    ylo == yhi /\ ¬ ylob /\ ¬ yhib /\ IntegerQ[ylo],
    (* Finite nonzero numbers to the power 0 equals 1. *)
    Which[ylo == 0,

      If[ (0 < xlo \/ (0 == xlo /\ xlob)) /\ (xhi < ∞ \/ (xhi == ∞ /\ xhib)) \/
          (-∞ < xlo \/ (-∞ == xlo /\ xlob)) /\ (xhi < 0 \/ (xhi == 0 /\ xhib)),
           1      1   , NaNg],
         closed closed
      (* Positive even power is like square function;
      test for zero straddle. *)
      EvenQ[ylo] /\ ylo > 0, Transpose[Which[
        (* Range is strictly
          negative. Order of endpoints reverses. *)
        xhi < 0 \/ (xhi == 0 /\ xhib),
        {{xhi^ylo, xhib}, {xlo^ylo, xlob}},
        (* Range is strictly
          positive. Endpoints preserve ordering. *)
        xlo > 0 \/ (xlo == 0 /\ xlob),
        {{xlo^ylo, xlob}, {xhi^ylo, xhib}},
        (* Range straddles zero. Closed
          zero is lower bound. Larger x^y is upper bound,
        but beware of ties between open and closed. *)
        True, Module[{t1 = xlo^ylo, t2 = xhi^ylo}, Which[
          t1 < t2, {{0, closed}, {t2, xhib}},
          t1 > t2, {{0, closed}, {t1, xlob}},
          True, {{0, closed}, {t1, xlob /\ xhib}}]]]],
```

```
(* Negative even power includes +∞ if zero straddle. *)
EvenQ[ylo] ⋀ ylo < 0, Transpose[Which[
  (* Range is strictly
     positive. Order of endpoints reverses. *)
  xlo > 0 ⋁ xlo == 0 ⋀ xlob,
  {{xhi^ylo, xhib}, {If[xlo == 0, ∞, xlo^ylo], xlob}},
  (* Range is strictly
     negative. Endpoints preserve ordering. *)
  xhi < 0 ⋁ xhi == 0 ⋀ xhib,
  {{xlo^ylo, xlob}, {If[xhi == 0, ∞, xhi^ylo], xhib}},
    (* Range straddles zero. Closed infinity is upper bound. smaller x^y
       is lower bound, but beware of ties between open and closed. *)
  True, Module[{t1 = If[xlo == 0 ⋀ ylo < 0, ∞, xlo^ylo],
    t2 = If[xhi == 0 ⋀ ylo < 0, ∞, xhi^ylo]}, Which[
    t1 > t2, {{t2, xhib}, {∞, closed}},
    t1 < t2, {{t1, xlob}, {∞, closed}},
    True, {{t1, xlob ⋀ xhib}, {∞, closed}}]
  ]]],
(* That leaves odd integer
 powers. Preserves ordering if positive. *)
ylo > 0,  xlo^ylo  xhi^ylo
          ─────    ───── ,
           xlob     xhib
(* Negative odd power. Reverses ordering. *)
True,  If[xhi == 0, -∞, xhi^ylo]  If[xlo == 0, ∞, xlo^ylo]
       ───────────────────────    ───────────────────── ,
               xhib                        xhib
xlo < 0, NaNg, (* Otherwise, negative x not allowed. *)
```

```
(* Non-integer exponent, and x is nonnegative. Find candidates. *)
  True,
  (* Lower left corner is in upper right quadrant,
  facing uphill: *)
  If[xlo ≥ 1 ⋀ ylo ≥ 0,
    lcan = Union[lcan, {powposleft[{xlo, xlob}, {ylo, ylob}]}]];
  (* Upper right corner is in lower left quadrant,
  facing uphill: *)
  If[(xhi < 1 ⋁ (xhi == 1 ⋀ xhib)) ⋀ (yhi < 0 ⋁ (yhi == 0 ⋀ yhib)),
    lcan = Union[lcan, {powposleft[rec[{xhi, xhib}], {-yhi, yhib}]}]];
  (* Upper left corner is in upper left quadrant, facing uphill: *)
  If[(xlo < 1 ⋁ (xlo == 1 ⋀ ¬ xlob)) ⋀ (yhi > 0 ⋁ (yhi == 0 ⋀ ¬ yhib)), lcan =
      Union[lcan, {rec[powposright[rec[{xlo, xlob}], {yhi, yhib}]]}]];
  (* Lower right corner is in lower right quadrant,
  facing uphill: *)
  If[(xhi > 1 ⋁ (xhi == 1 ⋀ ¬ xhib)) ⋀ (ylo < 0 ⋁ (ylo == 0 ⋀ ¬ ylob)),
    lcan = Union[lcan, {rec[powposright[{xhi, xhib}, {-ylo, ylob}]]}]];
  (* Upper right corner is in upper right quadrant,
  facing downhill: *)
  If[(xhi > 1 ⋁ (xhi == 1 ⋀ ¬ xhib)) ⋀ (yhi > 0 ⋁ (yhi == 0 ⋀ ¬ yhib)),
    rcan = Union[rcan, {powposright[{xhi, xhib}, {yhi, yhib}]}]];
  (* Lower left corner is in lower left quadrant,
  facing downhill: *)
  If[(xlo < 1 ⋁ (xlo == 1 ⋀ ¬ xlob)) ⋀ (ylo < 0 ⋁ (ylo == 0 ⋀ ¬ ylob)),
    rcan = Union[rcan, {powposright[rec[{xlo, xlob}], {-ylo, ylob}]}]];
  (* Lower right corner is in upper left quadrant,
  facing downhill: *)
  If[(xhi < 1 ⋁ (xhi == 1 ⋀ xhib)) ⋀ ylo ≥ 0, rcan =
      Union[rcan, {rec[powposleft[rec[{xhi, xhib}], {ylo, ylob}]]}]];
  (* Upper left corner is in lower right quadrant,
  facing downhill: *)
  If[xlo ≥ 1 ⋀ (yhi < 0 ⋁ (yhi == 0 ⋀ yhib)),
    rcan = Union[rcan, {rec[powposleft[{xlo, xlob}, {-yhi, yhib}]]}]];
```

$$\frac{powleft \quad powright}{openleft \quad openright} = \frac{lcan_{[\![1,1]\!]} \quad rcan_{[\![-1,1]\!]}}{lcan_{[\![1,2]\!]} \quad rcan_{[\![-1,2]\!]}};$$

```
  If[Length[lcan] > 1, If[
      lcan_{[1,1]} == lcan_{[2,1]} ⋀ (¬ lcan_{[1,2]} ⋁ ¬ lcan_{[2,2]}), openleft = closed]];
  If[Length[rcan] > 1, If[rcan_{[-1,1]} == rcan_{[-2,1]} ⋀
        (¬ rcan_{[-1,2]} ⋁ ¬ rcan_{[-2,2]}), openright = closed]];
```

$$\left.\frac{powleft \quad powright}{openleft \quad openright}\right]\!\!]$$

The *u*-layer version of the power function also tallies bits moved and numbers moved:

```
(* Power function in the u-layer, with data motion tracking. *)
powu[u_ /; uQ[u], v_ /; uQ[v]] := Module[{w = g2u[powg[u2g[u], u2g[v]]]},
  ubitsmoved += nbits[u] + nbits[v] + nbits[w]; numbersmoved += 3; w]
```

The exponential function **expg[g]** computes e^g where e is the base of the natural logarithm, $2.718281828\cdots$ and g is a general interval.

```
(* Exponential function in the g-layer. *)
expg[g_] := Module[{g1 = g[[1,1]], g2 = g[[1,2]], b1 = g[[2,1]], b2 = g[[2,2]],
    lo = Log[smallsubnormal], hi = Log[posbig], glo, ghi},
  glo = Which[
    g1 === NaN \/ g2 === NaN,  NaN
                               open ,
    g1 == -∞ /\ ¬ b1,  0
                       closed ,
    g1 == 0 /\ ¬ b1,  1
                      closed ,  g1 == ∞ /\ ¬ b1,  ∞
                                                  closed ,
    g1 < lo,  0
              open ,  g1 > hi,  posbig
                                open ,  True,  Exp[g1]
                                               open  ];
  ghi = Which[
    g1 === NaN \/ g2 === NaN,  NaN
                               open ,
    g2 == -∞ /\ ¬ b2,  0
                       closed ,
    g2 == 0 /\ ¬ b2,  1
                      closed ,  g2 == ∞ /\ ¬ b2,  ∞
                                                  closed ,
    g2 < lo,  smallsubnormal
              open          ,  g2 > hi,  ∞
                                         open ,  True,  Exp[g2]
                                                        open  ];
  If[g1 === g2, f2g[Exp[g1]],  glo[[1,1]]  ghi[[1,1]]
                               glo[[2,1]]  ghi[[2,1]] ]]
```

The corresponding ubound function is **expu[u]** where u is a unum or ubound; it tracks the data motion.

```
(* Exponential function in the u-layer; tracks data motion. *)
expu[u_] := Module[{i = nbits[ub], vb = g2u[expgint[ u2g[ub]]]},
  ubitsmoved += nbits[ub] + nbits[vb]; numbersmoved += 2; vb]
```

C.16 Absolute value, logarithm, and trigonometry functions

Unary operations are those that take one input and produce one output. The unary operations needed for examples in the text include logarithm, absolute value, and some trigonometric functions.

The logic for absolute value is surprisingly complicated since it must account for the endpoint ordering if the input spans zero, a little like what happens when squaring a general interval or ubound. The g-layer version of absolute value is **absg[g]**:

```
(* Absolute value in the g-layer. *)
absg[g_ /; gQ[g]] :=
  Module[{g1 = g[[1,1]], g2 = g[[1,2]], b1 = g[[2,1]], b2 = g[[2,2]]}, Which[
    g1 === NaN ⋁ g2 === NaN, f2g[NaN],
    g2 ≤ 0,   Abs[g2]  Abs[g1]
               b2       b1      ,
    g1 ≤ 0, Which[
                              0      Abs[g2]
      Abs[g1] < Abs[g2],   closed     b2     ,
                              0      Abs[g1]
      Abs[g1] > Abs[g2],   closed     b1     ,
                      0      Abs[g2]
      True,        closed   (b1 ⋀ b2) ],
             Abs[g1]  Abs[g2]
      True,    b1       b2      ]]
```

The ubound version is **absu[ub]**, which tracks data motion. As explained in Part 1, it may seem odd that there is any data motion associated with the absolute value of a number since with conventional floats only one bit is affected. With unums and ubounds, there is a very real possibility that the absolute value interchanges the left and right endpoints, which clearly requires moving the entire unum representation.

```
(* Absolute value in the u-layer. Tracks data motion. *)
absu[u_ /; uQ[u]] := Module[{i = nbits[u], v = g2u[absg[ u2g[u]]]},
  ubitsmoved += nbits[u] + nbits[v]; numbersmoved += 2; v]
```

The logarithm functions return NaN if the argument includes any negative values or is a NaN. The *g*-layer version is `logg[g]`.

```
(* Natural logarithm in the g-layer. *)
logg[g_ /; gQ[g]] :=
  Module[{g1 = g[[1,1]], g2 = g[[1,2]], b1 = g[[2,1]], b2 = g[[2,2]], glo, ghi}, Which[
```

$$g1 === NaN \lor g2 === NaN, \quad \begin{matrix} NaN & NaN \\ open & open \end{matrix},$$

$$g1 < 0 \lor (g1 == 0 \land b1), \quad \begin{matrix} NaN & NaN \\ open & open \end{matrix},$$

```
    True,
```

```
      glo = Which[
```

$$g1 == 0 \land \neg b1, \quad \begin{matrix} -\infty \\ closed \end{matrix}, \quad g1 == 1 \land \neg b1, \quad \begin{matrix} 0 \\ closed \end{matrix}, \quad g1 == \infty \land \neg b1, \quad \begin{matrix} \infty \\ closed \end{matrix},$$

$$True, \quad \begin{matrix} Log[g1] \\ open \end{matrix}];$$

```
      ghi = Which[
```

$$g2 == 0 \land \neg b2, \quad \begin{matrix} -\infty \\ closed \end{matrix}, \quad g2 == 1 \land \neg b2, \quad \begin{matrix} 0 \\ closed \end{matrix}, \quad g2 == \infty \land \neg b2, \quad \begin{matrix} \infty \\ closed \end{matrix},$$

$$True, \quad \begin{matrix} Log[g2] \\ open \end{matrix}];$$

$$If[g1 === g2, f2g[Log[g1]], \begin{matrix} glo_{[[1,1]]} & ghi_{[[1,1]]} \\ glo_{[[2,1]]} & ghi_{[[2,1]]} \end{matrix}]]]$$

The *u*-layer version is `logu[u]`. It tracks data motion, like most *u*-layer operations.

```
(* Natural logarithm in the u-layer. Tracks data motion. *)
logu[u_ /; uQ[u]] := Module[{i = nbits[u], v = g2u[loggint[ u2g[u]]]},
    ubitsmoved += nbits[u] + nbits[v]; numbersmoved += 2; v];
```

The prototype has the usual circular trigonometric functions other than secant and cosecant; it does not have the hyperbolic trig functions. The prototype does not provide inverse trig functions like arc sine, arc tangent, etc., though it is possible to use the try-everything solution method to obtain their values.

Remember that the trigonometric functions treat their arguments as being in degrees, not radians. This allows exact answers for special angles, like $\cos(60°) = \frac{1}{2}$ exactly. The routine leans heavily on the built-in transcendental functions of *Mathematica*, which can manipulate "π" as a symbol and not just as a value.

Each of the following trig functions is given as a *g*-layer and a *u*-layer version, where the *u*-layer tracks bits moved and numbers moved.

```
(* Cosine in the u-layer. *)
    cosg[g_ /; gQ[g]] :=
        Module[{g1 = g[[1,1]], g2 = g[[1,2]], b1 = g[[2,1]], b2 = g[[2,2]]},

        Which[

            g1 === NaN || g2 === NaN, {{NaN, NaN}, {open, open}},
            g1 === ∞ ∨ g2 === ∞ ∨ g1 === -∞ ∨ g2 === -∞,
            {{-1, 1}, {closed, closed}},
            g2 - g1 > 360, {{-1, 1}, {closed, closed}},
            (* Translate g1 and g2 so g1 is in [0,360). This assures
             tg2 is in [0,720], because of the previous If test. *)
            {g1, g2} -= ⌊ g1/360 ⌋ 360;

            g1 < 180,
            Which[g2 < 180 ∨ (g2 == 180 ∧ b2),
              (* Cos is monotone decreasing on (0°,180°) *)
              {{Cos[g2 °], Cos[g1 °]}, {b2, b1}},
              g2 < 360 ∨ (g2 == 360 ∧ b2),
              (* Angle spans 180 degrees; break tie for the high bound *)
              Which[Cos[g1 °] < Cos[g2 °], {{-1, Cos[g2 °]}, {closed, b2}},
                Cos[g1 °] > Cos[g2 °], {{-1, Cos[g1 °]}, {closed, b1}},
                True, {{-1, Cos[g1 °]}, {closed, b1 ∧ b2}}],
              True, {{-1, 1}, {closed, closed}}],
            (* g1 is in interval [180°,360°): *)
            True,
            (* Cosine is monotone increasing on [180°,360°): *)
            Which[g2 < 360 ∨ (g2 == 360 ∧ b2),
              {{Cos[g1 °], Cos[g2 °]}, {b1, b2}},
              True,
              ( * Angle spans 360 degrees; break the tie for the low bound *)
          Which[Cos[g1 °] > Cos[g2 °], {{Cos[g2 °], 1}, {b2, closed}},
            Cos[g1 °] < Cos[g2 °], {{Cos[g1 °], 1}, {b1, closed}},
            True, {{Cos[g1 °], 1}, {b1 ∧ b2, closed}}],

          True, {{-1, 1}, {closed, closed}}]

    ]];
```

```
(* Cosine in the u-layer, with data motion tracking. *)
cosu[u_ /; uQ[u]] := Module[{i = nbits[u], v = g2u[cosg[ u2g[u]]]},
    ubitsmoved += nbits[u] + nbits[v]; numbersmoved += 2; v];
```

The sine is derived from the cosine simply by using the complement identity $\sin(x) = \cos(90° - x)$. That preserves exactness, as long as the unum environment can represent the integers representing special angles in degrees.

```
(* Sine in the g-layer. *)
sing[g_ /; gQ[g]] := cosg[{{90 - g[[1,2]], 90 - g[[1,1]]}, {g[[2,2]], g[[2,1]]}}];
```

```
(* Sine in the u-layer, with data motion tracking. *)
sinu[u_ /; uQ[u]] := Module[{v = g2u[sing[ u2g[u]]]},
    ubitsmoved += nbits[u] + nbits[v]; numbersmoved += 2; v];
```

The cotangent is conceptually easier to write than the tangent since it repeats from 0 to 180° with a singularity at each endpoint of that period.

```
(* Cotangent in the g-layer. *)
cotg[g_ /; gQ[g]] :=
  Module[{g1 = g[[1,1]], g2 = g[[1,2]], b1 = g[[2,1]], b2 = g[[2,2]]},

    If[g1 === NaN || g2 === NaN, {{NaN, NaN}, {open, open}},

     If[Abs[g1] == ∞ ⋁ Abs[g2] == ∞, {{NaN, NaN}, {open, open}},

      If[g2 - g1 > 180, {{NaN, NaN}, {open, open}},

       (* Translate g1 and g2 so g1 is in [0,180). This also assures
          that g2 is in [0,360], because of the previous If test. *)
       {g1, g2} -= ⌊g1/180⌋ 180;

       Which[
         g1 == 0 ⋀ ¬ b1, {{NaN, NaN}, {open, open}},
         (* Cot[0°] is treated like 1/0 *)
         g1 == 0 ⋀ b1, Which[
           g2 < 180, {{Cot[g2°], ∞}, {b2, open}},
           g2 == 180 ⋀ b2, {{-∞, ∞}, {open, open}},
           True, {{NaN, NaN}, {open, open}}],
         g1 < 180, Which[
           g2 < 180, {{Cot[g2°], Cot[g1°]}, {b2, b1}},
           g2 == 180 ⋀ b2, {{-∞, Cot[g1°]}, {open, b1}},
           True, {{NaN, NaN}, {open, open}}],
         True, {{NaN, NaN}, {open, open}}]]]]];

(* Cotangent in the u-layer, tracks data motion. *)
cotu[u_ /; uQ[u]] := Module[{v = g2u[cotg[ u2g[u]]]},
  ubitsmoved += nbits[u] + nbits[v]; numbersmoved += 2; v];
```

Based on the cotangent routine, the tangent routine again uses a complementary angle identity: $\tan(x) = \cot(90° − x)$:

```
(* Tangent in the g-layer. *)
tang[g_ /; gQ[g]] := cotg[{{90 - g[[1,2]], 90 - g[[1,1]]}, {g[[2,2]], g[[2,1]]}}];

(* Cotangent in the u-layer; tracks data motion. *)
tanu[u_ /; uQ[u]] := Module[{v = g2u[tang[ u2g[u]]]},
  ubitsmoved += nbits[u] + nbits[v]; numbersmoved += 2; v];
```

This is clearly only a partial set of elementary functions for scientific and engineering applications. The above suffice as examples from which to code a more complete library, and they suffice to run all the examples in the text.

C.17 The unum Fast Fourier Transform

The unum version of the Fast Fourier Transform (FFT) looks exactly like the conventional one shown in Section 14.5, except that here we have not defined complex arithmetic operations with unums. So whereas the conventional form can write something like `gg*twiddle` where `gg` and `twiddle` are complex numbers, we have to expand the "*" to the operations on the real and imaginary parts of `gg` and `twiddle`. This routine could be greatly improved by application of some of the fused operations.

```
(* Conventional power-of-
  two Fast Fourier Transform performed with unum arithmetic. *)
cfftu[rru_List, n_Integer, iflg_Integer] :=
 Module[{ggu = rru, k = n / 2,
   th = If[iflg ≥ 0, -π, π], twiddleu, wwu, tru, tiu, i, j},
  While[k ≥ 1,

   wwu = {{-2 (Sin[tĥ / (2 k)])²}, { Sin[tĥ / k]}};

   twiddleu = {{1̂}, {0̂}};

   For[j = 0, j < k, j++,
    For[i = 1, i ≤ n, i += 2 k,
     tru = (ggu⟦i+j,1⟧ ⊖ ggu⟦i+j+k,1⟧);
     tiu = (ggu⟦i+j,2⟧ ⊖ ggu⟦i+j+k,2⟧);
     ggu⟦i+j,1⟧ = ggu⟦i+j,1⟧ ⊕ ggu⟦i+j+k,1⟧;
     ggu⟦i+j,2⟧ = ggu⟦i+j,2⟧ ⊕ ggu⟦i+j+k,2⟧;
     ggu⟦i+j+k,1⟧ = (twiddleu⟦1⟧ ⊗ tru) ⊖ (twiddleu⟦2⟧ ⊗ tiu);
     ggu⟦i+j+k,2⟧ = (twiddleu⟦1⟧ ⊗ tiu) ⊕ (twiddleu⟦2⟧ ⊗ tru)
     ];
    tru = ((twiddleu⟦1⟧ ⊗ wwu⟦1⟧) ⊖ (twiddleu⟦2⟧ ⊗ wwu⟦2⟧)) ⊕ twiddleu⟦1⟧;
    twiddleu⟦2⟧ =
     ((twiddleu⟦1⟧ ⊗ wwu⟦2⟧) ⊕ (twiddleu⟦2⟧ ⊗ wwu⟦1⟧)) ⊕ twiddleu⟦2⟧;
    twiddleu⟦1⟧ = tru];
   k = k / 2];
  For[i = j = 0, i < n - 1, i++,
   If[i < j, {
     tru = ggu⟦j+1,1⟧; ggu⟦j+1,1⟧ = ggu⟦i+1,1⟧; ggu⟦i+1,1⟧ = tru;
     tiu = ggu⟦j+1,2⟧; ggu⟦j+1,2⟧ = ggu⟦i+1,2⟧; ggu⟦i+1,2⟧ = tiu;}];
   k = n / 2; While[k ≤ j, {j = j - k; k = k / 2}]; j = j + k];
  ggu]
```

Appendix D Algorithm listings for Part 2

D.1 ULP manipulation functions

To manipulate multidimensional ubounds (that is, uboxes), we need support functions that promote and demote the fraction size and exponent size. These must be used with care since they affect the ULP size, but they are useful when we can determine which end of the ULP-wide interval moves and which end stays put. Since the first two routines require *exact* unums as input, nothing will happen to the value they represent; they simply decompress the fraction field or exponent field by one bit. The reason it is more complicated to promote the exponent field is because subnormals behave differently from normals, and there is also the possibility that exponent promotion causes a subnormal to be expressible as a normal.

```
(* Append 0 bit to the fraction of an exact unum if possible. *)
promotef[u_ /; unumQ[u] ⋀ exQ[u]] := If[fsize[u] < fsizemax,
  2 BitAnd[floatmask[u], u] + BitAnd[utagmask, u] + 1, u]
```

```
(* Increase length of exponent of an exact unum, if possible. *)
promotee[u_ /; unumQ[u] ⋀ exQ[u]] :=
  Module[{e = expo[u], es = esize[u], f = frac[u],
    fs = fsize[u], s = BitAnd[signmask[u], u],
    ut = BitAnd[utagmask, u] + fsizemax, nsigbits},
   Which[
    (* If already maximum exponent size,
    do nothing. This also handles NaN and ∞ values. *)
    es == esizemax, u,
    (* Take care of u=0 case,
    ignoring the sign bit. It's simply the new utag. *)
    e == 0 ⋀ f == 0, ut,
    (* If normal (nonzero exponent),
    slide sign bit left, add 2^{es-1}, increment esize. *)
    e > 0, 2 s + (e + 2^{es-1}) hiddenmask[u] +
     BitAnd[hiddenmask[u] - 1, u] + fsizemax,
    (* Subnormal. Room to shift and stay subnormal? *)
    fs - (⌊Log[2, f]⌋ + 1) ≥ 2^{es-1}, 2 s + frac[u] 2^{2^{es-1}} ulpu + ut,
    (* Subnormal becomes normal. Trickiest case. *)
    (* The fraction slides left such
    that the leftmost 1 becomes the hidden bit *)
    True, nsigbits = ⌊Log[2, f]⌋ + 1;
    2 s + (2^{es-1} + 1 - fs + nsigbits) hiddenmask[u] +
     (f - 2^{nsigbits}) 2^{fs-nsigbits+1} ulpu + BitAnd[utagmask, u] + fsizemax]]
```

With floats, addition and subtraction are done by first lining up the binary point. An analogous operation is useful with unums, where two exact unums with different utag contents are promoted such that they both have the same exponent size and fraction size, using the previous two routines. That makes it easy to compare bit strings.

```
(* Promote a pair of exact unums to the same esize and fsize. *)
promote[{u_, v_} /; unumQ[u] ⋀ unumQ[v] ⋀ exQ[u] ⋀ exQ[v]] :=
  Module[{eu = esize[u], ev = esize[v], fu = fsize[u], fv = fsize[v],
    ut = u, vt = v}, While[eu < ev, {ut = promotee[ut]; eu += 1}];
  While[ev < eu, {vt = promotee[vt]; ev += 1}];
  While[fu < fv, {ut = promotef[ut]; fu += 1}];
  While[fv < fu, {vt = promotef[vt]; fv += 1}];
  {ut, vt}]
```

Some ubox routines need to intentionally coarsen a result. Demoting (reducing the length of) the bit fields for fraction or exponent can easily turn an exact unum into an inexact one, but it will remain a valid bound for the original value. The more complex field demotion routine is the one for the exponent, **demotee**:

```
(* Demote exponent of a unum if possible. *)
demotee[u_ /; unumQ[u]] :=
  Module[{es = esize[u], mask = signmask[u] / 4, fm = floatmask[u],
    ut = BitAnd[u, utagmask], left2, ibit,
    s = BitAnd[signmask[u], u], e, f = frac[u]},
  left2 = BitAnd[u, 3 mask] / mask;
  Which[(* Cannot make the exponent any smaller: *)
    es == 1 ⋁ u == posinfu ⋁ u == neginfu ⋁ u == qNaNu ⋁ u == sNaNu, u,
    (* Subnormal; decreasing es means shifting frac by 2^(2^(es-2)) bits. *)
    expo[u] == 0,
    f = f / 2^(2^(es-2));
    ibit = If[FractionalPart[f] > 0, ubitmask, 0];
    BitOr[ibit, s / 2 + IntegerPart[f] ulpu + ut - fsizemax],
    (* If left two exponent bits are 00, result switches to subnormal.
       Exponent after the first two bits joins the fraction like a
       fixed-point number before shifting frac to the right. *)
    left2 == 2^^00,
    f = (2^(fsize[u]) + f) / 2^(-expo[u]+2^(es-2)+1);
    ibit = If[FractionalPart[f] > 0, ubitmask, 0];
    BitOr[ibit, s / 2 + IntegerPart[f] ulpu + ut - fsizemax],
    (* If left two exponent bits are 01 or 10, squeeze out second bit;
       if result is subnormal, shift hidden bit fraction bits right *)
    left2 ≤ 2^^10,
    e = BitAnd[expomask[u] - 3 mask, u] + BitAnd[u, 2 mask] / 2;
    If[e == 0, f = (2^(fsize[u]) + f) / 2];
    ibit = If[FractionalPart[f] > 0, ubitmask, 0];
    BitOr[ibit, s / 2 + e + IntegerPart[f] ulpu + ut - fsizemax],
    (* If the first two exponent bits are 11,
       always get an unbounded unum, all 1s for fraction: *)
    True, BitOr[(BitAnd[u, signmask[u]] + (fm - signmask[u])) / 2, ut] -
      fsizemax]]
```

By comparison, fraction demotion (`demotef`) is very simple:

```
(* Demote the fraction of a unum if possible,
even if it makes it inexact. *)
demotef[u_ /; unumQ[u]] := If[(*Cannot make the fraction any smaller*)
  fsize[u] == 1 \/ u == posinfu \/ u == neginfu \/ u == qNaNu \/ u == sNaNu,
  u, (*Else shift fraction right one bit.*)
  BitOr[BitAnd[u, floatmask[u]] / 2, BitAnd[utagmask, u] - 1]]
```

D.2 Neighbor-finding functions

Before finding neighbors on either side of a unum, we need three support functions
The **ulphi** function finds the width of the ULP just to its right of an exact unum,
taking into account its sign:

```
(* Finds size of ULP neighboring an exact unum. *)
ulphi[u_ /; unumQ[u] /\ exQ[u]] := Module[{s = If[ltuQ[{u}, {ô}], -1, 1]},

  u2g[u + s ubitmask][[1,2]] - u2g[u + s ubitmask][[1,1]]]
```

The **favore** and **favorf** helper functions prevent confusion stemming from unums
that have more than one legitimate (shortest bit string) form.

```
(* Look for alternative unum string that favors the exponent. *)
favore[unum_ /; unumQ[unum]] := Module[{u = demotef[promotee[unum]]},
  If[inexQ[u] \/ esize[u] == esizemax \/ fsize[u] == 1 \/ inexQ[u], u,
  While[fsize[u] > 1 /\ exQ[demotef[u]], u = demotef[u]]; u]]

(* Look for alternative unum string that favors the fraction. *)
favorf[unum_ /; unumQ[unum]] := Module[{u = demotee[padu[unum]]},
  If[inexQ[u], u,
  While[fsize[u] > 1 /\ exQ[demotef[u]], u = demotef[u]]; u]]
```

The **nborhi** function is surprisingly complicated, mainly because it allows a preferred
minimum ULP size. If it were not for that, an unwanted level of detail might ensue,
say, when crossing zero. In future versions, it should be possible to express
preferred ULP sizes as relative widths, not just absolute widths. The following
version suffices for the examples in this text. (Continues on next page.)

```
(* Find the right neighbor of a unum. *)
nborhi[u_ /; unumQ[u], minpower_] :=
  Module[{up, vp, s, t, ulpminu, ulpmin, overflow = False,

    ut = If[u == BitAnd[utagmask + signmask[u], u] /\ exQ[u], ô, u]},
    s = (-1)^sign[ut];
    ulpminu = Which[

      minpower < Log[2, smallsubnormal], smallsubnormalu,

      minpower > Log[2, maxreal], overflow = True; 2^⌊Log[2,maxreal]⌋,
```

```
True, 2^minpower]; ulpmin = u2g[ulpminu][[1,1]];
Which[

  u == posinfu ∨ u == sNaNu ∨ u == qNaNu,
  qNaNu, (* Values without nbors on the high side *)
  u == neginfu, If[

    overflow, If[

      utagmask == 1, -2 + ubitmask, (* Warlpiri environment *)
      -3 + ubitmask],

    negbigu + ubitmask], (* If -∞,
  use the (-∞,x) unum with the most negative x,
  unless the requested minpower caused overflow. *)
  inexQ[u], u2g[û][[1,2]], (* If inexact,
  always use the exact upper value *)
  overflow ⋀ u == 2̂ ⋀ utagmask == 1, 2̂ + ubitmask, (* Warlpiri "many" *)
overflow ⋀ u == 3̂ ⋀ utagmask ≠ 1, 3̂ + ubitmask,
  (* Also OK to use (3,∞) *) (* OK to overflow to
  ∞ if the minpower overflowed the environment. *)
  True, (* Reduce ULP until it equals ulpmin,
  or we run out of exponent and fraction bits *)
  t = u2g[ut][[1,1]]; ut = t̂;
  While[¬ IntegerQ[t / ulpmin], ulpmin /= 2];

  While[ulphi[ut] < ulpmin ⋀ ut ≠ favorf[ut], ut = favorf[ut]];

  While[esize[ut] < esizemax ⋀ ulphi[promotee[ut]] ≥ ulpmin,
   ut = promotee[ut]];
  While[fsize[ut] < fsizemax ⋀ ulphi[promotef[ut]] ≥ ulpmin,
   ut = promotef[ut]];

  ut + s ubitmask]]
```

Fortunately, the only thing that has to be done to create the counterpart function **nborlo** is to negate the argument, use **nborhi**, and negate again. The real number line is conveniently symmetrical. Unfortunately, "negative zero" introduces a tiny asymmetry in the logic, so that exception is handled here:

```
(* Find the left neighbor of a unum. *)
nborlo[u_ /; unumQ[u], minpower_] :=
  (* Watch out for negative zero, otherwise use nborhi *)
  If[sameuQ[{û}, {0̂}] ⋀ minpower < Log[2, smallsubnormal],
    smallsubnormalu + signbigu - ubitmask,
    negateu[{nborhi[negateu[{u}][[1]], minpower]}][[1]]]
```

The `findnbors` routine works in any number of dimensions to find *all* the nearest neighbor unums in the dimensions of a ubox *set*, using preferred minimum ULP sizes customized to each dimension from the `minpowers` list:

```
(* Find the neighbors, in all dimensions, of a unum. *)
findnbors[set_, minpowers_] := Module[{i, nbors, s, tempset = {}},
  For[i = 1, i ≤ Length[set], i++,
    s = set⟦i⟧;
    tempset = Union[tempset, Tuples[Table[{nborlo[s⟦j⟧, minpowers⟦j⟧],
        s⟦j⟧, nborhi[s⟦j⟧, minpowers⟦j⟧]}, {j, 1, Length[s]}]]]];
  Complement[tempset, set]]
```

D.3 Ubox creation by splitting

The easiest way to create a list of one-dimensional uboxes that tile a range is to use `uboxlist`, which accepts a ubound and a preferred minimum ULP width expressed as an integer, the log base 2 of the preferred width. They will be larger than this request if the minimum ULP size is larger than the request, and they will be smaller near any endpoint that does not fall on an even boundary of $2^{minpower}$.

```
(* Create a list of 1-dimensional uboxes from a ubound. *)
uboxlist[ub_ /; uboundQ[ub], minpower_Integer] :=
 Module[{utest = If[ub⟦1⟧ === neginfu,
     neginfu, nborhi[nborlo[ub⟦1⟧, minpower], minpower]],
   list1 = {}, list2 = {}, infinitybits, nbortemp},
  Which[ub⟦1⟧ === qNaNu ⋁ ub⟦-1⟧ === qNaNu ⋁
    ub⟦1⟧ === sNaNu ⋁ ub⟦-1⟧ === sNaNu, {qNaNu},
   sameuQ[ub, {neginfu}], {neginfu},
   sameuQ[ub, {posinfu}], {posinfu},
   True, (* Go up until we exceed the ubound range. *)
   infinitybits = ubitmask + expomask[utest] + fracmask[utest];
   nbortemp = nborhi[utest, minpower];
   If[BitAnd[infinitybits, utest] == infinitybits ⋀ esize[utest] <
       esizemax ⋀ nequQ[intersectu[ub, {nbortemp}], {nbortemp}],
    utest = promotee[exact[utest]] + If[inexQ[utest], ubitmask, 0]];
   While[sameuQ[intersectu[ub, {utest}], {utest}] ⋀ (utest =!= posinfu),
    list1 = AppendTo[list1, utest]; utest = nborhi[utest, minpower];
    If[BitAnd[infinitybits, utest] == infinitybits ⋀
        esize[utest] < esizemax ⋀ nequQ[intersectu[ub, {utest}], {utest}],
     utest = promotee[exact[utest]] + If[inexQ[utest], ubitmask, 0]]];
   (* Start from right, go left until utest equals last of list1. *)
   utest = If[ub⟦-1⟧ === posinfu,
     posinfu, nborlo[nborhi[ub⟦-1⟧, minpower], minpower]];
   While[gtuQ[{utest}, If[Length[list1] == 0,
       { (u2g[ûb])⟦1,1⟧}, {list1⟦-1⟧}]] ⋀ utest ≠ neginfu,
    PrependTo[list2, utest]; utest = nborlo[utest, minpower]];
   Join[list1, list2]]]
```

For some purposes (like the display of pixels on a screen), we can leave out the unums that are zero in any dimension. The **uboxlistinexact** routine is like **uboxlist**, but only produces unums that have nonzero width.

```
(* Create a 1-dimensional ubox list that leaves
   out the exact unums. *)
uboxlistinexact[ub_ /; uboundQ[ub], minpower_Integer] :=
  Module[{set = uboxlist[ub, minpower], temp = {}, i},
    For[i = 1, i ≤ Length[set], i++,
      If[inexQ[set[[i]]], AppendTo[temp, set[[i]]]]]; temp]
```

The next two routines are so similar that they probably should be consolidated. They both seek to split a range of values into smaller values representable in the *u*-layer, with a preference for the coarsest ULP resolution. The **splitub** routine prioritizes splitting off exact endpoints, so that repeated calls to it will eventually be given open intervals larger than an ULP. It then prioritizes the coarsest-possible ULP.

```
(* Split a ubound into three parts, if possible. *)
splitub[ub_ /; uboundQ[ub]] := Module[{b1, b2, g1, g2, gm},
  {{g1, g2}, {b1, b2}} = u2g[ub];
  Which[
    {{g1, g2}, {b1, b2}} ===
      {{Indeterminate, Indeterminate}, {True, True}}, {ub},
    ¬ b1 ∧ ¬ b2,
    If[g1 == g2, {ub}, (* Cannot split exact single values *)
      {{ub[[1]]}, g2u[ g1      g2   ], {ub[[-1]]}}],
              open    open
    (* else cleave off exact endpoints *)
    b1 ∧ ¬ b2, {g2u[ g1      g2   ], {ub[[-1]]}},
                     open    open
    (* cleave off exact right endpoint *)
    ¬ b1 ∧ b2, {{ub[[1]]}, g2u[ g1      g2   ]},
                                open    open
    (* cleave off exact left endpoint *)
    g1 == -∞, If[g2 == -maxreal, {ub},
      (* Cannot split the negative "many" region *)

      {{negbigu + ubitmask}, {negbigu}, g2u[ -maxreal   g2   ]}],
                                              open      open
    g2 == ∞, If[g1 == maxreal, {ub},
      (* Cannot split the positive "many" region *)
      {g2u[ g1    maxreal   g1    maxreal ],
            open  open      open  open

        {maxrealu}, {maxrealu + ubitmask}}],
```

```
(* See if open interval contains
   a unum different from either endpoint: *)
gm = u2g[(g1 + ĝ2) / 2];
```

$$gm_{[\![1,1]\!]} > g1, \left\{ g2u\left[\begin{array}{cc} g1 & gm_{[\![1,1]\!]} \\ open & open \end{array}\right], \{gm_{\hat{[\![1,1]\!]}}\}, g2u\left[\begin{array}{cc} gm_{[\![1,1]\!]} & g2 \\ open & open \end{array}\right] \right\},$$

$$gm_{[\![1,2]\!]} < g2, \left\{ g2u\left[\begin{array}{cc} g1 & gm_{[\![1,2]\!]} \\ open & open \end{array}\right], \{gm_{\hat{[\![1,2]\!]}}\}, g2u\left[\begin{array}{cc} gm_{[\![1,2]\!]} & g2 \\ open & open \end{array}\right] \right\},$$

```
True, {ub}]] (* Cannot split; must be smallest ULP size. *)
```

The bisect routine is a *g*-layer function that checks for infinite endpoint values, but for finite-valued endpoints it looks for the coarsest-possible integer power of 2 near the mean value.

```
(* Bisect an inexact general interval along a coarsest-
   possible ULP boundary. *)
bisect[g_ /; gQ[g]] := Module[{gL = g[[1,1]], gM, gR = g[[1,2]], m},

  gM = Which[
    gL < 0 ∧ gR > 0, 0,
    gL == -∞ ∧ gR > -maxreal, -maxreal,
    gL < maxreal ∧ gR == ∞, maxreal,
    m = 2^⌊Log[2,gR-gL]⌋;
    IntegerQ[gL / m], If[gR - gL == m, (gL + gR) / 2, m ⌊gL / m + 1⌋],
    True, m ⌈gL / m⌉];
```

$$\left\{ \begin{array}{cc} gL & gM \\ g_{[\![2,1]\!]} & open \end{array}, \begin{array}{cc} gM & gR \\ open & g_{[\![2,2]\!]} \end{array} \right\} \right]$$

D.4 Coalescing ubox sets

There are several ways to put sets of uboxes back together, and the prototype has a few primitive approaches for doing this in one and two dimensions. They are the most in need of algorithmic efficiency improvement of any of the routines here, since they are doing things in an order n^2 way that could be done in order n time. They also need to be more robust in handling overlapping uboxes. For example, the following pair of uboxes would not be coalesced.

The `coalescepass` routine examines a set for the occurrence of adjacent inexact-exact-inexact unum triples, since those can sometimes be replaced by a single inexact unum of twice the ULP width.

```
(* Make a single pass at a set of 2-dimensional uboxes to
   coalesce subsets that form a coarser single ULP-wide unum. *)
coalescepass[set_] := Module[{gset, g, g1, g2, ndim, udim, width, i, j},
  ndim = Length[set[[1]]];
  gset = Table[u2g[set[[i,j]]], {i, 1, Length[set]}, {j, 1, ndim}];
  For[j = 1, j ≤ ndim, j++,
   For[i = 1, i ≤ Length[gset], i++, g = gset[[i]];
    If[g[[j,2]] == {open, open},
     width = g[[j,1,2]] - g[[j,1,1]];
     If[g[[j,1,1]] ≥ 0,
      If[EvenQ[g[[j,1,1]] / width] ⋀ Length[
         g2u[{{g[[j,1,2]] - width, g[[j,1,2]] + width}, {open, open}}]] == 1,
       g1 = Insert[Delete[g, j], {{g[[j,1,2]], g[[j,1,2]]},
         {closed, closed}}, j];
       g2 = Insert[Delete[g, j], {{g[[j,1,1]] + width, g[[j,1,2]] + width},
         {open, open}}, j];
       If[MemberQ[gset, g1] && MemberQ[gset, g2],
        gset = Complement[gset, {g, g1, g2}] ⋃ {Insert[Delete[g, j],
          {{g[[j,1,1]], g[[j,1,2]] + width}, {open, open}}, j]}]],
      If[EvenQ[g[[j,1,2]] / width] ⋀ Length[g2u[
          {{g[[j,1,1]] - width, g[[j,1,1]] + width}, {open, open}}]] == 1,
       g1 = Insert[Delete[g, j], {{g[[j,1,1]], g[[j,1,1]]},
         {closed, closed}}, j];
       g2 = Insert[Delete[g, j], {{g[[j,1,1]] - width, g[[j,1,2]] - width},
         {open, open}}, j];
       If[MemberQ[gset, g1] && MemberQ[gset, g2],
        gset = Complement[gset, {g, g1, g2}] ⋃ {Insert[Delete[g, j],
          {{g[[j,1,1]] - width, g[[j,1,2]]}, {open, open}}, j]}]]
    ]]]];
  Table[g2u[gset[[i,j]]][[1]], {i, 1, Length[gset]}, {j, 1, ndim}]]]
```

The `coalesce` routine (it really should be called `coalesce2D`) calls coalescepass until no more consolidation is possible.

```
(* Make multiple passes at a set of 2-dimensional uboxes to coalesce
   subsets that form a coarser single ULP-wide unum until all
   possible coalescing is done. *)
coalesce[set_] :=
 Module[{newset = set}, While[coalescepass[newset] ≠ newset,
   newset = coalescepass[newset]]; newset]
```

The next two routines are for one-dimensional ubox sets. The `coalesce1D` routine makes fewer assumptions about an existing ubox list **ubset**, attempting to merge an input argument ubound **ub** with that set if its ranges overlap or touch.

```
(* Add a ubound or unum to a 1-dimensional set of ubounds,
coalescing if possible. *)
coalesce1D[ubset_List, ub_ /; uQ[ub]] :=
 Module[{i, newset = {}, ut = If[uboundQ[ub], ub, {ub}]},
  If[ubset == {}, {ub},
   (* First look for any overlapping or touching ubounds,
   and merge them with ut. If disjoint, append to newset. *)
   For[i = 1, i ≤ Length[ubset], i++,
    If[ubset⟦i⟧ ≈ ut ⋁
      nborlo[ubset⟦i,1⟧, -∞] ≈ ut ⋁ nborhi[ubset⟦i,-1⟧, -∞] ≈ ut,
     ut = {(minub[ubset⟦i⟧, ut])⟦1⟧, (maxub[ubset⟦i⟧, ut])⟦-1⟧},
     AppendTo[newset, ubset⟦i⟧]]];
   AppendTo[newset, ut]; newset]]
```

The **ubinsert** routine works in conjunction with the try-everything approach, and works very well when it starts out empty and can preserve ordering with each new candidate to add to the solution set. The commented **Print** statements can be uncommented to produce a monitoring of the progress of the solution-finding algorithm.

```
(* The ubinsert insertion routine assumes ubset is a
  sorted set of disjoint ubounds, and has no NaN ubounds. *)
ubinsert[ubset_, ub_ /; uboundQ[ub]] :=
 Module[{i, j, k = Length[ubset], lefttouch = False,
   righttouch = False, newset = ubset},
  If[k == 0, (*Print["   First element. ",{ view[ub]}];*) {ub},
   j = k; While[j > 0 ⋀ gtuQ[newset⟦j⟧, ub], j--];
   (*Print["insertion point = ",j]*);
   If[j > 0, lefttouch = ({nborhi[newset⟦j,-1⟧, -∞]} ≈ ub)];
   If[j < k, righttouch = ({nborlo[newset⟦j+1,1⟧, -∞]} ≈ ub)];
   Which[
    lefttouch ⋀ righttouch, newset = Join[Drop[newset, {j, k}],
      {{newset⟦j,1⟧, newset⟦j+1,-1⟧}}, Drop[newset, {1, j + 1}]];
     (*Print["   Joined on both sides. "]*),
    lefttouch ⋀ ¬ righttouch, newset = Join[Drop[newset, {j, k}],
      {{newset⟦j,1⟧, ub⟦-1⟧}}, Drop[newset, {1, j}]];
     (*Print["   Joined on left side. "]*),
    ¬ lefttouch ⋀ righttouch, newset = Join[Drop[newset, {j + 1, k}],
      {{ub⟦-1⟧, newset⟦j+1,-1⟧}}, Drop[newset, {1, j + 1}]];
     (*Print["   Joined on right side. "]*),
    True, newset = Join[If[j + 1 > k, newset, Drop[newset, {j + 1, k}]],
      {ub}, Drop[newset, {1, j}]];
     (*Print["   Inserted new ubound, not touching. ",newset]*)
   ]; newset]]
```

The `hollowout` routine is a little like consolidation; it removes interior points and leaves only the boundary. Sometimes that can be the most efficient way to operate with a multidimensional shape.

```
(* Remove interior points from a set of uboxes,
leaving the boundary only. *)
hollowout[set_] := Module[{gnbors, gset, i, j, n = Length[set],
    ndim = Length[set⟦1⟧], nbors, nnbors, boundary = {}, ub},
    gset = Table[u2g[set⟦i,j⟧], {i, 1, n}, {j, 1, ndim}];
  For[i = 1, i ≤ n, i++,
    ub = set⟦i⟧;
    nbors = findnbors[{ub}, Table[0, {ndim}]];
    nnbors = Length[nbors];
    gnbors = Table[u2g[nbors⟦i,j⟧], {i, 1, nnbors}, {j, 1, ndim}];
    j = 1;
    While[j ≤ nnbors ⋀ gsetintersectQ[gnbors⟦j⟧, gset], j++];
    If[j < nnbors, AppendTo[boundary, {ub}]]];
  boundary]
```

D.5 Ubox bounds, width, volumes

The `minub` and `maxub` functions find the leftmost and rightmost points on the real number line described by two ubounds, taking into account their open-closed states to break ties and to record the result.

```
(* Find the minimum of two endpoints,
considering open-closed state. *)
minub[xub_ /; uboundQ[xub], yub_ /; uboundQ[yub]] :=
  Module[{xg = u2g[xub], yg = u2g[yub]},
    If[xg⟦1,1⟧ < yg⟦1,1⟧ ⋁ (xg⟦1,1⟧ == yg⟦1,1⟧ ⋀ ¬ xg⟦2,1⟧), xub, yub]]

(* Find the maximum of two endpoints,
considering open-closed state. *)
maxub[xub_ /; uboundQ[xub], yub_ /; uboundQ[yub]] :=
  Module[{xg = u2g[xub], yg = u2g[yub]},
    If[xg⟦1,2⟧ > yg⟦1,2⟧ ⋁ (xg⟦1,2⟧ == yg⟦1,2⟧ ⋀ ¬ xg⟦2,2⟧), xub, yub]]
```

The `width` and `volume` routines do just what they sound like they should do, with `width` returning the (real-valued) `width` of a unum or ubound, and `volume` returning the total multidimensional volume of a set of uboxes (following page):

```
(* Find the real value of the width of a unum or ubound. *)
width[u_ /; unumQ[u] ⋁ uboundQ[u]] := Module[{glo, ghi},
   {glo, ghi} = If[unumQ[u], u2g[u]⟦1⟧, u2g[u]⟦1⟧];
   ghi - glo]


(* Find the total n-dimensional volume of a ubox set. *)
volume[set_] := Module[{n = Length[set], ndim = Length[set⟦1⟧]},
   ∑ⁿᵢ₌₁ ∏ⁿᵈⁱᵐⱼ₌₁ width[set⟦i,j⟧]]
```

D.6 Test operations on sets

The **setintersectQ** function has other uses, though; it tests if a ubound *ub* intersects any ubound in a set of ubounds, *set*:

```
(* Check a ubound for intersection with a set of other ubounds. *)
setintersectQ[ub_, set_] := Module[{intQ = False, ndim = Length[ub], n},
   For[i = 1, i ≤ Length[set], i++, n = 0;
    For[j = 1, j ≤ ndim, j++, n += Boole[{ub⟦j⟧} ≈ {set⟦i,j⟧}]]
     If[n == ndim, intQ = True]];
   intQ]
```

The **gsetintersectQ** test does the equivalent for general intervals instead of ubounds:

```
(* Check a general interval for intersection with a set
 of other general intervals. *)gsetintersectQ[gs_, gset_] :=
Module[{intQ = False, n, ndim = Length[gs]},
   For[i = 1, i ≤ Length[gset], i++, n = 0;
    For[j = 1, j ≤ ndim, j++, n += Boole[nneqgQ[gs⟦j⟧, gset⟦i,j⟧]]];
    If[n == ndim, intQ = True]];
   intQ]
```

For safe programming practice, a function requiring a set of general intervals can be checked with **glistQ** to make sure every member of the set is a general interval.

```
(* Check if an argument is a list of general intervals. *)
glistQ[g_] := Module[{q = True}, If[Head[g] == List,
   For[i = 1, i ≤ Length[g], i++,
    If[¬ gQ[g⟦i⟧], q = False]], q = False]; q]
```

Since all it takes is one NaN to make an entire calculation a NaN, **glistNaNQ** can be called to return True if *any* entry is a NaN, otherwise False. The **ulistNaNQ** does the equivalent test for *u*-layer values.

```
(* Check if a list of general intervals contains a NaN anywhere. *)
glistNaNQ[g_] := Module[{q = False},
  For[i = 1, i ≤ 1 Length[g], i++,
    If[g⟦i,1,1⟧ === NaN ⋁ g⟦i,1,2⟧ === NaN, q = True]]; q]
```

```
(* Check if a list of unums or ubounds contains a NaN
   anywhere. In the prototype, signalling NaN is tested for,
but in a real environment would halt the calculation. *)
ulistNaNQ[u_] := Module[{i, q = False},
  For[i = 1, i ≤ 1 Length[u], i++,
    If[u⟦i⟧ == qNaNu ⋁ u⟦i⟧ == sNaNu, q = True]]; q]
```

The `ulistQ` test returns True if every member of the set is a unum or ubound.

```
(* Check if an argument is a list of unums or ubounds. *)
ulistQ[u_] := Module[{i, q = True}, If[Head[u] == List,
  For[i = 1, i ≤ Length[u], i++,
    If[¬ uQ[u⟦i⟧], q = False]], q = False]; q]
```

D.7 Fused polynomial evaluation and acceleration formula

The routines for loss-free evaluation of polynomials use several support functions. The **polyTg** function uses a finite Taylor expansion to relocate the argument from powers of x (as a general interval) to powers of $(x - x_0)$:

```
(* Polynomial helper function that uses powers of x -
   x0 instead of x. *)
polyTg[coeffsg_, xg_, x0g_] :=
  Module[{bi, i, j, k = Length[coeffsg], coeffstg = coeffsg, pg, xmg},

    If[x0g⟦1,1⟧ == -∞ ⋁ x0g⟦1,2⟧ == ∞, {{-∞    ∞}
                                          closed closed}',

      For[j = 0, j ≤ k - 1, j++,
        bi = Binomial[k - 1, j];
        pg = timesg[coeffsg⟦-1⟧, {{ bi     bi  }
                                   closed closed}];

        For[i = k - 2, i ≥ j, i--,
          bi = Binomial[i, j];
          pg =
            plusg[timesg[coeffsg⟦i+1⟧, {{ bi     bi  }
                                         closed closed}], timesg[x0g, pg]]];

        coeffstg⟦j+1⟧ = pg];
      xmg = minusg[xg, x0g];
      pg = coeffstg⟦k⟧;
      For[i = k - 1, i ≥ 1, i--,
        pg = plusg[timesg[pg, xmg], coeffstg⟦i⟧]];
      pg]]
```

The `polyinexactg` function evaluates a polynomial using each endpoint as an x_0 in the `polyTg` function. The correct answer lies in the intersection of the resulting general interval; if this intersection is between the values of the polynomial evaluated at the exact endpoints, then there is no need for further refinement.

```
(* Polynomial helper function that evaluates
 a polynomial at the endpoints of an inexact unum,
and intersects them to tighten the result. *)
polyinexactg[coeffsg_, xg_] :=
```
$$\text{intersectg}\left[\text{polyTg}\left[\text{coeffsg, xg, } \begin{matrix} xg_{[1,1]} & xg_{[1,1]} \\ \text{closed} & \text{closed} \end{matrix}\right],\right.$$
$$\left.\text{polyTg}\left[\text{coeffsg, xg, } \begin{matrix} xg_{[1,2]} & xg_{[1,2]} \\ \text{closed} & \text{closed} \end{matrix}\right]\right]$$

The `polyexactg` function takes an exact general interval xg and evaluates it exactly in the *g*-layer, using Horner's rule. A fused multiply-add would work as the loop kernel also, but since the routine takes place entirely in the *g*-layer, there is no difference between fused operations and separate ones.

```
(* Polynomial evaluation of an exact general interval,
using Horner's rule. *)polyexactg[coeffsg_, xg_] :=
 Module[{i, k = Length[coeffsg], pg = coeffsg[-1]},
  For[i = k - 1, i ≥ 1, i--,
   pg = plusg[coeffsg[i], timesg[pg, xg]]]; pg]
```

The `polyg` routine evaluates the polynomial of a general interval xg, using a set of general interval coefficients *coeffsg* (one for each power of xg from 0 to *n*), without information loss at the *u*-layer level of accuracy. The technique here should work for functions other than polynomials, if the equivalent of a `polyTg` function can be derived. (Continues to following page.)

```
(* Polynomial evaluation of a general interval without u-
 layer information loss. *)
polyg[coeffsg_ /; glistQ[coeffsg], xg_ /; gQ[xg]] := Module[

 {gL, gR, gM, k = Length[coeffsg], min, minQ, max, maxQ, pg, trials},
 Which[

 (* Dispense with NaN cases. *)
```
$$xg_{[1,1]} === \text{NaN} \lor xg_{[1,2]} === \text{NaN} \lor \text{glistNaNQ[coeffsg]}, \begin{matrix} \text{NaN} & \text{NaN} \\ \text{open} & \text{open} \end{matrix},$$
```
 (* Constant case. Just
  return the first (and only) coefficient. *)
 k == 1, coeffsg[1],
 (* Linear case is a fused multiply-add; no dependency problem. *)
 k == 2, fmag[coeffsg[2], xg, coeffsg[1]],
 (* Exact argument is also easy, since no dependency problem. *)
 xg[1,1] == xg[1,2], polyexactg[coeffsg, xg],
```

```
(* Quadratic or higher requires finesse. Intersect tbe two endpoint-
   based evaluations. *)
True,
trials = {xg};
```

$$gL = \text{polyexactg}\left[\text{coeffsg}, \begin{array}{cc} xg_{[1,1]} & xg_{[1,1]} \\ \text{closed} & \text{closed} \end{array}\right];$$

$$\text{If}\left[xg_{[2,1]}, gL_{[2]} = \{\text{open, open}\}\right];$$

$$gR = \text{polyexactg}\left[\text{coeffsg}, \begin{array}{cc} xg_{[1,2]} & xg_{[1,2]} \\ \text{closed} & \text{closed} \end{array}\right];$$

$$\text{If}\left[xg_{[2,2]}, gR_{[2]} = \{\text{open, open}\}\right];$$

$$\text{If}\left[gL_{[1,1]} < gR_{[1,1]} \lor (gL_{[1,1]} == gR_{[1,1]} \land \neg gL_{[2,1]}),\right.$$
$$\quad (* \text{ then } *) \{\text{min, minQ}\} = \text{Transpose}[gL]_{[1]},$$
$$\quad (* \text{ else } *) \{\text{min, minQ}\} = \text{Transpose}[gR]_{[1]};$$

$$\text{If}\left[gL_{[1,2]} > gR_{[1,2]} \lor (gL_{[1,2]} == gR_{[1,2]} \land \neg gL_{[2,2]}),\right.$$
$$\quad (* \text{ then } *) \{\text{max, maxQ}\} = \text{Transpose}[gL]_{[2]},$$
$$\quad (* \text{ else } *) \{\text{max, maxQ}\} = \text{Transpose}[gR]_{[2]};$$

$$\text{While}\left[\text{Length[trials]} \geq 1,\right.$$
$$\quad pg = \text{polyinexactg}[\text{coeffsg, trials}_{[1]}];$$

$$\quad \text{If}\left[\text{intersectg}\left[\text{u2g}[\text{g2u}[pg]], \text{u2g}\left[\text{g2u}\left[\begin{array}{cc} \text{min} & \text{max} \\ \text{minQ} & \text{maxQ} \end{array}\right]\right]\right] === \text{u2g}[\text{g2u}[pg]],\right.$$

```
    trials = Rest[trials],
    trials = Join[bisect[trials[[1]]], Rest[trials]];
```

$$\quad gM = \text{polyexactg}\left[\text{coeffsg}, \begin{array}{cc} \text{trials}_{[1,1,2]} & \text{trials}_{[1,1,2]} \\ \text{closed} & \text{closed} \end{array}\right];$$

$$\quad \text{If}\left[gM_{[1,1]} < \text{min} \lor gM_{[1,1]} == \text{min} \land \neg gM_{[2,1]},\right.$$
$$\quad \{\text{min, minQ}\} = \text{Transpose}[gM]_{[1]}];$$

$$\quad \text{If}\left[gM_{[1,2]} > \text{max} \lor gM_{[1,2]} == \text{max} \land \neg gM_{[2,2]},\right.$$
$$\quad \{\text{max, maxQ}\} = \text{Transpose}[gM]_{[2]}];$$

$$\quad \begin{array}{cc} \text{min} & \text{max} \\ \text{minQ} & \text{maxQ} \end{array} = \text{u2g}\left[\text{g2u}\left[\begin{array}{cc} \text{min} & \text{max} \\ \text{minQ} & \text{maxQ} \end{array}\right]\right]$$

$$\left.\left.\right]\right]; \begin{array}{cc} \text{min} & \text{max} \\ \text{minQ} & \text{maxQ} \end{array}\right]\right]$$

The **polyu** function is the *u*-layer version of **polyg**. Unlike most *u*-layer functions, it does not tally the numbers and bits moved since its iterative nature makes it difficult to compare to a conventional polynomial evaluation. It clearly does more operations than the naïve method of evaluating the polynomial of an interval, but it is worth it to produce a result that completely eliminates the dependency error.

```
(* Polynomial evaluation of a unum or ubound without u-
   layer information loss. *)
polyu[coeffsu_ /; ulistQ[coeffsu], u_ /; uQ[u]] :=
  Module[{coeffsg = Table[u2g[coeffsu[[i]]], {i, 1, Length[coeffsu]}]},
   g2u[polyg[coeffsg, u2g[u]]]]
```

The **accub** function is shown here as an example of how to craft library functions that eliminate dependency error by fusing the operations. It was used in the chapter on gravitational dynamics. Because x and y appear in both the numerator and denominator of the formula for the acceleration,

$$\text{acc}[\{\mathbf{x\_, y\_}\}] := \frac{-Gm}{\left(\sqrt{x^2+y^2}\right)^3}\ \{\mathbf{x, y}\}$$

there is some information loss when x and y are inexact. While *numerical* calculus is "considered evil" in Chapter 21, symbolic calculus is incredibly useful for finding the minima and maxima of functions in library function design. We can evaluate functions of inexact values by evaluating them at their endpoints **and at any local extrema that lie within their range of uncertainty**. By differentiating the **acc** function above with respect to x and y, we discover extrema along the lines $y = \pm\sqrt{2}\ x$ and $x = \pm\sqrt{2}\ y$. So if we have a two-dimensional ubox for x and y, it always suffices to calculate **acc** at its corners and at any points in the range that intersect the lines $y = \pm\sqrt{2}\ x$ and $x = \pm\sqrt{2}\ y$.

```
(* Acceleration bound that avoids the dependency
   problem between numerator and denominator. *)
accub[{xub_, yub_}] := Module[{i, j, x, y, a, b, s,
    amin = ∞, amax = -∞, bmin = ∞, bmax = -∞, tests = {}},
    {x, y} = {u2g[xub]〚1〛, u2g[yub]〚1〛};
    tests = Tuples[{x, y}];
    For[i = 1, i ≤ 2, i++, For[j = 1, j ≤ 2, j++, For[s = 1, s ≥ -1, s -= 2,
        If[x〚1〛 < s y〚j〛/√2 < x〚2〛, AppendTo[tests, {s y〚j〛/√2, y〚j〛}]];
        If[y〚1〛 < s x〚i〛/√2 < y〚2〛, AppendTo[tests, {x〚i〛, s x〚i〛/√2}]];
        If[x〚1〛 < s y〚j〛 √2 < x〚2〛, AppendTo[tests, {s y〚j〛 √2, y〚j〛}]];
        If[y〚1〛 < s x〚i〛 √2 < y〚2〛, AppendTo[tests, {x〚i〛, s x〚i〛 √2}]]
    ]]];
    For[i = 1, i ≤ Length[tests], i++,
        {a, b} = acc[tests〚i〛];
        {amin, amax} = {Min[a, amin], Max[a, amax]};
        {bmin, bmax} = {Min[b, bmin], Max[b, bmax]}];
    If[amin == 0 ⋀ inexQ[xub〚1〛] ⋀ inexQ[xub〚-1〛],
        amin += smallsubnormal / 2];
    If[bmin == 0 ⋀ inexQ[yub〚1〛] ⋀ inexQ[yub〚-1〛],
        bmin += smallsubnormal / 2];
    If[amax == 0 ⋀ inexQ[xub〚1〛] ⋀ inexQ[xub〚-1〛],
        amax -= smallsubnormal / 2];
    If[bmax == 0 ⋀ inexQ[yub〚1〛] ⋀ inexQ[yub〚-1〛],
        bmax -= smallsubnormal / 2];
    {{amîn, amâx}, {bmîn, bmâx}}]
```

D.8 The try-everything solver

The `solveforub[`*domain*`]` solver requires a `conditionQ[`*ub*`]` Boolean function to be defined by the user. Since it takes some practice to get the question right (see Chapter 17 for examples), the following code has some debugging-style `Print` statements that have been commented out. They can be un-commented to help discover why a routine is running too slowly or producing trivial-looking results because the wrong question was asked by `conditionQ`. A better version of `solveforub` would include a *minpower* parameter that prevents the solver from examining the real number line below a certain ULP size of $2^{minpower}$, similar to `nborhi` and `uboxlist`. The function below will go to the limits of the settings of `setenv`, and will produce the *c*-solution for an initial search space of *domain*, where *domain* is a list of ubounds. In the worst case, it will iterate as many times as there are representable values in the environment, so it is best to set the environment low initially.

```
(* The try-everything solver. Has commented-out Print statements
   that can be uncommented to see why long running solvers are
   taking too long. *)
solveforub[domain_] :=
 Module[{i, new, sols = {}, temp, trials = domain, ub},
  While[Length[trials] > 0,
   new = {};
   For[i = 1, i ≤ Length[trials], i++, ub = trials⟦i⟧;
    If[conditionQ[ub],
     temp = splitub[ub];
     (*Print["Splitting ", view[ub]," into ",
      Table[ view[temp⟦j⟧],{j,1,Length[temp]}]]*);
     If[Length[temp] == 1, (* unsplittable. Join to
       existing region or start new one. *)
      sols = ubinsert[sols, temp⟦1⟧] (*;
      Print["Joined ", view[ub]," to sols: ",
       Table[ view[sols⟦j⟧],{j,1,Length[sols]}]]*),
      (*Print["New split to examine: ",
       Table[ view[temp⟦j⟧],{j,1,Length[temp]}]];*)
      new = Join[temp, new];
      (*Print["The 'new' list: ",
       Table[ view[new⟦j⟧],{j,1,Length[new]}]]*);
     ]
    ]
   ];
   trials = new
  ];
  sols]
```

D.9 The guess function

As described in Chapter 18, the `guessu[x]` function turns a unum, ubound, or general interval into an exact value that is as close as possible to the mean of its endpoints. When there is a tie between possible values that are as close as possible, the routine selects the one with the shorter unum bit string, which is the equivalent of "rounding to nearest even" for floats. This enables unums and ubounds to duplicate any algorithms that are known to be successful with floats, such as Gaussian elimination and stable fixed-point methods. If the guess is to satisfy a condition, the guess can be checked by testing the condition using exact unum arithmetic. If it passes, then it is a provable solution, but there is no guarantee that it is a complete solution of all possible values that satisfy the condition. If the solution set is known to be *connected*, then the paint bucket method of Chapter 15 can be used to find a *c*-solution, using the guess as a seed.

```
(* Turn a unum or ubound into a guess,
for imitation of float behavior. *)

guessu[x_ /; (uQ[x] ⋁ gQ[x])] := Module[{a, b, g, gu},

  (* If not already a general interval with exact endpoints,
  make it one. *)
  g = If[uQ[x], u2g[x], u2g[g2u[x]]];
  {a, b} = g[[1]];
  Which[

    (* Take care of trivial cases: NaN or an exact interval. *)
    a === NaN ⋁ b === NaN, qNaNu,
    a == b, â,
    a == -∞ ⋀ b == ∞, 0̂,
    True, (* Average the endpoint values and convert to a unum. *)
    gu = (a + b̂) / 2;
    If[exQ[gu], gu, (* If exact, we're done. That's the guess. *)
      (* else round to nearest even. *)
      If[BitAnd[ulpu, gu] == 0, gu = gu - ubitmask, gu = gu + ubitmask];
      (* Compress away the last zero in the fraction,
      since it is even. *)
      If[fsize[gu] == 1, gu, BitShiftRight[BitAnd[floatmask[gu], gu], 1] +
        BitAnd[efsizemask, gu] - 1]]]]
```

Remember: There is nothing floats can do that unums cannot. ∎

For further reading

A conventional "Bibliography" seems like an anachronism in an era where powerful web searching is ubiquitous. The purpose here is to point out some important (or at least entertaining) work that might be missed by a cursory web search.

Acton, Forman S., *Numerical Methods that Work,* Harper & Row, 1970, updated and reissued by the Mathematical Association of America, 1990.
http://onlinepdfbooks.blogspot.com/2013/12/
numerical-methods-that-work-by-forman-s.html

Bailey, David H., Barrio, Roberto, and Borwein, Jonathan M., "High precision computation: *Mathematical* physics and dynamics," *Applied Mathematics and Computation,* volume 218 (2012), pages 10106–10121.
http://www.davidhbailey.com/dhbpapers/hpmpd.pdf

Bailey, David H., and Borwein, Jonathan M., "High-precision arithmetic: Progress and challenges," manuscript, 8 Aug 2013.
http://www.davidhbailey.com/dhbpapers/hp-arith.pdf

Goldberg, David, "What every computer scientist should know about floating-point arithmetic," *Computing Surveys*, Mar 1991.
http://docs.oracle.com/cd/E19957-01/806-3568/ncg_goldberg.html

Gustafson, John L., and Snell, Quinn O., "HINT: A new way to measure computer performance," *Proceedings of the 28th Annual HICSS,* IEEE Computer Society Press, Jan 1995.
http://hint.byu.edu/documentation/Gus/HINT/ComputerPerformance.html

Kahan, William, "How futile are mindless assessments of roundoff in floating-point computation?" web-published document, Jan 2006. http://www.cs.berkeley.edu/~wkahan/Mindless.pdf

Kahan, William, "Ellipsoidal error bounds for trajectory calculations," web publication of abbreviated lecture notes, Nov 1999.
http://www.cs.berkeley.edu/~wkahan/Math128/Ellipsoi.pdf

Kahan, William, "How Java's floating-point hurts everyone everywhere," originally presented at the *ACM 1998 Workshop for High-Performance Network Computing,* Mar 1998.
http://www.cs.berkeley.edu/~wkahan/JAVAhurt.pdf

Knuth, Donald E., *The Art of Computer Programming Volume 1: Fundamental Algorithms* and *Volume 2: Seminumerical Algorithms*, Second Edition, Addison-Wesley Publishing, Reading MA, 1973.

Kulisch, U. W., and Miranker, W. L., *Computer Arithmetic in Theory and Practice*, Academic Press, New York NY, 1981.

Lynch, Thomas, and Schulte, Michael J., "A high radix on-line arithmetic for credible and accurate computing," *Journal of Universal Computer Science*, volume 1, no. 7, Jul 1995.
http://www.jucs.org/jucs_1_7/a_high_radix_on/Lynch_T.html

Muller, Jean-Michel et al., "Floating-point arithmetic," web publication 2009.
http://perso.ens-lyon.fr/jean-michel.muller/chapitre1.pdf

Rump, Siegfried M., "Algorithms for verified inclusion," in R. Moore, editor, *Reliability in Computing, Perspectives in Computing,* pages 109–126. Academic Press, New York NY, 1988.

Sterbenz, Pat H., *Floating-Point Computation*, Prentice-Hall, Englewood Cliffs NJ, 1974.

Index

2001: A Space Odyssey and "beyond infinity," 28.

absg, **absu**, 155, 341, 380.
absolute value (non-unum), 12.
absolute value (unum), examples, 177, 213; general approach, 155; prototype, 341, 380.
acc, defined, 289, 399; examples, 292, 297, 399.
acceleration calculation in physics simulations, 274, 274–280, 282–285, 288–309, 312–315.
accub, defined, 297, 347, 399; examples, 298.
accuracy, argument reduction for trig functions, 75, 157; automatic management of, 2, 41, 122–126, 180–181, 221, 297, 343–344; C approach to, 58–59; coprocessor, 46, 59–60; cost trade-off, 86–88; dot product, 61–62, 162–163, 172, 252, 271–272, 342, 370; excessive, 3, 38; fused multiply-add, 60–61; *g*-layer, 56, 59–62, 162–163, 165, 252, 271–272; guard digits, 3, 56; *h*-layer, 74–78, 185; IEEE Standard floats, 37–38, 45–46; improvement by intersection, 108, 166, 169; inconsistent, 2, 3, 5, 18, 20, 46, 56–60, 62, 68, 84, 160, 166–167, 264, 331; input values, 6, 75–77, 189, 313; insufficient, 3, 111, 117–126, 127, 182; Java approach to, 3, 158; output result, 6, 41, 77–78, 124, 192, 209, 301; pi, 75, 157, 195–198, 207–210; scratchpad, 56, 59–62, 162–163, 165, 252, 271–272; significance arithmetic, 72; speed, vs., 62, 86, 88, 94; ubound tracking, 72, 150, 195; unum range of, 41–43, 45–46; unums vs. floats, *xv*, 1, 3, 51, 165, 173–192, 193, 248–249, 291–309.
ACRITH, 225.
Acton, Forman S., 245, 337.
addition (non-unum), biased integer, 11; cumulative rounding error, 111, 117–121; fixed-point, 16; fixed-size integer, 9–10; float, 21–22; floats with a ubit enabled, 34; rational arithmetic, 14; sign-magnitude, 12–13.
addition (unum), automatic error control example, 121–126; data motion measurement, 114–115; examples, 112, 114–115, 117, 126; hardware for, 116–117; prototype, 341, 364–365, 370, 371, 372, 396, 397; special symbol ⊕ for, 111, 365; table, 111–113; visualization, 115–116; "unbiased rounding" counterexample, 117–121.
"almost infinite," and "almost nothing," 28–31, 90, 112, 130, 202, 369.
argument reduction, 75, 156–158.
Archimedes, 195–196.
Ariane rocket launch disaster, 83.
ASCI Program, *xv*.
ASCII characters, 36, 328.
associative law of algebra, 58, 164–167, 194, 329.
automatic accuracy control, 2, 41, 122–126, 180–181, 221, 297, 343–344.
autoN, 78, 339, 374.
Avogadro's number, 3, 40, 62, 318.

backward error analysis, 76.
Bailey, David H., 42, 135, 184–187, 337.
bandwidth, 1–3, 5–6, 38, 54, 59, 62, 88, 92, 101, 160, 193, 210, 212, 216, 331.
baseball analogy, 69.
bias, *esizesize* and *fsizesize*, 39; exponent, 18, 34, 53, 94, 339, 340, 354, 357; rounding, 117–121, 271; signed integers, 10–11.
bias, function to extract exponent bias, 339, 354.
big, **bigu**, 47, 52–53, 339, 354, 355.